POWER, THE PRESS AND THE TECHNOLOGY OF FREEDOM

Freedom House

Freedom House is an independent nonprofit organization that monitors human rights and political freedom around the world. Established in 1941, Freedom House believes that effective advocacy of civil rights at home and human rights abroad must be grounded in fundamental democratic values and principles.

In international affairs, Freedom House continues to focus attention on human rights violations by oppressive regimes, both of the left and the right. At home, we stress the need to guarantee all citizens not only equal rights under law, but equal opportunity for social and economic advancement.

Freedom House programs and activities include bimonthly and annual publications, conferences and lecture series, public advocacy, ongoing research of political and civil liberties around the globe, and selected, on-site monitoring to encourage fair elections.

Focus on Issues
General Editor: James Finn

This publication is one in a series of Focus on Issues. The separate publications in this series differ in the method of examination and breadth of the study, but each focuses on a single, significant political issue of our time. The series represents one aspect of the extensive program of Freedom House. The views expressed are those of the authors and not necessarily those of the Board of Freedom House.

About the Author

Leonard R. Sussman illuminates this book with forty years of personal observations of journalists, their censors and their publics. As a journalist, briefly as a government spokesman and a long-time freedom advocate, Sussman has met with media people and their controllers in forty-five countries. He was for twenty-one years executive director of Freedom House. He has served on U.S. delegations to Unesco, the Conference on Security and Cooperation in Europe, on special African aid, and on a congressional panel on Cooperation and Competition in Outer Space. He has written extensively for newspapers, magazines and journals and contributed to textbooks on international communications. His earlier books: *Mass News Media and the Third World Challenge* and *Glossary for International Communications: Warning of a Bloodless Dialect*. Mr. Sussman is presently senior scholar in international communications of Freedom House.

POWER, THE PRESS AND THE TECHNOLOGY OF FREEDOM

The Coming Age of ISDN

LEONARD R. SUSSMAN

Focus on Issues, No. 9

FREEDOM HOUSE

Copyright © 1989 by Freedom House

Printed in the United States of America. All rights reserved. No part of this book may be used or reproduced in any manner without written permission except in the case of brief quotations embodied in critical articles and reviews. For permission, write to Freedom House, 48 East 21st Street, New York, N.Y. 10010

First published 1989.

Cover design by Emerson Wajdowicz Studios, N.Y.C.

Distributed by

National Book Network
4720 Boston Way
Lanham, Md. 20706

Library of Congress Cataloging-in-Publication Data

Sussman, Leonard R.
    Power, the press and the technology of freedom : the coming age of ISDN / Leonard R. Sussman.
       p.   cm. — (Focus on issues ; no. 9)
    Includes index.
    ISBN 0-932088-39-2 : $24.95
    1. Freedom of the press—Developing countries. 2. Journalism--Political aspects—Developing countries. 3. Integrated services digital networks. 4. Developing countries—Political aspects. 5. Communication—International cooperation. I. Title. II. Series: Focus on issues (Freedom House (U.S.)) ; 9.
PN4748.D44S87   1989
323.44'5'091724—dc20   89-16806
                                                CIP

Contents

Dedication

Acknowledgments

Glossary

Historic Digital Photo

Preface   1

Introduction: **Technology of Freedom**   5

Chapter I: **The Word is Power**   21

Chapter II: **Revolution Without Losers**   53

Chapter III: **The Too-Easy Equivalence of Liberty**   93

Chapter IV: **Fear of Information Power in "Socialist" Countries**   99

Chapter V: **Power and Press in the Developing Countries**   125

Chapter VI: **Count the Ways of Censorship**   159

Chapter VII: **How the News Was Reported and Distorted—A Survey of State versus Information Power in 74 Countries**   189

    Table 1: Kind and degree of influences over editorial content
    Table 2: Formal government influences on media
    Table 3: Informal government influences on media
    Table 4: News "management" practiced by government
    Table 5: Harsh measures to control news on news content
    Table 6: Private ownership influence on news concern
    Table 7: Self-censorship and other pressures

Chapter VIII: **The Information-Power Struggle: 88 Country Reports**   207

Chapter IX: **The Reagan/Press Conflict**   341

Chapter X: **Networks of Freedom**   357

Chapter XI: **Don't Fear the Slippery Slope, and Other Recommendations**   419

Epilogue: **Will ISDN Facilitate Peace and the Human Imagination?**   439

Moral   451

Appendices   453

   A. Journalism Morbidity Table—1988

   B. Roster of Government-run News Agencies

   C. Letter of Violeta Barrios de Chamorro to Daniel Ortega, President of Nicaragua

   D. Listing of Countries by Degree of Change, 1987

   E. Correspondence: Amadeu Mahtar M'Bow, Director-General of Unesco

   F. Questionnaire: On Impediments to the Free Flow of Information in the News Media of the World in 1987

   G. News Media Control by Countries in 1988; also showing countries included in our Press Freedom Survey, and our country descriptions

   H. Government Press Ownership, Licensing, Censorship, and Guidance

   I. Proposals made at the London Information Forum of the Conference on Security in Europe

Notes   477

Index   487

Dedication

To the independent-minded journalists
    of the developing countries, who are often

Undervalued and intimidated
    by Third World governments,

Under-represented in Unesco's debates
    on communication development, and

Under-employed by Western news media.

They deserve better.

May they receive it in the Age of ISDN.

Other books by the author

Mass News Media and the Third World Challenge

Glossary for International Communications:
Warning of a Bloodless Dialect

Introduction and editing of books by the author

Big Story: How the American Press and Television Reported and Interpreted the Crisis of Tet 1968 in Vietnam and Washington, by Peter Braestrup (two volumes)

Three Years at the East-West Divide, by Max M. Kampelman

Today's American: How Free?, with James Finn

Acknowledgments

NEARLY 100 JOURNALISTS and academics in seventy-four countries served as correspondents for this book. Over many months in 1987 they responded to a lengthy questionnaire. Most also wrote extensive descriptions of the clash between state power and information in their countries. Except for two who have recently died, all but one still live in their homelands. For this reason, more than half asked to remain anonymous. Most seeking anonymity reside in Marxist-oriented countries or in authoritarian nations, politically right as well as left of center. In fairness, we provide anonymity for all our living correspondents. Clearly, however, the request of so many to remain nameless is a revealing commentary on the state of journalism in three-quarters of the world.

We shall name two correspondents. Both died while their work for us was under way: Percy Qoboza, the leading black editor of South Africa (whom we describe later), and Dr. Tim Matthews, a respected white observer of Zimbabwe, whose last writing appears here. We cherish the memories of both men, and are deeply indebted to all our correspondents.

For their support during more than two decades when I served as executive director of Freedom House, including the year producing this book, my appreciation to the board of trustees of the organization, particularly the present officers, Max M. Kampelman, chairman, John W. Riehm, president, and Leo Cherne, honorary chairman. A bow, too, to R. Bruce McColm, my talented successor.

My thanks, too, to the foundations whose generous grants helped make possible the continuing Freedom House program on international communications of which my work and this book is a part: the Cowles Charitable Trust, the Earhart Foundation, the Forbes Foundation, the Esther A. and Joseph Klingenstein Fund, Inc., the New York Times Company Foundation, Inc., and the Fondación Angel Ramos, Inc. The late Angel Ramos, publisher of *El Mundo* and the *World-Journal* in San Juan, P.R., hired me as cable editor in December 1941. I have long been grateful for that rapid immersion in wartime journalism.

Special thanks are due the Joyce Mertz-Gilmore Foundation and the late Robert Wallace Gilmore.

Portions of the book which discuss the philosophy and politics of international communications, and particularly the application of technologies to human freedom were extensively informed by the writing and personal encouragement of the late Ithiel de Sola Pool, professor at Massachusetts Institute of Technology. He combined uniquely the perspective, past and future, of a scholar with the precision and creativity of the technologist.

I am indebted to Bert Cowlan, specialist in all aspects of telecommunications and related technologies, for carefully vetting the chapter on the freedom networks. He has long been a mentor to me, and to many U.S. delegations doing battle on the international communications barricades.

I appreciate the insight on high-definition television and related foreign trade issues which John Diebold provided.

For permission to reprint portions of her article on Ethiopia which appeared in *Index on Censorship*, I am grateful to Roberta Cohen-Korn. Hers is the best reporting of the appalling events in that country.

The Bush inaugural photograph, historic in its own right, was provided by the Associated Press. My thanks to Harold G. Buell and Nate Polowetzky, assistant general managers for news photos and foreign news, respectively.

The report from Algeria has been prepared with the assistance of l'Observatoire de l'Information, Montpellier.

Dr. Kim Sang Man, South Korea's distinguished journalist and publisher, was extraordinarily helpful during my visit to his country in 1988.

My appreciation to David Berley for his translations from the Russian and French.

I acknowledge the helpful interest of Enrique Zileri, chairman of the International Press Institute (London), and editor of *Caretas*, one of Latin America's most diverse magazines. Early in this project I was encouraged by Peter Galliner, IPI's director.

For her careful, imaginative copy-editing of these pages I thank Molly Finn.

Finally, I profited from the opportunity to serve for two weeks in May 1989 on the U.S. delegation to the London Information Forum of the Conference on Security and Cooperation in Europe. The proposals I submitted for the future action of the thirty-five nation CSCE appear in Appendix I.

While all this assistance has been valuable, the judgments are, for better or worse, mine.

—L.R.S.

Glossary of Acronyms, Terms and Organizations

AFP Agence France-Presse.

AP Associated Press.

ASEAN Association of Southeast Asian Nations.

CATV Broad-band Cable Television.

CCITT Consultative Committee on International Telephone and Telegraph, ITU.

Common Carrier Communications system which offers its service to the public but does not control the messages.

DBS Cross-border broadcasting by satellite.

Downlink/Uplink The earth station for receiving or sending satellite signals.

EEC European Economic Community, presently twelve countries.

Fiber Optics A pencil-thin bundle of glass-fiber filaments which can move tens of thousands of phone calls simultaneously in bursts of laser beams.

FoI Freedom of Information Act.

GATT General Agreement on Trade and Tariffs.

Geostationary Orbit Satellite path 22,300 miles above the equator. The satellite's position remains constant because its speed is synchronized to the earth's rotation.

HDTV High definition television—many-lined clarity.

Intelsat International Telecommunications Satellite Organization created by the U.S., now owned and managed cooperatively by 115 member states. It carries most of the satellite traffic.

Intersputnik International System and Organization of Space Communications. The satellite system created by the Soviet Union and used by East-bloc countries.

IPDC International Program for the Development of Communications, Unesco.

IPI International Press Institute, London.

ISDN Integrated Systems of Digital Networks, substituting "systems"—national and international—for the present usage of ISDN, Integrated *Services* Digital Networks.

ITU International Telecommunications Union.

Laser Stands for light amplification by stimulated emission of radiation. Sends an extremely narrow beam of light.

Modem Lets computers talk to one another (a "modulator-demodulator").

NWICO New World Information and Communication Order.

OECD Organization for Economic Cooperation & Development.

OTA Office of Technology Assessment (Congress).

OTP Office of Telecommunications Policy (White House).

PBS Public Broadcasting Service (U.S.).

PTT Post, Telephone & Telegraph—the government agency in many countries which traditionally owns and operates communications systems.

QUBE A two-way, interactive cable system which enables the user to shop, bank, vote or receive other services by cable.

SIG Senior Interagency Group (U.S.).

Switching Center The link between computer or telephone lines.

Transponder A receiver and relayer of transmissions from satellites.

Teletext A one-way medium which delivers words and pictures using the TV system entering the home or office.

Unesco United Nations Educational, Scientific & Cultural Organization.

UPI United Press International (formerly United Press).

USIA U.S. Information Agency.

VANS Value Added Network Services—electronic mail, data interchange and consumer services.

VCR Video cassette recorder.

Videotex A two-way, interactive medium that connects the home TV or personal computer to a central data base, usually by telephone circuits.

WPFC World Press Freedom Committee, Washington.

History, Twice Over: Inaugural Photo Sent "Live," U.S.-Wide in 40 Seconds

This swearing-in of the forty-first president was historic for a publicly *un*recognized achievement: *40 seconds* after President George Bush completed his oath, this photograph was transmitted live and instantaneously received by 1,000 newspapers across the United States. It was sent by the Associated Press directly from the booth facing the president. This was the first photo of a public event sent instantaneously via telephone from a magnetic disc in an electronic camera. No film or processing was needed. Newspapers thousands of miles away received the photo, and published it in editions already entering the pressroom. "We are on the way to a digital world," said Hal Buell, AP's news photo assistant manager.

Digitalization and telephony are vital cores of the coming Age of ISDN. Significantly, too, the camera was manufactured in Japan.

Preface

How It Began

THIS BOOK HAS many origins. It may have begun during a sad conversation I had in Ghana with a victimized journalist; or speaking to a Czech writer caught in a bureaucratic web, but nostalgic over his days as a sportswriter when his country was free; or while detained briefly in Paraguay; or during an afternoon in Moscow with the editor, then fresh out of the gulag, who was harassed and imprisoned anew for taking glasnost seriously; or during conversations in many countries with men and women who labor in journalism, suffering quietly, often expectedly. Visiting some forty-five countries during the past decade also strengthened the vein of optimism running through this book. I was present when long-endured oppression was turning to hope in Spain, Portugal, Zimbabwe, El Salvador, China, Chile, South Korea and the Soviet Union. I shuddered at the burden borne by journalists in such countries. Their daily struggle to secure and share information makes them heroes, all too often martyrs. They inspire the rest of us to use freedom with deep respect for that right and the responsibility it entails.

I did not always do so. As a working journalist in the Caribbean at the outset of World War II, I reported a small ship running the U.S. naval blockade of Nazi-controlled warships in Martinique. I shouldn't have (as the naval censor was quick to advise). I learned about Third World journalism as cable editor rewriting "pro-Prensa" wire copy of United Press destined for Latin America. Editing raw radio reports for the Foreign Broadcast Intelligence Service of the FCC, I received a crash education in propaganda. And as press secretary to Puerto Rico's governor I saw the government's side of public information.

I came, however, to share the views of my boss, John Lear, who had been the government's information coordinator after writing for the Associated Press. Twenty-five years later he recalled, "I was awfully young in those days," and "believed that anything was possible for those who wished for it hard enough." He added, "But I have come to wonder whether it actually is possible to orchestrate information about the government of a democracy." It is possible, and to a limited degree, essential; but as latter-day spin-doctors prove, it can be misinformative.

By 1967, I believed that there was a third side to the story—not only the government's position, say, and the press's version. There should also be a regular critique of the performance of the news media. The Vietnam war had so divided Americans that President Johnson did not seek reelection. Just twenty-four hours after he withdrew he reminded the nation that for journalists or presidents, "where there's great power there must also be great responsibility." It appeared sometimes that the public knew less and less about more and more. The communications technologies could transmit a far greater volume of raw information, conveying more undigested fact than ever before. Too often, the focus of interpretative material was on the unknowable future rather than on a thorough examination of the past and present.

I enlisted Peter Braestrup to study the performance of the American print and broadcast media in covering the Vietnam war. That seven-year study produced the definitive two-volume work, *Big Story: How the American Press and Television Reported and Interpreted the Crisis of Tet 1968 in Vietnam and Washington*.[1]

On 10 November 1967 I invited eleven distinguished communicators to evaluate the role of the mass media in America. Participants included William Benton, who had been the principal proponent of the free flow of information when he served on U.S. delegations to the United Nations and Unesco organizing meetings, and was later a U.S. Senator; Jack Gould, broadcast editor of the *New York Times*; Harold D. Lasswell, a leading analyst of the sociology of communication and a member of the 1947 Hutchins Commission which proposed "a new and independent agency to appraise and report annually upon the performance of the press"; and Harry D. Gideonse, president of Freedom House, former president of Brooklyn College and a colleague of Senator Benton on the nationally broadcast radio program "University Round Table of the Air."

We were concerned about the growing number of countries in which press freedom was diminishing, and also about America, where we feared sensational reporting was undermining the credibility of the press. Freedom House asked: "...must free institutions be overthrown because of the very freedoms they sustain?" We were wary of the rapidly rising importance of the television camera as the principal "analyzer" of the world to its citizens. We fretted over the great influence of a handful of newspapers in determining not only what their readers, but television viewers as well, would consider today's reality. Everything left unreported would, in effect, not have happened.

We worried, too, about the overreaction of some journalists to criticism of their difficult work. Many see hidden agendas in even the valid analyses of their products. They too often regard criticism by a governmental official, whether justified or a matter of personal pique, as a step toward outright censorship. It reminds me of my undergraduate days. I learned from philosopher Sidney Hook to reject the logical fallacy of the slippery slope. (It holds that once a step is taken, no matter how small, it leads eventually to the worst possible conclusion, at the bottom of the slide.) That fallacy, as I mention in Chapter XI, influences how U.S. journalists regard their relationship to the government. An American editor said recently that when you look at all the media that surround us, you realize why people feel they're living in a "press state." He added, journalists "may recognize that as the diametrical opposite of a police state, but," he said, "a lot of people think they're similar."

We concluded—twenty years ago—that the unease with American journalism was a threat to the democratic system. We thought representative audiences should be invited regularly to see television newscasts to determine how audiences react, what they understand, and what they feel they have missed. Funding could not then be secured for such "consumer" research. It was never undertaken. The idea is still on the table.

During the past sixteen years, however, Freedom House has regularly assessed the level of political and civil rights in every country in the world. Many of the criteria affecting civil liberties include the performance of the mass news media in each nation. For the past eight years, I have particularly examined the status of the print and broadcast media in every country and have published the results in the yearbook *Freedom in the World: Political Rights and Civil Liberties*.[2] The survey we devised for this book (see pp. 207-340) elaborates on those studies. Questionnaires containing some 200 points were completed by correspondents in seventy-four countries. Many responded at some risk to their persons and their employment. I have therefore omitted the names of all.

Names *not* to be omitted:

Jessie Miller, for twenty-two years my associate at Freedom House. For this study, she helped construct the survey, select correspondents, and maintain a large research archive. She placed the entire manuscript, several times, into a word processor. More than that, Jessie provided the sly questions hinting that an idea was not what it seemed to the author. Her contribution, especially to this work, has been considerable, not the least because of her great humor. After handling so much futurist communication technology, she tells this story:

A man at an airport sees another, carrying two very large and heavy bags. He puts them down every few steps to talk to his wrist. The observer stops him to inquire, and is told he is speaking into a wrist-telephone.

"Where can I get one?," he asks.

"In Switzerland, for $300," is the reply.

"Here, take $400 for it."

The new owner happily rushes off, amazed at how much telephony is crammed into a small space.

"Wait," the seller shouts, laboriously handing over the two bags, "you forgot the batteries!"

That's Jessie.

James Finn, editorial director of Freedom House, has created a growing book and magazine production program of which this volume is a part. His editorial judgment has markedly improved this book. Mark Wolkenfeld also labored mightily through this text. My gratitude to both.

Marianne, my wife, remained understandingly patient for many months, even as I saw power in a gentle stream, and a slippery slope in the cornfields of Vermont.

Mark J. Sussman, my son, brilliantly shared the interviewing and research chores in Moscow in 1987. David W. Sussman, my son, coauthored with me a study of the legal aspects of the Unesco Mass Media Declaration referred to in the text. Lynne Sussman-Hyde, my daughter, took time from her professing to provide helpful suggestions.

Rebecca, Benjamin and Jane, in their first, second and third years, citizens of the twenty-first century, will, I believe, see the Age of ISDN, and wonder how their grandparents could have survived in a world so unfree, yet with such promise.

—July, 1989

Introduction

Technology of Freedom

[O]nly as men are brought into counsel, and state their own needs and interests, can the general interests of a great people be compounded into a policy suitable to all. So, at this opening of a new age, in this its day of unrest and discontent, it is our part to clear the air, to bring about common counsel; to set up the parliament of the people.
 —Woodrow Wilson in a presidential campaign address, 1912.

THE PRESS IS a metaphor for *all* forms of delivering information and transmitting ideas to and from the people. In the 1990s and the century beyond, the technologies of communication will serve to enlarge human freedom everywhere, to create inevitably a counsel of the people, if not the voting parliament Wilson foresaw. Directly ahead is the creation of the Integrated Systems of Digital Networks (ISDN). ISDN will provide diverse kinds of information and varied forms of delivery such as we can now begin to imagine, and, imagining, create. Nearly every man and woman on earth will be able to communicate in a few moments with someone continents away. Everyone will have immediate access, at home or at the workplace, or through a nearby communal telephone, to a vast volume of diverse information—a volume such as even the world's finest libraries or news services cannot provide today. The cultures of even the smallest, least familiar peoples will be preserved, and made accessible to everyone, everywhere. New communications technologies will induce the human mind to think more clearly, to test new possibilities, to gain confidence and even exhilaration from the process of idea-discovery. Perhaps most responsive to President Wilson's dream, people now silent will have their voices heard. Whether this serves to create a parliament of the people will remain a political determination. But so pervasive will communications become, accessible to all, that the technology will tend to overcome political restrictions.

Since the first written word appeared, however, freedom has often succumbed to state power. That power has strictly controlled communications. In the new era, there will be voice and picture linkage to virtually every living person—and interaction between sender and receiver. Will the new systems avoid censorious state controls? At the outset, that will depend on whether the new communications linkages are put in place by believers in democratic communications. Political democrats can assure the free flow of information. The democratic communicators are not likely to be primarily ideologues on the side of political freedom. They may well be inventors and entrepreneurs who profit from the opportunities of a relatively free market, and the political system that makes possible individual risk-taking and small-scale inventiveness—two major aspects of technological development in a free society. Major communications developments these days come from small, creative risk-takers, not large sluggish corporations.

Some corporate fears may stimulate greater American inventiveness and productivity. The current assumption that U.S. electronics are inescapably second to Japanese technology, particularly in the crucial field of high-definition television (HDTV), may be overcome by U.S. creators spurting ahead, leapfrogging to produce HDTV via the Integrated Systems of Digital Networks (ISDN). HDTV is the key to many industrial advances beyond the field of television entertainment. HDTV matters. The digital networks of ISDN may replace the vaunted analog or voltage systems of HDTV, in which the Japanese presently lead in research and development. By developing "smarter" computer chips based on binary digits rather than voltage changes, and carried entirely by fiber optic cable rather than less certain broadcast transmissions, U.S. technology may bypass even the still-to-be-perfected Japanese analog (voltage) HDTV systems. The integrated systems of ISDN are carriers of data, voice and pictures over networks linked to other networks.

Technologists, too, will influence the freeness of the flow. The new technologies will have their own momentum, even in less-than-democratic societies. Leaders in the Soviet Union and the People's Republic of China now recognize that their social, economic and educational systems must be revolutionized, not just reformed, if their peoples are to avoid bankruptcy and their political systems not succumb to crucial challenges. The core of the problem and the key to its solution is the need to computerize information, make it accessible to the broad public, and put hundreds of millions to work in the post-industrial information era.

The developing countries have also come to this realization. They demanded an amorphous "new world information and communication order." That demand was based on two premises:

(1) a few Western countries today determine the content of the world's information flow, and (2) with vast, new communications linkages Third World news and views will be even more completely overwhelmed. But recognition is also slowly gaining in the Third World that past panaceas

have failed to develop the economies of the new nations, let alone feed, clothe or house their peoples. Consequently, these nations are perilously unstable, and their political futures uncertain.

In scores of Third World countries after World War II, decolonialization and self-determination did not automatically produce the promised viable governments and humane political systems; quite the opposite. Early Western theorists predicted that economic development would lead to freer, democratic governance and more stable societies. After four decades, that has not happened in most of these countries, though they received billions in loans and grants. Nor, as was forecast by others, has utopian ideology generated either substantial development or political freedom. For diverse reasons, Third World authoritarians harshly restricted or censored the flow of information to, from and within their countries. But this, like other panaceas, has not advanced social or economic development, or enhanced citizens' political rights.

The communications revolution is not a panacea. It is becoming a reality with fresh potentials anticipated every day. In the waning years of this millennium, that new source of idea-power—the new technology of communications—is beginning to alter every form of human relationship and activity. This technology, like older forms, has the potential to enhance or constrain human freedom. But there is a major difference: the new communications tools are technologies of scale and diversity. Their very variety, number and speed defy the controller who would ban or censor from a central point. Censors in the past relied on central control of the medium (the printing house or broadcast studio) to inhibit dissent or diversity. Control was secured from the center, from which spokes of information radiated to the periphery. The new communications technologies need no "center." The new instruments are linked at the periphery as well as the center, often bypassing the center. Networks of information join other networks at a speed beyond the capacity of the human mind to monitor before a message has been delivered. And then it is unlikely the center will know that the message has been conveyed.

The new technologies are the conduit for generating vast information-power. Almost simultaneously world-wide, they convey, store or retrieve current speech, text, data or pictures, and information from all of human history. They also

- facilitate problem-solving in all human disciplines
- convey news and information needed to determine governmental policies as well as citizen judgments
- broaden the scope of scientific investigation, and the development of still newer technologies
- enable questioners to confirm data, verify policies, and cross-check information instantaneously

- widen the horizon of individuals through far greater cultural and educational opportunities
- enable men and (finally in many societies) women to examine shared problems from diverse cultural, historical and political traditions
- encourage the user to develop greater electronic literacy and the power of logical thinking

The new communications technologies are still only imperfectly understood even in highly industrialized countries. Video games that attracted American youth in the early 1970s helped prepare today's university students for elaborate gambits in computer logic. Today, the Japanese video games called Nintendo are present in many homes of eighteen-year-old males in the U.S. Some spend four hours a day at these games. In Tokyo, young men wait in street lines for twenty hours to buy the latest video game.

By mastering intricacies of the computer we train our biologic brains to think. And perhaps one day we will program the computer to develop artificial intelligence. We can only do so after we understand ourselves better. Meanwhile, that same machine presses us to ponder alternatives for human survival on an increasingly problem-strewn planet.

The common instruments of "information power" are the telephone and telefax, the computer and its value-added terminal, the television and VCR, domestic and cross-border radio, the recorder and the copier, and the fiber optic cable and satellite dish. Combining these instruments produces seeming miracles. A high-speed computer in London linked by satellites and telephone to a terminal enables Manila to call up a manuscript in the Library of Congress. All the books in that great library can be stored in a computer no larger than a home refrigerator. Every eight years, it is now estimated, computer science doubles the entire volume of information available to all humans everywhere.

There are restraints to linear progress. National security will provide some valid, but mostly invalid, reasons for limiting access to information. Proprietary interests of writers and publishers will create copyright controls. Professionals in science and other fields will similarly want to patent their developments. A new set of legal protections will be needed to guard the privacy of individuals as well as the security of their property. And Third World leaders must be persuaded that their fragile cultures will not be overwhelmed by the pervasive new communications flow.

Some fears are well founded. In the Age of ISDN (Integrated Systems of Digital Networks)—the network of networks—computer literacy will be even more important for the individual's lifetime development than ordinary reading capability in the age of books and newspapers. Indeed, the distribution of computer terminals, in itself, can be a democratizing force in every country. But the withholding of terminals, as many authoritarians today deny their citizens access to telephones, will inhibit such democratic potential. In many places, the cost of installing terminals will result in uneven access to the new world of information.

This revolutionary information era is not the "new world information and communication order" (NWICO) imprecisely but vociferously demanded at Unesco meetings for more than a decade. The new era is not ordered. If anything, it will be disorderly, as befits a democratic mode. It will provide "small" communications technologies so that even the most distant, the poorest, the least heard peoples and cultures can be brought readily on line and received anywhere in the world. All will be able to interact. Small domestic networks will link rural areas to urban centers. These will connect with the centers and peripheries of neighboring states; and these, in turn, with the most distant places. All networks will have access to a vast array of choices provided by all other networks in the linkage. Perhaps thirty or forty years hence, the world-wide network of networks will be in place. The early phase of the Age of ISDN is already operative.

Some fear it as too mechanistic. They confuse the information revolution with automation or robotics. But humans need not be enslaved by information machines. Other humans are the traditional slave-masters. The censors described in this book are redblooded humans. They short-circuit today's information technology in order to restrict the power of the nongovernmental news and information media. Communications technology transmits human ideas. Electrical impulses are idea-neutral.

Others, particularly in the Third World, envision ISDN as the ultimate denial of information diversity by its ability to tie together vast national and international networks. This reflects the fear that a few Western entrepreneurs will dominate the electronic networks. Yet the very nature of the diverse linkage should enable Third World voices to be heard, as they are not now.

Perhaps the most compelling fear is expressed by those who foresee an Orwellian conclusion: a political tragedy on a vast scale. An elite in one country or several would come to dominate not only the mass media, but the masses as well. Ironically, however, the Marxist-Leninists who have crudely dominated their societies for up to seventy years now fear the introduction of the new communications technologies into their countries. This fear is based partly on the linkage to free-enterprise systems overseas, and the attraction this may have for citizens of "socialist" countries. But mainly the Marxist fear is based on the democratizing nature of the new communications technologies, and the implication of this for sharing future control in their own society. There is validity to this Marxist fear. Democratic systems are based on real choices for citizens. The new technologies provide a diversity of communications systems, and of varied news, data and information within those systems. Such variety is bound to alter the political and economic landscape, as well as the policies and persona of the political leadership.

The technology also provides for heightened interaction between the user and the sender—a democratizing force to be feared by the authoritarian. On a more profound level, the magnification of sources and volume of information must inevitably raise new questions about the nature of the society and the role of information in all aspects of life, including governance.

In all societies, the new communications tools are ending the false separation between technology and other aspects of human intercourse: culture, general education, domestic politics and geopolitics. Communications technology influences the imagination, and therefore all forms of human endeavor. It is a pervasive social process. There are no guarantees that technology per se will democratize the society. Some technologies support and some undermine democracy. The test is whether communications technologies provide diverse contents, can readily meet changing needs of the users, and are inexpensive, permitting wide distribution. Louis Mumford warned, "those who are concerned with maintaining democratic institutions" must ensure that their efforts "include technology itself."[1] This is in the early American tradition. Benjamin Franklin spoke of the interdependence of philosophy, democracy and technology. That concept was never more appropriate than in assessing the broad social implications of the new communications technologies for all societies. While the communications revolution was still in its infancy, there were both intellectual scoffers and supporters. Several have exchanged positions from being unfriendly to friendly to technology, and vice versa.

Since the printing of the Gutenberg Bible, political elites have feared each new information system. For ages, books were kept out of the hands of peasants lest they get revolutionary ideas. When they were permitted to read, their texts were carefully censored. Mass printing has never produced a revolution, nor did it create Nazism, fascism or communism. Guns and political cruelty produced all three, aided afterward by the propaganda channels of the totalitarians. It is far more likely that the explosive diversity of the new communications technologies will prepare the way for democratic governance. There are likely to be systems within systems. Radio has become far more diversified since television entered most American homes. Today, 188 million Americans listen to 505 million radio sets and more than 10,000 radio broadcasters. Cable TV provides news, information and entertainment not generally found on the large television networks or affiliates. VCRs offer classical film, documentaries and other subjects not regularly provided on cable or the networks. Computer terminals carry a myriad of news and information: statistical data, encyclopedic references, journalistic archives, current and past financial information, and much more. Future terminals will carry printouts and videotext of Third World news, information and cultural offerings now available only in distant lands. All of this provides information-on-demand. The individual user becomes the center, not a speck on the periphery. The mind of the singular man or woman can roam distant horizons, not only of geography but of meanings and potentials. Most important, he can interact. He can express himself, and respond to others. He can even move back in time and space, and find new perspectives for himself and his future. He can become active, rather than simply reactive (far different from the couch potato tuning only to tonight's entertainment fare). There will, indeed, still be the Rambos for those who seek them. But there will be much, much more. More than we can even envision as the Age of ISDN opens.

We may even see the end of ideology, as Daniel Bell predicted somewhat prematurely in 1960. Marxism began foundering in the Soviet Union decades before Gorbachev's appearance, but glasnost provided the coup de grace. Ideologues of whatever stripe are subverted by idea diversity and interaction among the citizenry, and the structure of quest and analysis which the new communications technologies establish. This alone will not produce a democratic society, but it will provide the structure for political forces to create one.

Nor will these technologies in themselves guarantee that a nation already democratic will support the democratization of the world's cultures. The availability of limitless film, news and information from the Third World cannot assure the calling up of this rich assortment on the terminals of the industrialized nations. With easy access to this world of diversity, however, new interests can readily develop in the Age of ISDN. No utopian era is promised. At least, however, wherever in the world a telephone line runs, or shortwave radio is accessible—almost everywhere by 2050—global citizens can acquire some greater sense of interdependence. But first political will must permit the linkages.

* * *

A DEMOCRATIC SOCIETY is not a slumbering, undisturbed aggregation. It is vibrant, testing, even tense. The tension comes from the testing of the possibilities, even the limits of law and custom. Out of that, perhaps, comes new law, new custom. Certainly in an age of vast change, such as communications technologies provide, there must be regular testing. The miniaturization of voice and data carriers opens new vistas for transmitting knowledge—and for unwanted surveillance. The networking of information makes possible both highly centralized and infinitely decentralized switching points—control points, if one is authoritarian, free-flow points, if libertarian. The technologies will not solve political problems. They will create new ones. But for the first time in history the great interlinks will make possible the empowering of entire publics through access to diverse information systems. It will take political will to allow such democratization to happen.

If and when it does occur, the censors will be overwhelmed, and finally made superfluous. In the great array of information made available at the touch of a keyboard or a telephone, the primitive concept of press controls and other information blockages will be seen for what they already are: anachronistic. They have an ancient, if not honorable, history. Censorship began in Germany in 1529. A half-century later in England the infamous Star Chamber restricted printing rights to certain presses. They did this in order to prevent "greate enormities and abuses" practiced by "dyvers contentyous and disorderlye persons professinge the arte or mystere of Printinge or sellinge of bookes."[2] We may be seeing the last massive gasp of the public censors. When thirty-eight journalists are killed in one year, as in 1988, hundreds more arrested, and still larger numbers harassed, as they were; when political systems in three-quarters of the nations of the world are still afraid of what their citizens may read, see, or hear—as they are—the race toward the democratization of communication is far from over, as Chapter VII attests.

But it is quite possible that this is, indeed, the last great surge. And with ISDN perhaps forty years away, we can be optimistic, even as we are realistic about today's plague of censorious nations. It is the purpose of this book to make that contrast: to describe the present press- and idea-control states, and make clear that no nation, not even this most democratic, with the freest press in history, is free of the tensions which information power stimulates when it challenges state power.

In a democratic society, there are irresponsible journalists and corrupt officials. Both rend the democratic fabric. Inevitably, some will seek legislative or administrative remedies. Often these will not be necessary. Public rejection of journalistic sensationalism, and more vigilant monitoring of officials will be sufficient to avoid further damage—until other aberrations occur. And they will, in a free society. They can be minimized. Yet the freedom of the press cannot be absolute. No political or civil right can be absolutely granted in a democratic society. There are slippery slopes at every important crossroad. In the clash of rights —free press v. fair trial, for example—a winner one day in court may be a loser next time. The law and the judge are the mediators. Journalists win some, when they invoke shield laws to protect themselves from revealing sources of information. But sometimes it is more important to protect a defendant than support a journalist's credibility with his source, important as it is to support the public's right to know next time.

Journalists losing such decisions often claim a "chilling" attack on their First Amendment rights. They feel they are sliding down the slippery slope. But many fail-safe guards protect the press before it is truly hampered. Indeed, during a normal clash of rights it can be damaging to cry "Emergency."

Even during the Reagan years, when the administration sought to stop official leaks and reduced access to government documents under the Freedom of Information Act, no serious reduction in press freedom occurred (see Chapter IX). There was anti-conservative bias and anti-administration commentary, along with approval of the "Great Communicator" by those in the media who favored the president. The administration was also attacked for its information restriction policies. Some administrative restrictions were posed and not put in place. Others were only minimally applied. The tension between independent information power-wielders and the state minimized the controls.

* * *

A COUNTRY IN transition, moving toward democracy and a free press, is South Korea. In Seoul in May 1988 I interviewed journalists and the minister of information. That was just hours after elections which made three opposition parties the majority in the South Korean legislature. This historic vote followed the 1987 presidential election which signaled a sea change, a democratization, in South Korean affairs. For the first time in many years, journalists felt free of domination from the Blue House. The information minister confirmed his role in the democratization of information. The American ambassador told me he was encouraged.

Radical students, however, were stepping up ritualistic street demonstrations, though their earlier demands had been largely realized with the elections. A viable political opposition was in place, and part of governance.

Yet street power was heady, particularly when it could easily be projected onto television screens the world over. During the 1988 Olympics, there were regular, thoroughly staged happenings. A CBS reporter called a toll-free line in Seoul and asked the student headquarters when their next demo would occur.[3] "Five o'clock." At the appointed hour, the students arrived holding Molotov cocktails made from small amounts of kerosene in a bottle. They were warned by helmeted, shield-bearing police not to cross the white line in the middle of the street. The police, for their part, would not cross it either. But if the students did, the police would chase them. Automobile traffic went through while rocks and cocktails were thrown overhead. Sometimes the demo would stop to allow the traffic to pass, and the students would take time off for ice cream. The understanding was that nobody would get hurt. One block away, in crowded down-

Power, the Press and the Technology of Freedom

town Seoul, all was quiet, business as usual. Meanwhile, the students shouted carefully designed slogans which only the cognoscenti would understand, even in Korean. "This," said a *Newsweek* photographer who had been watching the demos for two years, "is like a Cecil B. De Mille staging—there are flames, and action, but nobody gets hurt." One could see that some of the players, police as well as students, had developed ballet-like forms. They move in swiftly, engage the "enemy" and pull back, unscathed. Satiated. Recorded on film for worldwide attention.

In Seoul those sights and sounds—Molotov cocktails bursting into orange flames on populated streets or youths shouting slogans, chased by helmeted police with protective shields—became a force unto itself: information power challenging state power. The Korean demos, by deft use of information media and street theater, symbolize the clash between the power-seekers and power-sharers. With obvious Korean cultural overtones, these demos were reminiscent of the "happenings" in Berkeley, Chicago and a hundred streets of America in the 1960s. The scenario, however, is still watched carefully by today's rulers in Kuala Lumpur, Moscow, Pretoria, Jerusalem and wherever independent information power is cause for alarm or suppression.

How frightening is information power? And to whom? How useful? And to whom? And how does the interaction play out, for all the actors?

That is the subject of this book. We provide some documentation of that interaction in half the countries of the world. We detail, as never before, the scores of ways the flow of information is influential, in widely varying styles and degrees, in all countries. We also test some answers to the questions posed above. Mainly, we show why there are no easy answers.

We pose the harsh challenges to American communications systems —journalists, manufacturers of electronic hard- and software, and the U.S. information carriers—all of whom face immediate competition from Japan and Western Europe. The free enterprise system, exemplified best in the United States, is not working well here at the moment. Indeed, America's communications technology may be over the hill before the end of the '90s—the "hill" being the learning curve. America's major communications entrepreneurs have missed a generation of experience by their slow response to the potentials of crucial new technologies. They have sacrificed long-term profits, market share and, most important, revolutionary development because they are oriented to short-term considerations [see Chapter X].

But this is an optimistic book.

The world is witnessing changes—the information revolution—recorded with similar impact only twice since man first walked the earth: the agricultural and industrial revolutions. No one can predict where the new changes will lead. That they will alter every aspect of human relationship is certain.

Domination of the information channels by television dramatically assisted the civil rights movement in the 1960s, altered the presidential political process in the '70s, popularized hospices, malpractice suits, feminism and the plight of the homeless in the '80s. Significant social change is already attributable to the pervasiveness of the electronic media. And we are just at the beginning of such social change. Indeed, the new media may transform the home and office by the information products they make accessible. A new social environment may be developed. The individual personality, too, may be altered by the seemingly limitless choices the new media provide, and the process of selection and testing that this implies. The impact will vary from country to country, and from class to class, different for the highly educated and the illiterate, but a significant change for both. As each has access to a similar keyboard, however, the dividing lines between them will blur and eventually diminish. It is highly unlikely that, once installed, the new information systems can ever again become segregated.

The information revolution is already producing the key technologies for the Age of ISDN, some forty years off. That has the capability of providing, for the first time in human history, the democratization of information. IDSN represents not only the interlocking deployment of all manner of communications services, but the empowering of the individual through access to communications: a giant step toward the democratization of politics and governance, toward human freedom.

* * *

HOW SHALL WE take that step—how do we get from here to there? There will be as many ways as there are countries to be linked. But there may be some common lines of approach in the development of ISDN.

We assume that political/ideological considerations will influence the early stage of ISDN. The greatest initial advances will be made by societies which already assure their citizens a high level of choice among policies and leadership, and which couple such political choices with the freedom to make economic decisions (whether called free enterprise or some other name). We can already see the groundwork being laid today for such early advances in ISDN in the United States, Japan and Western Europe.

It is likely that the first "universal" installation of terminals tied to telephone lines will come in France, Japan, West Germany and the United States, possibly in that order. There may be more terminals in use in the U.S. than in the other three countries combined within five years.

But the American "Third World"—the several million below-poverty-line families—may remain unlinked to ISDN for some time.

In this regard, the poorest Americans may share the ISDN time-lag with hundreds of millions of citizens in developing countries. During that time lag, it is likely that the information flows will provide an enormous volume of news, data and other ideas in words, sound and pictures. These flows, however, will be fashioned principally for those presently using the services. While every marketing effort will be made to secure maximum penetration of all potential user groups, there will probably be less linkage by citizens with low incomes. Ultimately, the steadily reducing cost of ISDN will make it possible, even economically profitable, not to say morally imperative, for everyone in a politically free society to have access to a terminal. It is likely that terminals and telephone linkage will be provided free of charge to low-income families. This could be rationalized in anticipation of user expenditures for value-added services. This arrangement has already proved successful in France for a large urban population.

When universal linkage in a politically free country occurs it is likely the information flow will be expanded to include data and services not previously believed of general interest. This may include services of special use to the urban and rural poor—the updating of services, for example, that have been widely disseminated since the U.S. government dispatched county agents to teach farming and health practices in the nineteenth century. Some of these programs, for urban and rural dwellers, and based on the most advanced late-twentieth-century practices, will be immediately useful in many Third World countries. At that stage, perhaps some years after ISDN is already universally available in the politically free industrialized states, the leaders of developing countries will begin to drop their resistance to tying into ISDN.

Third World resistance will be based on the assumption that the flows of information will be principally oriented to Western interests, and to the style of governance in Western democratic societies. Indeed, since some of the first "universal" linkages will appear in North America, Western Europe and Japan, this assumption has some validity. But it would be self-defeating for Third World leaders to hold their citizens aloof from the earliest possible ISDN linkage. For it will be a two-way channel. The sooner developing countries permit their people to participate, the earlier will the cultures, the news, the development process of the Third World become known and accessible to more than five billion inhabitants of the planet. It is likely, therefore, that the more politically democratic developing countries will take the lead—Costa Rica and Barbados, for example—having as models recently developed South Korea and Taiwan. As all such countries join ISDN, the worldwide information flows will change. As current data, cultural elements, news and histories of these countries are added to the flows, Third World interests and accessions of information will broaden the data bases of all users, everywhere.

So far we have assumed that the democratic and democratically-inclined states—developed and developing—will be among the earliest large users of ISDN. Whatever the time period needed for this development, it is likely that the less-free states will be experimenting with growing but still largely restrictive use of ISDN, domestically and internationally. Indeed, it seems probable that even highly restrictive societies, using the Soviet Union as an example, will begin to link their specialized citizens to international news and data flows. Control will be maintained initially by permitting only a single "gateway" for all computerized data entering the USSR. Most of the flow comes from the United States through the San Francisco-Moscow teleport; a second link goes through Vienna.

For many years, a small elite in the Kremlin has had access to some highly secret daily Tass news bulletins specially prepared for them and no other citizens. Tass produced for this elite a relatively objective report of the world. That circle has probably been expanded under Gorbachev; more important, with the policy of glasnost, a somewhat more accurate picture of the world is available to the average Soviet citizen. The circle of those receiving objective data from abroad in the sciences and technology has been expanded. At some point, Soviet citizens through their institutions will be permitted to interact, on line, in real time, with men and women in other countries. It is only several years, after all, since automatic telephone switching for international calls was restored after the Kremlin shut it down. It may take some years before telephones are permitted tens of millions of Soviet citizens now denied them. Though microcomputers will enter homes, printers may only be available at government offices. Restrictive states, however, have disadvantages: financial and educational systems that do not encourage typing, programming and the keeping of accurate records, an inefficient telephone system, and little relationship between buyers and sellers which stymies innovation. Yet technology may drive policy in the USSR as elsewhere.

As great, new advances are made in science and technology through computerization and ISDN linkages, the Kremlin will probably find it necessary to acknowledge the democratizing factor of ISDN. The growth in the volume of phone calls already makes surveillance difficult. These linkages will, indeed, permit Soviet citizens in any part of the country to interact in real time with fellow citizens across the Asian land mass. Video cassette recorders are already adding to the horizontal communication provided in the USSR by ham radio, tape recording and photography. The next stages are telefax and electronic mail—all difficult to monitor. What will this mean for restive nationality groups in the Baltic

States, Armenia, and the Tatars displaced a thousand miles from their ancestral lands?

The threat to restrictive information systems will not come mainly from dissident messages or anti-state electronic samizdat, but from the erosion of government controls over great volumes of information. After balancing the risk of decentralized power against the advantages of ISDN, one of the Soviet Union's leading communications specialists unequivocally opts for computerization: Academician Andrei Ershov describes his country's urgent need for "informatization." He calls the use of the new communications technologies "an inescapable phase in the development of human civilization." This, he says, requires "the democratization of society's informational structure" as the "precondition of its vitality and its capacity for development." Ershov sees the United States completing its informatization by 2020, but he warns that the Soviet Union is presently lagging far behind. The USSR, he adds, "must rehabilitate the concept of individual initiative."

Marxist leaders in Eastern Europe will be watching such developments as clues to their own internal problems. And authoritarian rulers in Africa and elsewhere will face similar trade-offs. Should they join ISDN and secure the information needed to advance their countries' development? Should they permit only limited access, by having a small elite go online? The elite might monitor data which could create political problems if trickled down to the entire population. But would not undue restriction, in turn, hamper national development? For even restrictive domestic communication policies may not long keep knowledge of the potentials of ISDN from even the most remote countries, or from closed sectors within nations.

Communications systems such as shortwave radio will be increasingly difficult to control from the authoritarian center. The USSR and Eastern Europe have discovered that. They ended shortwave jamming of incoming broadcasts in 1988. Such attempted closure had reduced the credibility of the nation's leaders and its political system. Slowly, authoritarian elites will be pressed to install the domestic communications systems which provide vital developmental information, and perhaps only incidentally stimulate democratic instincts.

Numerous political and economic variables make it impossible to forecast when ISDN may become truly universal. One may consider the goal of the International Telecommunications Union—the first and most experienced intergovernmental agency—long predating the League of Nations. The ITU hopes to persuade governments to make a telephone accessible to every person on earth early in the twenty-first century. One may assume that the information linkage of ISDN would accompany that timetable. Decades before, however, there will become apparent the massive effects of ISDN in the democratic industrialized countries, the democratic Third World, and in other politically flexible states. This remarkable thrust forward into a new epoch of peaceful development should soon persuade the still-authoritarian states to recognize that they are failing their citizens, and their nations' development, by providing less than full linkage to the Integrated Services of Digital Networks. Sooner or later, even the authoritarians will have to take their chances in societies transformed by the democratization of information. No aspect of humankind can any longer escape the influence of the new communications technologies. They will soon alter all the natural and social sciences, all levels of education, all forms of cultural activity, all geopolitics.

Everywhere.

1.
The Word is Power

Knowledge and human power are synonymous.
   —Francis Bacon, *Novum Organum*, Aphorism i, 1620

To an earlier age knowledge was power, merely that and nothing more: to us it is life and the summa bonum.
   —Charles S. Pierce, Annual Report, Smithsonian Institution, 30 June 1900.

In the last century strong governments neither loved nor respected the press. They scorned it and felt it was better done away with...But now, a totalitarian government does not suppress the printing press but reduces it to a single voice...Totalitarians want to force thoughts, consciences, words and type into one channel.
   —Alberto Gainza Paz, "Freedom of the Press" speech at Northwestern University, 1 October 1951.[1]

Government is by nature allergic to criticism. It would be ludicrous not to expect it to interfere with editorial policies. There is an inherent conflict between the media and government.
   —Andrew Sesinyi, Director of News, Radio Botswana, September 1987.

Most reporters are simply intrigued by the human activity known as power; they like to observe it and describe its...effects.
   —James Deakin, *Straight Stuff*, 1984.

Even as today's journalists assert their right to greater interpretive freedom, they find it difficult to admit to the partiality all interpretation entails...The danger lies in identifying a partial viewpoint with the common good and the whole truth.
   —S. Robert Lichter, Stanley Rothman, Linda S. Lichter, *The Media Elite*, 1986.

THE WORD IS power—from the beginning of time till now—power influencing the interplay between *nations,* power driving the social and political dynamics *within nations*, and power to fulfill the human potential of the individual *citizen*. On all three levels, the new technologies of communications will vitally affect the fate of international affairs, individual nations, and all citizens of the world.

The Age of ISDN will provide the maximum use of "intellectual technology" created by and for the "postindustrial society." Daniel Bell forecast this more than a decade ago.[2] He saw preindustrial and industrial societies in most of the world. The former have economies such as farming or mining based mainly on processing natural resources. The latter mainly fabricate products using natural resources. The postindustrial society—such as the U.S. is becoming—is organized around knowledge. For such a society, intellectual technology is essential. It examines and codifies the nature of life, things and social interactions. It gathers, stores, retrieves and manipulates information and theoretical knowledge—the keystone of still broader scientific, economic, technological and social development.

In the broadest sense the new communication networks will inspire —require—the most definitive socio-political decisions. In making knowledge (if not yet wisdom) easily accessible to nearly everyone, long-accepted assumptions and traditions will be challenged. Indeed, the open-endedness of information sources and the ease of posing questions will inevitably undermine elitism and enhance democratic forms in social as well as political relationships. The age-old conflict facing philosophers and politicians will arise with greater force: the challenge which freedom poses to equality. The new communication systems will epitomize the freedom to choose: first, kinds of information, then, options to solve problems, large and small. But the utopian goal of human equality will be no closer to achievement, even as institutions are made more accessible, ideas more diverse, and human and material potentials more exhilarating.

A meaningful flow of information alone will not increase knowledge or wisdom, but it will alter the forms of language, and ways of thinking and reasoning. The computer will reorganize bodies of knowledge in accordance with its own programs. New education policies will be as necessary as new industrial and communications policies.

Communications are the nervous systems of human agencies— international, national and individual. The new linkages are far more necessary to those agencies than electricity, for ISDN will carry information, not only electrical impulses. These "nerves" will be the facilitators of action, the catalysts of social change.

That change will alter all power structures everywhere. It will require adaptation to new interpersonal relationships, and therefore new industrial, social and political policies. For the United States, which has *no* industrial policy, *no* communications policy (except the First Amendment), there will be an urgent need to devise a deliberate policy for social change.

For example, the communication-poor in the U.S. will be as disabled as today's employees paid poverty-level salaries. Muddling through will no longer serve, nor will reliance solely on market forces. Where matters of international sovereignty are deeply involved, as in international communications in the Age of ISDN, corporate power by itself will no longer be able to dominate foreign or domestic policy-setting without governmental participation. There must at least be a social framework in which the new communications impact on society and the world is systematically and not haphazardly addressed.

Below the level of domestic macroplanning (involving government and public representation) should be the regulators, who insure that the agreed-upon constraints are followed, and the market forces, which will primarily reflect the objectives and choices of individual citizens and their agencies. Bad planning or no planning can result in the dehumanizing that many fear in the prominence of computers. If we are to secure the great gains which ISDN can provide, we shall have to plan not only for the liberating values but for avoidance of the tyrannies as well.

No less than a major social revolution is involved. Governments other than the U.S. already regard "information sovereignty" as their vital domain. Other countries see the great concentration of information power in the U.S. as a threat to *their* sovereignties. The European Economic Community (EEC) in 1992 will drop tariff and other barriers among its twelve member nations. This, in effect, will form a cartel of America's Western European competitors. The Japanese, in association with Asian producers, form another major communications giant. All countries, however, must eventually negotiate the linkages of competitive systems to form a truly universal network of networks (ISDN). For that, America will need far more concentration on information policy than it has ever displayed.

This will require econometric studies of the implications of ISDN systems: the effects of computerization on unemployment and on closing down of major industries as information "plants" replace traditional production; the retention in seemingly uneconomic terms of national-security industries; the rapid loss to the trade union movement of previously organized workers; and the undermining of the strike weapon as "professionals" in widely separated workplaces replace factory hands and office workers.

The traditional games of power-playing will be thoroughly altered from the interpersonal to the international levels. Power will be at stake at every turn. Ten years ago the French government began preparing for leadership in European communications, and challenges to U.S. dominance. The French regarded the International Business Machines Corporation (IBM) as the *most* important competitive entity on the world scene. At that time, IBM had 80 percent of the computer manufacturing business and was going beyond data processing and planning to enter telecommunications—the use of satellites as well as landlines to transmit information. "No firm, and no government either, has so mastered the chain extending from component to satellite," said the French government study of IBM.[3] "As a controller of networks," the French report added, IBM "would take on a dimension extending beyond the strictly industrial sphere: it would participate, whether it wanted to or not, in the government of the planet. In effect, it has everything it needs to become one of the great world regulator systems."[4] Sooner or later, said the French, IBM must open a dialogue with national governments.

There have since been such dialogues in international meetings. Now the time has come for the U.S. to develop a long overdue communications policy by joining other governments as they enter into serious dialogue not only with IBM, but with all major manufacturers and users of the new communications technologies. That need is the direct consequence of the power of the word in the new communications environment.

* * *

DOMESTIC IMPLICATIONS MUST also be examined against *today's* realities for both press and government if we are to understand the movement toward the Age of ISDN.

Who speaks or writes the word, and to whom; what it claims for itself, and when; what it means, and why—these are the variables, the uncertainties, no matter how clearly, ardently, authoritatively the word is presented.

This chapter discusses the uses and abuses of both state and information power; considers whether, indeed, it is accurate to distinguish between two power centers or regard both as simply parts of a single monopoly; and begins to examine the fragmenting but democratizing potential in the coming technologies of communication. We continue to weigh, then, today's realities, heavily tinged with skepticism and even pessimism, and the coming decades' potentialities, which support the broadest optimism.

We recognize the gap between the present and the coming Age of ISDN—the network linkage of tens of thousands of domestic and global communications channels through the Integrated Systems of Digital Networks. Recall the distance between the Dark Ages and today; remember, progress has been speeded several thousand-fold. Admittedly, startling advances in agricultural, industrial and social development have not led to liberation for all humankind, but there has been significant improvement. Human rights have become codified, and violations recognized, if not immediately rectified. The imminent revolution—the networking of all communications capabilities—will require the yielding of central power, indeed the fragmentation of many central controls, in order to *maintain* national security and citizen peace. In the past and present, protection of national security is the rationale most national leaders use for coalescing all power in the hands of a few. Soon, the sharing of communications power will be essential to achieve domestic progress, and to participate in universal discourse and international life.

Further consequences:

• A proposed "new world information and communication order" (NWICO) will have become meaningless in the terms propounded in the 1970s and 1980s. Third World countries' news will be heard instantly on the new communications networks accessible everywhere, in homes as well as news services worldwide. The democratizing programs of ISDN will replace the nebulous NWICO.

• Totalitarian or authoritarian regimes—whether Communist, Socialist, or right-wing—by yielding communications power, will find it increasingly difficult to retain absolute political power.

• Communicators in many fields—journalists, data processors, and countless marketers of other services—will face national and global entrepreneurs who seek to bring ever larger communications systems under single corporate control. It is likely, however, that communications monopolies will be avoided, first, by governmental regulatory acts, and second, by the very fragmentation induced by the variety of new communications forms and channels, and the low cost of access to the new media.

Whatever the new forms, the principal conveyor of news and news-related information to publics everywhere will be the journalist. The basic challenge facing the journalist and the news media will remain. There will continue to be distortions in reporting fact, and bias accompanying analysis. Absolute truth may evade even the most earnest truth-seeker, for—recognize it or not—"truth must be stalked from a point of view," in the words of Will Irwin, a popular journalist early in this century. The charge of bias—pro-Western, anti-Third World—is insistently levelled at American journalism by the developing countries and the Soviet bloc. They regard the U.S. press as politically conservative, U.S. government-dominated, and informed by Western training, experience and objectives. Western Europeans often criticize the U.S. press as less than serious, somewhat irresponsible, politically provincial, yet admirably independent.

Press power, people power

Four months before leaving office in 1989, George Shultz recalled Ronald Reagan's private meeting with corporate communications leaders in 1980. To help candidate Reagan in the primaries, Shultz had assembled the group in San Francisco. Reagan gave his general speech, and then asked, "What can I do for you in Washington?"

There was a long, embarrassing silence, Shultz recalled eight years later. Then, finally, a communications leader spoke up:

"Leave us alone!"

The 1988 audience of telecommunications specialists exploded in laughter, and understanding. "Get the government off our backs," is a familiar plea of corporate leaders, and especially regulatory-prone communications managers. Secretary Shultz, however, almost immediately demonstrated the dilemma of both government and private communicators. Without acknowledging a dilemma he next took credit for recent U.S. efforts to secure more favorable universal regulatory measures at GATT and ITU for American telecommunications and space activities. Agreement of government to "leave us alone" is a power-sharing implement, no less than is governmental assistance in securing global rights-to-the-road for U.S. communicators. Often those who demand laissez-faire also want government assistance when domestic or foreign problems arise.

So far in American experience, neither regulation nor deregulation has supported the slippery slopers' fear that personal freedom would suffer from sensible governmental activity, in a field as endowed with public interests as communications. The power of state and of press have always been in contention in the U.S. Those who favor nearly absolutist freedom for the news media cite Thomas Jefferson *before* he became president:

> The basis of our government being the opinion of the people, the very first object should be to keep that right; and were it left to me to decide whether we should have a government without newspapers, or newspapers without a government, I should not hesitate a moment to prefer the latter.

Press supporters rarely cite Jefferson *after* he became president, and suffered the angry barbs of the press:

> Even the least informed of the people have learnt that nothing in a newspaper is to be believed. This is a dangerous state of things, and the press ought to be restored to its credibility if possible. The restraints provided by the laws of the states are sufficient for this if applied. And I have therefore long thought that a few prosecutions of the most prominent offenders would have a wholesome effect in restoring the integrity of the presses.

Even Jefferson, the constitutionalist, did not consider prosecution of the press as a slope that would lead inevitably to the endangering of the republic. Yet, two centuries later, prosecution is not the answer.

Power is an inescapable element of human existence. Every man or woman, no matter how "powerless" in political or economic terms, exerts some power at some time, at least over family or friends. And, in fact, even the "powerless" have latent power in their numbers. Indeed, the first act of revolutionaries is to employ the power they decried when exerted by their adversaries. Yet, in truth, there is no power in "the masses," but only in individuals who join together for limited purposes, and limited times, and exalt a leader. The instinct for power is present to some degree in everyone,

Power is the essential organizing force of any society. The more complex the society, the more diverse the channels of power. In such a complex society, the interaction of different power centers helps produce a political and social consensus. This defines the rules of the road (stopping automobiles at red lights, and directing airplanes in flight); the norms of behavior (murder and thievery are punishable; public nakedness and bawdiness, proscribed); and the regulating of political and commercial acts which may lead to the monopolization of power.

For power itself is without moral stigma. It can be used for good or evil. In a democratic society, power is essential. Power provides the arithmetic of representative governance. Power does not always produce an exact replication of the ideas and decisions in the proportion in which they are held by individuals in society. But wealth and influence, derived from family or business associations, can temporarily alter the power equations. In as complex a society as the United States, however, even the wealth of a Rockefeller or Harriman could not assure them the White House, while far humbler backgrounds did not preclude the Nixon and Carter presidencies.

They succeeded by coalescing diverse centers of power sufficient to gain office, but their effective power did not extend to the other influential power centers in the Congress, the full executive bureaucracy and the judiciary. If Machiavelli were alive today, he would have to write a far more complex guide than the one he wrote about how to deal with princes in the simple Italian city-states. Machiavelli would particularly have to understand that no president, even the "great communicator," has been able to control that other large source of power after governance itself—the print and broadcast news organizations. The role of the news media as power players is typical of all other actors on the scene. All wield power, though most decry it in others, particularly when new combinations of political or economic power threaten to tip the scales of influence in a new direction.

Power is more acutely felt when it is exerted by a new actor. Before the Wagner Act, trade unions were weak, and unorganized labor was entirely at the mercy of employers. Federal labor regulations created new power in government, bolstered that of unions, and forced business into new relationships with employees. Power existed before the Wagner Act. It was mainly in the hands of business; since then, it has been shared. Labor-management relations today reflect the fragmentation of power apparent throughout U.S. society. Large power centers are opposed by other large power centers. The real power of one group wanes as another waxes. But power remains, not as a hidden, nefarious force, but as a highly visible projection of the will of groups, individuals—not a class, not a hereditary elite, not even an establishment secure for all time.

Since power is a dominant factor in all social organizations, this can be said:

• Power begins with the individual, who attracts like-minded individuals to support an idea, business or political act.

• Power is ultimately centered in the corridors of the nation-state, but only after it flows through journalistic, cultural, religious, corporate, political and other power centers.

• Power, as governance, reflects a definable system of ideas and objectives, to be realized through specific institutions.

• Power requires the continuous understanding and support of those already committed to the central ideas and objectives, and the increasing adherence of the society generally. For such understanding and persuasion, mass communications are essential.

In a democratic, market society, therefore, the independent news and information media, therefore, are most essential to the operation of government and to the continuation in power of the present governors. Direct power is the ability to mobilize human and property resources to attain statist goals. But direct power today is so fragmented that officials from the president down have changed from being rulers to administrators. They must plead or persuade, rather than demand or command. They must particularly rely on the indirect power of journalism. Indirect power is the capability of the mass media to set agendas, reveal facts, and challenge ideas and goals of those seeking or exerting power. By such acts, the mass media themselves exhibit power which confronts the direct-power structure of governance and the related political apparatus.

Power, thus, is not an abstraction residing in a class, a corporation, or a newspaper. Power is the use of the instruments necessary to generate change, or, in the case of the mass media, to influence the movers and shakers. The power of an editor is not in the control he has over his relatively small staff, but in his ability to select the news and views he prints or broadcasts. Those views may be taken up by other groups —business, corporate, political, religious, intellectual—and relayed to the actual power-wielders of the city or nation.

That indirect power—to set agendas for national discussion, support competition in public places, interpret favorably or critically the policies of the administration—is often as potent as the direct power of governance; sometimes, the media are more powerful by delaying or stalemating government policies and activities.

All governments thus regard mass communications as a rival force. Authoritarian and totalitarian governments have an easy answer. The governments either own the mass media, or influence their content by overt or covert controls.

But how does the press in a free society use its power? Does it provide intelligence for a free people to make important decisions about their society, or at least provide adequate information to both the governed and their governors about how representative governance is functioning? Do the news media also keep the bureaucrats informed of one another's activities, and the rapidly changing world beyond U.S. borders (for news reports of foreign affairs often arrive in Washington before official accounts of the same events)? Do the mass media join together to form a coalescence of power directed for or against a policy, official or activity?

The quick answer: "The press" is a very mixed bag. Lincoln Steffens, the muckraking New York journalist put it cynically:

> When [a newspaper] is young, honest, and full of ideals, it is attractive, trusted, and full of the possibilities of power. Powerful men see this, see its uses, and so seek to possess it. And some of them do get and keep it, and they use, abuse, and finally ruin it.[5]

Included in the press are large, cosmopolitan daily newspapers with staffs in many cities in the U.S. and abroad; thousands of smaller dailies and weeklies which focus mainly on news and comments closer to their readers' homes and primary interests; two great global news services reporting for many thousands of publications and broadcasters; local broadcasters' and networks' news departments, including cable news; several weekly news magazines; and thousands of columnists, commentators and specialized magazines which cover segments of the news field. These are "the press."

Newspapers compete with one another, and all the media with the others. They compete for public attention. That translates into the purchasing of their product, or the advertising space or time sold within the product. That competition should lead to a healthy diversity of news coverage and viewpoints. It can also lead to a distortion of the competition by less-than-responsible treatment of news. Distortion may take the form of "pack journalism"—a horde of reporters overwhelming a newsworthy person or event so as not to be left behind by the competition. Or it may take the form of new-fashioned "scoops"—stories which no other writer publishes because many of the "facts" are highly imaginative. Or by distortion instead of sound investigative reporting. At its best, this form provides deep examination of a subject which would not otherwise come to light, and certainly not as pointedly. Distortion may set in where investigative reporting stretches the knowable to include the unknowable in order to fit a preconceived premise. Then, fiction parodies journalism, and the public is misled.

Power in the field of global news reporting and dissemination is full of paradoxes. The major services headquartered in the West (AP, UPI, Reuters, AFP) produce every day 50 to 80 percent of the world news flow. This is bound to have great influence not only on the public everywhere, but on the policymakers and intelligence services in every government in the world as well. There is no overt cohesion among the four global news services. They actively compete for access to news, clearer analysis, and some assessment of the implications of today's news for tomorrow's developments. Thus they compete not only for coverage of news, but for the sale of their services to all manner of publishers, broadcasters and, eventually, electronic digital networks.

Competition leads to the packaging of some news, commentary, and information bases to meet special needs of some clients. Cricket or soccer may get more space, baseball, less. The Organization of African Unity (OAU) may be better covered by Reuters and AFP, and the Organization of American States (OAS) featured by AP and UPI. In the larger sense, the French and British services provide far greater coverage of the French- and English-speaking countries in Africa and Asia than the American ones do. Similarly, the American global services package special coverage for their clients in Latin America or within the United States.

That is not to say that any of these services alters the standards of news coverage as they move from country to country. One could consider this relatively universal editorial standard a mark of achievement. Yet this standard often provides a cudgel with which to strike the four Western services. It is said they act with uniformity; that is, they design their coverage to meet the best interests of their primary clients in the Western countries. The four services act with similar news standards and writing styles. A simple declarative report of knowable facts is the fairest, clearest form for reaching all the diverse nationalisms and cultures. The primary users of all four services have similar editorial values, and monetary and trade interests. The news services thus set separate editorial agendas, but only after they have demonstrated shared values.

Despite their power, however, the four news systems are surprisingly small and poorly financed. The total staffs number less than 10,000 employees. Fewer than half are assigned outside their home countries.

Consequently, foreign news coverage is greatly restricted. The "big four" thus become targets of Third World attacks twice over: first, for employing Western editorial values, and second, for rejecting the active promotion of "development news" as part of that standard.

Development news is a stated need of developing countries. They seek a journalistic commitment to the *promotion* of their economic and social development. The "big four" cannot make this commitment and still adhere to their editorial values which stress "objectivity." Implicit in this demand for development news is the commitment to a governmental information agenda. Such an agenda is quite understandable for governmental information agencies (such as the U.S. Department of Agriculture's farmers-aid services) but not for an independent news service. That is not to say that Western news media are justified in ignoring the major economic and social preoccupations of developing countries, and the processes needed for their development. But the independent media, at their best, will not accept as their own, official handouts or the information agendas of a government.

The "big four" rationalize their rejection of development news in practical terms. They say they do not have the staff, space and time to cover economic, agricultural and industrial training and development. Most crucial, however, is the basic definition of news employed by the "big four": the coverage of *exception*. News is that which is new, unusual, even exotic. In recent years, however, there has been a notable improvement in the coverage of Third World longer-term or "process" stories by the global services.

The paradox, then, is the fragility of the four great global news agencies; fragility cum power.

The great communicating power of television and the several nationally influential newspapers can, indeed, distort reality for a time. The "pack" tendency causes TV network editors to pick up spectacular stories floated in early editions of the *New York Times* or *Washington Post*. In turn, coverage by one TV outlet will inspire a competitive telecaster to seek a follow-on angle to the same theme. Thus, a continued story takes on a life of its own.

But other tendencies are also at play. There are news fads. One personality, organization or cause will be widely covered for awhile, then almost as spectacularly dropped, if not subject to anti-hero treatment in the second stage. And, of course, the great profusion of news and commentary on every side serves to diminish the longevity of every story. This may be a loss in the case of vital information which should be kept in view. But it is a protection for society in the case of information inadequately treated or distorted by the mass media. Profusion and diversity, nevertheless, tend to be the free society's built-in defense against error as well as monopolization of information. It is an untidy, often unsatisfactory arrangement. Yet it has the great advantage of fragmenting media power before it can overstep bounds which are poorly defined. Most hazards stem from the saturation of coverage and the prominence of display given to information with little social value, or even anti-social themes. Some bounds are partially incorporated in libel laws, antitrust laws and regulations preventing the domination of a U.S. geographic area by the cross-ownership of newspapers and television or radio outlets.

It can still be argued that in an era when government has become so big, diverse, and powerful, it is necessary to have great independent organs of communications—print and broadcast—serving as watchdogs over officials, and analysts of their policies. Power begets power, and both state and press must be effective as well as, in the end, limited. State power and information power are complementary, even when they are most adversarial.

Governor Mario Cuomo put it well in lecturing an audience of journalists in 1986:

> The founding fathers knew precisely what they were dealing with... The press of their time was not only guilty of bad taste and inaccuracy, it was partisan, reckless, sometimes vicious...They might have written [constitutional] amendments that never mentioned freedom of the press. Or they might have tried to protect against an imperfect press like the one they dealt with, with conditions, qualifications, requirements, penalties. But they didn't. They knew the danger. They knew that broad freedoms will be inevitably accompanied by some abuse and even harm to innocent people. Knowing all the odds, they chose to gamble on liberty.[6]

The absolutist case for the press is set forth by Herb Greer.[7] He says that *only journalists* can decide what is ethical journalism, what is in the public's interest to communicate, whether one who is reported has the right to reply, whether one has been libeled, and whether a reporter has a responsibility under law. Only journalists can decide. Greer concludes "there is no such thing as an abuse of press freedom."

The slippery slopers of the press regard *any* limitation, even the most sensible libel protection, as an immediate subversion of the First Amendment. This conception removes the press from responsibility under law, unless application of legal restraint is first approved by other journalists. This privilege is not granted presidents, Supreme Court justices or congressmen. This is the ultimate slippery slope argument. It holds that *any* law, no matter how reasonable, is bad because one law can lead to another, and the next may violate press freedom. It is as extreme to say there is no such thing as violation of the press as it is to assert there is no such thing as abuse of press freedom.

There is, indeed, no such thing as unlimited freedom—except for a Robinson Crusoe on an unpopulated island. Every law protecting one person is a restriction on another. "We have hidden from our critics. And worst of all, we have hidden behind the First Amendment," said Richard D. Smyser, president of the American Society of Newspaper Editors in 1985. The test of a functioning free press is not the absence of laws carried to an extreme, but whether minimal restrictions against libel and monopolization are reasonable in statute, and reasonably enforced, with clear and specific ground rules laid out in advance for all to understand.

The entire democratic fabric is just that fragile. And yet that limited interference is also the strength of the system.

That system must deal, however, with increasing challenges from within the communications industry itself. Because of its own power, the information industry challenges state power. Conceivably, information power could coalesce into a giant conglomerate that would, itself, limit diversity of news and views even while vastly increasing the flow of information. This is a *possibility,* not a present-day analysis, though some specialists regard the present information structure as highly centralized and increasingly tied to the military-industrial complex. That analysis has been repeatedly made by Herbert Schiller of the University of California at San Diego. In three volumes,[8] he argues that (l) American news and information media treat information as a saleable commodity, (2) they dominate the domestic and world information flow through ties to the Fortune 500 corporations, and (3) these information corporations are replacements for the lost U.S. dominance in manufacturing industries.

All of this, says Schiller, amounts to "cultural imperialism." He defines it as "the sum of the processes by which a society is brought into the Western world system and how its [elites are] attracted, pressured, forced and sometimes bribed into shaping social institutions to correspond to, or even promote, the values and structures of the dominating center of the system." The global market "imperative" of the U.S. and Western Europe, says Schiller, is to "organize the world system." There is "fury and bewilderment by those responsible for the informational apparatus in the core area [the U.S. particularly] when there is any indication of rejection."

Schiller sums up his opposition to the free-flow-of-information concept by saying, "When there is an uneven distribution of power among individuals or groups *within* nations or *among* nations, a free hand—freedom to continue doing what led to the existing conditions—serves to strengthen the already-powerful and weaken further the already-frail."

The remedies, however, are limited. Either one opts for equality of *result* or equality of *opportunity*. The former, egalitarianism (never attempted anywhere in the world, and probably not possible), means depriving some to favor others. The latter (attempted in democracies, but never adequately achieved) means opening the system so that those underrepresented may have the chance to participate. The guarantee of result must install, instead of a relatively free market, a system which puts a harsher centralized controller—the state and its police power—in place of the less fearsome power of diverse economic managers.

Another media critic, Ben Bagdikian,[9] maintains that fifty giant U.S. corporations, linked through their boards of directors, control not only pharmaceuticals, airlines, agribusinesses, banking, insurance, electronic manufacturing and defense contracting but also the majority of U.S. mass media: newspapers, magazines, radio, television, books and movies. The implication is clear: the press, in all its major forms, at least will not embarrass its financial partners; at worst, it will support the general interests of the affiliated corporations. If this financial linkage has the implied effect, certain big stories detrimental to affiliates would go unreported, or toned down.

The broadcast networks and the major newspapers cited by Schiller and Bagdikian are the same media which conservative critics say are dominated by the liberal elitists who serve as the new powerbrokers in the United States. These critics regard publicly supported radio and television, and the commercial "media elite"—three broadcast networks, ABC, CBS, and NBC; *New York Times, Washington Post, Time* and *Newsweek*—as dominated by liberal journalists. One of the most respected columnists, the late Joseph Kraft, criticized his colleagues for their attitudes:

> We no longer represent a wide diversity of views. We have ceased to be neutral in reporting events...The media have been taken in tow by the adversary culture...We are skeptical of established authority.[10]

The Lichter-Rothman studies[11] examined the media elite based on extensive interviews with them. While far from the majority of American journalism, the elite (as everywhere) sets the national agenda and the style of reporting and analysis. The wire services (AP, UPI) are *not* among the media elite. They cover the world for most of the press, and provide much of the coverage used by the Third World.

The media elite, say Lichter-Rothman, are distinctly liberal—by self-description, by their own voting records, and their attitudes toward society and governance—and perhaps most important, by using journalism to change events and social structures. They are homogeneous. They are "America's new powerbrokers." Some 54 percent of the journalists responding said they are left of center, and 19 percent right of center. Corroboratively, they described their journalist colleagues as 56 percent on the left. Their voting records follow this pattern. In four presidential elections, not less than 80 percent of the media elite voted for the Democratic candidates. The journalists went 87 percent and 81 percent for the losers against Richard Nixon, and 81 percent for Carter when the country gave him 51 percent. The attitudes of the elite media differed markedly from national norms on many other political, religious and social issues.

The media elite, say Lichter-Rothman, are more than watchdogs of government. The elite discover new realities—the day's event or personality—"by reintegrating" their own experiences to conform to the new situation. The elite journalists choose sources and select reportable information that generally matches their own liberal orientation. This is not usually a conscious decision, say Lichter-Rothman. That phenomenon is more complex than deciding to report with bias, or, worse still, conspiring with others to do so. The Lichter-Rothman analysis says in political science terms what Third World critics have bluntly charged either out of their own interests or biases, or, in many cases, determination to censor or control the flow of information: Journalists report out of subjective preconditioning.

Yet the same year of the Lichter-Rothman survey of media elites (1980), the Michael Robinson study found that liberal Ted Kennedy was treated more negatively in the primary broadcast coverage than Jimmy Carter, and Carter (the more liberal in that election) drew worse coverage than Ronald Reagan. In 1984, moreover, Reagan's coverage was far more favorable than Mondale's, who was liberal. Also in 1984, Jesse Jackson, the very liberal Democrat, received the most positive coverage of all contenders in his party. Again, in the 1988 primaries, the Lichter-Rothman survey found Jesse Jackson getting by far the most favorable press coverage. When they compared the coverage for party bias, however, Lichter-Rothman found barely a 2 percent difference after a half year of campaigning. They concluded: "Evidence for charges of partisan bias is less than compelling."[12]

The Lichter-Rothman warning is nevertheless appropriate: "Unless journalists take responsibility for their creative role in shaping the news, they will remain mired in a debate over bias that is misleading and acrimonious." That debate is increasingly polarized. It powered the Reagan administration to reduce government information flows, cut access to official data, increase secrecy, crack down on leaks, and restrain officials from revealing information after they have left government service.

American journalism's reaction to the challenge in the international arena supported the U.S. withdrawal from Unesco in 1985.[13] Uniquely, out of what each major player regarded as its self-interest, the U.S. government and the conservative as well as liberal media acted together to withdraw from the organization.

The media elite exemplify—for democratic and anti-democratic governors alike—the challenge to governance by wielders of alternative power. Elite journalists enter the profession with an "anti-authority outlook" in order to "exert moral power...against traditional restrictions and institutional authority," say Lichter-Rothman. That is quite different from investigative journalism, which seeks answers wherever the search leads, and different, too, from adversarial journalism, which uncovers another side of a particular story and poses it in perspective against the official version.

Often, the constructive engagement in investigative or adversarial journalism is, however, regarded by ideologues—whether or not with cause—as reportage inspired by the journalist's personal political or social objectives.

Given the increasing power of the U.S. media to attack or stalemate government, the charge of media bias has become an effective counter-attack. "In our business we think it is our duty to change the public's perception of a politician who is coming off better than we think he is," says Brit Hume, White House correspondent for ABC News, but "we feel no comparable responsibility to a politician who is coming off worse than we know him to be." Public understanding of the adversaries—state power v. information power—is not enhanced by the acrimony. Nor does bitterness improve the journalist's understanding of his role as actor in the play, not merely as observer or reporter.

Journalists, whatever their national or political background, are bound to reflect their personal and national cultures and attitudes (whether the dominant or the contrary attitudes). But journalistic training and editorial monitoring do reduce the likelihood of outright bias reflected in news reports. The Lichter-Rothman study concludes that further examination is needed to ascertain whether ingrained personal attitudes of the journalist influence his coverage of particular events or personalities.

The growing criticism, however, is further recognition of the power of the media, and the inclination of aspirants to enter journalism in order to seek and perhaps share that power. That may be achieved without bending the rules of journalism. A journalist who would bend these rules belongs in another field—politics, perhaps.

The *New York Times*' former executive editor, A. M. Rosenthal, spelled out in a memo to his staff his "beliefs" in the character of the newspaper:[14]

The belief that although total objectivity may be impossible because every story is written by a human being, the duty of every reporter and editor is to strive for as much objectivity as humanly possible.

The belief that no matter how engaged the reporter is emotionally he tries as best he can to disengage himself when he sits down at the typewriter.

The belief that expression of personal opinion should be excluded from the news columns.

The belief that our perjorative phrases should be excluded, and so should anonymous charges against people or institutions.

The belief that every accused man or institution should have the immediate right of reply.

The belief that we should not use a typewriter to stick our fingers in people's eyes just because we have the power to do so.

The belief that presenting both sides of the issue is not hedging but the essence of responsible journalism.

Rosenthal also observed that *good* journalism had become more complex, indeed, more cerebral:

Our business is facts. We have learned to search deeper for facts, to dig for the meaning of facts, to relate facts to each other, to analyze them and put them into perspective...

Newspapers were always quite good at reporting what people said, and then what people did. Now we are also reporting what people think—theology, scientific thought, the meaning of the law—because what people think influences what they say and do.

As long as the news media remain the most important channels of information, the media will continue to be locked in a struggle for power. That struggle will be diversified, however, by the installation in all countries of the networks of freedom. These are the liberating technologies of communication. They will fragment information power, and thereby democratize it. A government or a commercial consortium would not be able to sit astride most of the significant communications switch points.

That liberating network of freedom—the Age of ISDN—is the major theme of this book.

Only one "national power"?

The theory of the single national power[15] holds that there only *appears* to be press power as distinct from state power. "The principle of watchdoggery notwithstanding," this theory states, "the relationship between press and [state] power is less adversarial than symbiotic."

This not only diminishes the watchdog role of the press, but also its adversarial function and its agenda-setting role. In holding to this theory

Herbert Altschull, professor of journalism at Indiana University, denies the objectivity of the press. For to be a watchdog, an adversary, or an agenda setter the press would do more than simply record what is happening: it would alter events. In Altschull's view, the press is not an "independent actor," and not a disinterested, objective observer. The press is part of the larger establishment. The press needs the power of the political and economic leadership, just as those leaders need the power of the news media. Together, "they live and/or die," says Altschull, "and it is together that they disseminate information, or that they set the public agenda."

It should not be a surprise that "the press"—as "the church," "the military," "the university," "the workers," "the bankers," and many more—is an institution of society. That is not to say that any one is the controller of political power, though all influence it, sometimes more or less than the others. Teachers of our children, for example, may have a greater lifetime influence than the press. In a stable society, all these forces are in symbiosis. Indeed, even the most objective press, one more objective than presently exists, would still influence political and economic power merely by being the carrier of information of many kinds. To note that information is power, is by now a truism—not a dark threat, but a potential for more broadly informing the public.

Within a decentralized society such as the United States, the struggle to exert power and achieve more is widely differentiated. The press may use its considerable power to advance owners' interests, editorial/journalistic interests, regional interests, national interests, universal (environmental) interests or simply the interest of reasonableness in a given situation.

Altschull holds otherwise. He says the "content of the press is directly correlated with the interests of those who finance the press." He cites four different forms of financial control: the *official* pattern in which content is determined by government edicts, as in the Soviet Union; the *commercial* pattern reflecting the views of advertisers and commercial allies including publishers; the *interest* pattern, influenced by a political party or a religious group (or, he might have added, trade unions, feminists, gay liberation, or others); or the *informal* pattern in which content controls are applied by friends or relatives who provide funding.

While Altschull acknowledges that the press in the United States has greater political autonomy than it has in the USSR, he seems not to concede that the American press also has greater "economic autonomy, or cultural autonomy." Consequently, he believes it is wrong for Freedom House, in his words, "to classify the press of one country free and another controlled." He also sees variations within countries. He regards the *Literary Gazette* in Moscow as having more political autonomy than *Izves-*

tia. And in the U.S., the *Nation* magazine, he says, has greater autonomy for criticizing the government than *U.S. News and World Report.* Altschull's overriding conclusion: "the political system in a country derives from the economic power structure, and [thus] the press of that country will therefore at any given time by and large reflect the ends of those who manage the economy."

This generally Marxist analysis holds true for the Soviet Union. We are presently seeing the enormous economic failures and their political consequences for the Soviet system driving the content of the mass news media in a new direction. The objective is precisely to improve the central management of the entire political system. This may require some controlled sharing of limited power in the periphery (the lower levels of the bureaucracy, in provinces distant from Moscow, and in parts of press coverage and commentary which do not undermine but rather, sustain central power). The present purpose is not to open the society to critical analysis of fundamental ideology or structural questions. Press power in the USSR is not only in a symbiotic relationship with state power, but press power is indistinguishable from Soviet state power. Press power is completely subservient to the objectives of governmental leadership, while permitting certain marginal variations supporting the same centralized control and end result. Thus, *Ogonyok, Moscow News* or the *Literary Gazette* may experiment with some new subjects or forms of journalism, but only to the extent they serve the narrowly defined needs of perestroika and centralized management.

Altschull's suggestion of rough equivalence between the U.S. and the USSR, based on the thesis that economic power determines press content, undervalues the absolute political control over economic power in the Soviet Union, and the diversity of economic and other interests which vie for political power in the United States. By examining the degree to which state power—political, economic or other—controls or influences the news media, it is possible to form a meaningful judgment of press autonomy in any country. There are significant variations within less centralized countries. Some media will be more autonomous than others. But the fundamental determination in any country is the degree to which the center of political power—the state and the governors—yield power to the mass media to operate with autonomy. Political power is the dominant force in every society. Economic power is secondary in the U.S. as in those countries in which state power is so strong that power sharing outside the presently ruling elite is impossible.

Where centralized control is overwhelming, and no commercialism exists, as in the Soviet Union, there has always been factionalism in the bureaucracy. Power is narrowly held, and has been challenged and put down virtually in secret since the Stalin purges. It is meaningless, therefore, to assert that the press of the USSR "reflects the ends of those who manage the economy." Soviet leaders control the political as well as the economic functions of the entire country. It would be difficult, if not impossible, to demonstrate that they keep a constant eye on the economic "bottom line." Only after seventy years have they realized that something has been terribly wrong with the country's productivity, but that's quite different from showing that economic considerations have governed the content of the mass media in the Soviet Union for seven decades.

Differentiating the autonomy of the *Literary Gazette* and *Izvestia* also tells little about the ideology of central control which permeates the mind of every writer in the Soviet Union. Yet one can differentiate the degree of autonomy in other countries where political power is less inclusive, more fragmented, structurally pluralistic or under constitutional strictures limiting state power.

Interestingly, as the survey made for this book shows (see pages 189-204), countries with the least political freedom, but still having some functioning commercial interests, reflect greater economic pressures on journalists than nations with the most political freedom for citizens and the press.

It is therefore misleading to suggest that because of commercial economic forces in countries of relative political autonomy, the press has no greater autonomy than news media in highly centralized states. Altschull holds that state political power and press power are symbiotic. That at once suggests that press power is a recognizable force (as it is). It also acknowledges the distinction between the two power sources. But that does not support the implication that a relatively monolithic economic power (overtly or covertly) funds the press in order to exert political control. Power in decentralized societies is sought for diverse reasons, and there are distinct clashes of some economic power with other economic power.

The press, as we say elsewhere, is a very mixed bag. Altschull acknowledges this and sees a "higher level of political autonomy" in the U.S. than in the USSR, citing the *Nation* challenging the political order. But writers on the *Nation* advance the interests, ideology and power of the publisher, as exerted through editors thoroughly committed to an anti-establishment point of view. It is not possible to get a balanced hearing in the *Nation* for another view or a thoroughly objective view of any fundamental issue. It is useful to know where to turn for a consistent anti-establishment critique. It is no less useful to know that there are other publications as dogmatic on the other side, or still others that will be less dogmatic, and provide analyses based on different criteria. Many different criteria and approaches to public issues may be found in the myriad of large and small publications, and some broadcast and cable outlets in the United States.

To support Altschull's thesis one would have to show that there is monolithic financial support for existing political power, or its obverse, the use of financial power to upset existing political power. More than that, there would have to be demonstrated the explicit monolithic use of press power to sustain or alter the existing leadership or policies. But to see *both* forces, and many more, in play at once—as they are—is to call seriously into question the Altschull thesis.

Certainly state power clashes with information power. That is one of the main theses of this book. But to suggest that all information power in the United States is virtually monolithic, and that all economic power is not ranged against political power, but is merely a part of the same umbrella establishment, is to deny the continuous struggles for domination or, at best, power accretion at every level, in the political structure of America.

There are, indeed, economic interests of publishers and shareholders. These clash with the economic interests of other publishers and shareholders. And quite apart from such bald display of economic power through the press, are significant efforts in the most responsible news media to insulate working journalists from the known financial or political interests of owners and publishers.

The greatest protection of the public is the diversity of owners and the diversity of interests, and also the drawing of journalists from diverse social backgrounds and different political orientations. The concentration of ownership of mass media in fewer corporations is a distinct source of concern, particularly if new owners are attracted solely by the investment- or political-power potential, and not by the objective of building a useful information channel. Until the millennium arrives, however, there is some assurance of diversity in American journalism. It is revealed not only in the fact that owners generally come from one side of the political spectrum, and elite journalists, by their own descriptions, from the other side, a distinction demonstrated in the rough balancing of content in the newsrooms, if not always on the editorial pages.

The coalescing of state and press power in an earlier, less sophisticated era in America is described in Lincoln Steffens's autobiography. Long before the first White House press conference was ever held, Steffens had a daily date with President Theodore Roosevelt as he was being shaved by his barber. Steffens asked very direct questions, engaged in frank political discourse, and shared Roosevelt's private thoughts. These helped inform Steffens's readers in a prestigious New York newspaper. When Steffens wanted to query other officials of the government he was given this handwritten note signed by the President:

Jan 9th 1906
The White House
Washington

To any officer or employee of the Government;
   Please tell Mr. Lincoln Steffens anything whatever about the running of the government by or under officers of the Executive that you know (not incompatible with the public interests) and provided only that you tell him the truth—no matter what it may be—I will see that you are not hurt.
                                    T. Roosevelt [16]

That was more than a half century before the Freedom of Information Act!

Power surge

What people think is a clue to how they may act. Action is power. In all ages, the uses of power are well understood by the powerful, and only remotely observed—often negatively, physically felt—by the powerless.

Human history may be traced back five million years. Farming and some urban dwelling goes back a few thousand years. The agricultural revolution was far more recent. The industrial revolution appeared about 250 years ago. Modern communications, in its primitive state, entered a century ago. The most significant communications power surge—the information age—is happening today, before our eyes. By its very nature, the information age is different from the agricultural, industrial, and even communications revolutions. This age can be seen by billions of humans as it happens. On every continent, masses of people are familiar with radio and television, and understand that a world of change separates them from their ancestors. What these changes will yet mean to them as individuals, to their country and to the world is largely unknown. Never before have the wielders of power been so uncertain about their own institutions and their ability to exploit or withstand the driving forces of communications technologies. For there are positive and negative potentialities for the rulers and the ruled in the promises of the information age.

Information is a key to power. Placing that key in the hands of institutions or individuals outside government can become a threat to governance. Placing such pervasive power in the hands of government can overwhelm the individual citizen. Yet, we should act to assure that the instruments of communication will prove democratizing. Individual citizens living under repressive regimes will also get access, finally, to information that can reduce their degree of servitude.

Transnational corporations already acquire and deliver information across national borders faster than governments can monitor transactions. Banks and monetary traders using split-second information transmissions now supercede the central bankers of the world in setting monetary standards. "This enormous flow of data," says Walter Wriston, "has created an Information Standard which has replaced the Gold Standard and the Bretton Woods Agreements" on currency regulations.[17] The global market for news, money and capital has altered dramatically the traditional power of governments, not only the central banks. The binding thread is the worldwide communications network. This "electronic global market," says Wriston, has produced what amounts to a giant vote-counting machine, which conducts a running tally of what the world thinks of a government's diplomatic, fiscal and monetary policy. That opinion, he adds, "is immediately reflected in the value the market places on a country's currency."

Governments do not like power they cannot control. Not only monetary information, but news and information of every kind carries similarly uncontrollable power to and from countries. Those nations that do not live by universal suffrage intensely dislike this new challenge, but even the most democratic governments find outside influences difficult to acknowledge and manage. Yet all countries have no choice. The computer screens, the news teleprinters, the networked telephones and other information providers will increasingly spill out data and news, relatively uncontrolled by many who fear the information age.

This new surge of power, largely undirected by the state, is governed, instead, by the driving force of the technologically possible and usable. Indeed, some new electronic developments appeared well before the market found applications for the innovations. But sooner or later most innovations, added to existing applications, have become part of standard systems. Such seemingly anarchical surges are changing the relationship between the public and private sectors, and changing the institutions of both. These changes will require new thinking about the constitutional arrangements of democratic societies, as well as the autocratic institutions of authoritarian and even totalitarian governments. Not the least threatening to all systems of governance is the speed with which the information age is developing and the need—now—to accommodate.

Although the new age will affect everyone on the planet, the power surge will first reach the industrialized societies. That presents a dilemma to the poorer, developing countries. On the one hand they want to use the new communications technologies to speed their nations' economic development. On the other hand they fear not only the further loss of control but the possible overwhelming cultural changes that may accompany the new age.

The Third World recognizes that the major media of communications are centered outside the developing countries. These media, say Third World leaders, do not serve the interests, values or objectives of the majority of the world's people. This belief stems mainly from the soon-to-be archaic debates over "a new world information and communication order."

Apart from the perception by government officials and communicators that they are real or potential adversaries, they are also direct or indirect collaborators. Direct on-air questioning of Anwar el-Sadat by American independent television anchorpersons elicited the Egyptian president's pledge to visit Israel and break a decades-long Mideast deadlock. That visit made possible the Camp David Accords approved by President Sadat and Prime Minister Menachem Begin with President Jimmy Carter's active intercession.

In hostage-taking and other terrorism in the '70s and '80s, television too often became the cat's-paw of terrorists, making governmental countersteps still less productive than they had been.

In the late '80s the People's Republic of China and the Republic of Korea (South), which had no formal diplomatic ties, exchanged television news programs. Both participate in Asiavision, the new link of Asian national television systems modeled after Eurovision. Both cooperatives permit news and information pictures to cross national boundaries. This limited movement of extranational, extracultural information is presumed to have less influence over viewers than messages directly geared to domestic interests, policies and goals. In the case of China, however, the reception of television from South Korea has been credited with persuading the Chinese government to participate in the 1988 Olympics. And political and economic relationships may soon follow.[18]

Perhaps the largest estimate of information power in the late '80s was made by Charles Wick while head of the U.S. Information Agency (USIA). He said that military power has diminished in importance as an arbiter of international affairs, and in its place information power has dramatically increased. Public diplomacy is on the ascendancy, he added, citing his creation Worldnet, and international broadcasting extravaganzas such as the globally televised celebration in London in 1988 of the birthday of Nelson Mandela, the black leader imprisoned for several decades in South Africa. The Worldnet teleconferences enable reporters in countries abroad to interview live, in real time, top-level officials of the U.S. government. Wick foresaw "atrophy and inconsequence" for those societies which still do not allow the free flow of information.

It is likely that state controllers of the information flow—the majority of governments in 1989—will allow more carefully screened entry of foreign news, information and entertainment well before they permit their domestic communicators to make independent choices in critical content areas of news and information.

Statist information power

Every government seeks the most friendly reporting of its acts and aims; the freest states manage the news the least; the least free countries completely control the information flow.

The Soviet Union, in this era of Gorbachevian glasnost, permits the controlled public discussion of "current" Communist history (provided it does not shake Leninism), current ecological problems (disposal of hazardous waste after television discussions of specific dangers), and social issues (price of consumer goods and food-production lags, which generated embarrassed responses from ministers on television). But all of that is intended to enhance the present overweaning policy of perestroika (renewal). The crucial challenges to central authority in Azerbaijan, Armenia and the Baltic area are not subjects for unlimited public discussion in any Soviet media.

An intimate description of the changing uses of television power by the State is given by Leonid P. Kravchenko, first deputy chairman of Gostelradio, USSR.[19] Gostelradio is officially known as "state television and radio," and some may "wince" at the "qualifier, state," says Kravchenko. One may "doubt whether there can be free expression of opinions or pluralism" and "wonder whether I am really no more than a courier for the Kremlin who comes to work with an attaché case full of instructions for every editorial board as to what should be broadcast and when, who should be rebuked, and who, on the contrary, patted on the back." This, he says, is not the case. The scale of Soviet broadcasting—120 TV centers—"rules out dictating from Moscow," Kravchenko adds.

He acknowledges that "a mere several years ago Soviet television was rather conservative...the ears of the television were hard of hearing and the eyes of our television were hard of seeing." But, now, he says, "television is at the very center" of the country's official policies of perestroika and glasnost. In other words, Gostelradio is "independent," but committed to basic governmental policies. Broadcasters are "independent" to pursue news and commentary within the bounds of official policies, but hardly to challenge fundamental policies. At the historic nineteenth national conference of the Soviet Communist Party, in 1988, television was a major instrument for spotlighting for the public the individuals and policies obviously favored by the top leadership. While this give and take was historic and refreshing, it was still far from the open-endedness one expects from independent broadcasters in the West.

Similarly, the media are used selectively in all the Arab countries except Lebanon. And in Iran, the rules have subordinated electronic and other cultural communications to the Islamification of the entire society. There, innovations are not accepted for the sake of technological advancement, but in order to further the religio-ideological objectives of the government. The Friday Islamic prayers at mosques all over Iran are by far the most inclusive form of communication in the country. In addition to the prayers, a sermon provides the social and political message that is broadcast throughout the land. This is the total deployment of communication for stipulated central objectives. Thus, traditional and modern communications modes are intertwined, but for now the modern systems are thoroughly under the control of government and religious traditionalists.

A careful observer of Iran, his birthplace, is Hamid Mowlana, a communications specialist who returned there briefly in mid-1988. He concludes that "despite the growth and expansion of modern media of all kinds in Iran, the traditional channels of communication produce a more powerful stream of information and news than is even conceivable." He adds, "Book publishing, magazines, newspapers, radio, television and even video sets play lesser roles than the traditional channels of communication through mosques, bazaars, and literally thousands of traditional institutions given social and political legitimacy by religious and other opinion leaders." Moreover, Mowlana found, a new private university in Teheran has been created for national political leaders-to-be. The university has three colleges, all tying economics, political science and other studies to the propagation of Islam. This institution is expected to educate the elite youth who will move into controlling positions in the postwar governments of Iran.[20] Clearly, as in few other countries, domestic information power in its modern form has been thoroughly harnessed to state power for traditionalist purposes.

Yet a remarkable incident in Iran early in 1989 demonstrates the complex impact of mass communications on one of the most hermetically sealed theological cultures in the world. The government sought during the recent Iran-Iraq war to exploit an imported Japanese television "soap" drama as a mass morale builder. The tactic succeeded—far beyond official expectations. Following is part of the Associated Press report, 1 February 1989 (W1208):

> *Nicosia, Cyprus*—A Japanese soap opera which has become the most popular television program in Iran has landed four Tehran Radio executives in prison. They were sentenced by a court [and to be given fifty lashes] for broadcasting a phone-in program in which a caller said the leading character of the TV show, "Ushin," was a better role model for Iranian women than the daughter of the Prophet Mohammed, founder of Islam.
>
> The oriental soap opera about a young girl growing up in post-World War II Japan amid great suffering, has become a cult among Iranians. The drama, which began in 1988, was chosen by Iranian officials to serve as a role model for Iranians during the eight-year

war with Iraq, when they suffered hardships, Iraqi rocket attacks and the loss of, by some estimates, a million of their countrymen.

Thus, on the birthday of the Prophet's daughter Fatima Zahra, when the woman who telephoned in to the daily "Family" radio program was asked whether Fatima Zahra was a fitting model for Iranian women, she gave an emphatic no.

Fatima Zahra is not acceptable because she lived 1400 years ago, she said, "I prefer Ushin."

Several morals can be drawn: Television is a particularly powerful tool in a country enduring great stress and violence; government officials who fully control the media system cannot entirely cope with its consequences; human reactions, even in a totalitarian society, find their way onto mass communications and influence them. Perhaps such a pervasive influence penetrated the mind of the Ayatollah Khomeini. One day after condemning the four radio executives, he pardoned them in a broadcast announcement.

In India, in 1988, Prime Minister Rajiv Gandhi tried to impose legislative power over independent newspapers. He had a Defamation Bill rushed through one house of Parliament. The press mounted immediate opposition. All newspapers did not publish for one day. Journalists marched in the streets. Opposition party leaders joined the foray. After several days Prime Minister Gandhi capitulated and withdrew the Defamation Bill. It would have provided penalties for "grossly indecent or scurrilous" writing. Mainly, the bill was a threat to those newspapers which have for several years uncovered and widely publicized corruption in the Gandhi regime. The *Statesman* editorialized that "the civil and criminal law of libel, slander and defamation that already exists in this country is quite sufficient to defend the legitimate interests of the government or anyone else."[21]

Thus, the combined power of the press, projecting itself as the people's surrogate, and bolstered by a political opposition party, was sufficient to turn back executive power. This in the world's most populous democracy, and a developing country as well.

The Iranian decision—subordinating modern communications to traditional culture—seems to reflect at one extreme some demands heard at Unesco for a new world information order. Those demands are based on the premise that indigenous cultures are subverted or submerged by the wanted or unwanted inflow of modern technologies bearing foreign news, information and entertainment. Few Third World countries share that perception in this extreme form, and still fewer accept the Iranian model. But all developing countries believe themselves powerless before the present massive news and information flow mainly from the West. They fear even greater torrents of foreign messages as the information age proceeds.

But they have yet to consider formally and publicly the universally empowering potentials in the Age of ISDN—the network of networks to which all can contribute as well as receive.

Democratic response

Recognizing these potentials, the more democratic states are beginning to alter basic law and practice to maximize the advantages of pre-ISDN, as well as future deployment of global networking technologies. The traditional monopolies of communications systems are giving way to more competitive arrangements (the breakup of AT&T is a notable successful example). Public networks, particularly in Europe, are converting to open networks of mixed public and private ownership. Smothering centralized planning is changing to more pragmatic decision-making (even in China). Central controls generally are yielding to diversified implementation by various actors. This leads to the diminishing of formal, official controls, and the determination of content and channeling by the force of demands made upon the network from outside the official hierarchy. Truly, these are major steps in the democratization of the content and flow of news and information.

The twenty-first-century goal—democratization of information power—implies universal service (access for everyone, in some form), assurance of message inclusion for all, and at uniform and affordable rates.

All countries may be faced with the challenges to state power from the independent news media, particularly television. American television plays a particularly pervasive role in influencing U.S. attitudes toward foreign affairs. Third World critics argue that the U.S. media, particularly television, take their policy leads from the U.S. government. For a quarter-century, before the Shah of Iran was overthrown, the U.S. press generally conveyed through news coverage and commentary the U.S. policy on Iran. This assumed that the Shah was a democratizing modernizer with popular support. Reality was more complex, and quite different. After the Shah was ousted and U.S. hostages held, American television became a toy in the hands of the radical students inside the U.S. Embassy. Even the Ayatollah Khomeini became a political hostage to the radical hostage-takers. He joined them, exploited their deed, and then ousted them from the revolutionary movement, sometimes by execution. None of this appeared in the simplistic U.S. press coverage.[22]

The U.S. press, as well as the government, sought new approaches to Iran. White House operatives created the Iran-Contra linkage. The news media, however, relentlessly covered the Iran-Contra exposure. That coverage had clearly damaging consequence for U.S. relations with Iran, Nicaragua, Israel, Saudi Arabia and Lebanon. Television cameras and commentators in many capitals pried unpleasant facts and assumptions from often unwilling actors in the complex political drama. The lengthy congressional hearings were largely conceived for televised coverage. Television paraded its wares day after day, and influenced the events it covered. The medium was hardly a cat's-paw for U.S. policies.

Television, however, was absent from Cambodia in 1975-79. The Communist military prevented Western coverage. Neither the Western print or broadcast media probed the bamboo curtain. Meanwhile, in the silence, more than one million men, women and children were slaughtered. No one can be certain that timely disclosure would have saved lives. But obviously, silence abetted the horrors. The failure to cover the Cambodian holocaust was not television's, but the print media's—particularly the influential newspapers of record. Until late in the day, when massacres were occurring all around the correspondents on the scene, they excused the Marxist regime as merely responding to the intrusion of American bombers into a supposedly neutral country.

Television, forevermore, appears to be a part of the policy-making and unmaking process in democratic countries. Television applies new technology to the coverage of foreign affairs. The small, hand-held camera easily places the global eye at the scene of action. The satellite conveys word and image from distant places, including the eye of a hurricane. The hastily called press conference, or the exploitation of leaks, reduce the time for careful decision making by officials. Since television news is usually where the camera manages to be at the moment of action, that becomes the focus of the story, whether or not the central theme, or the moving personality, is somewhere else. The prejudging of an event by those assigning the camera crew often can determine where the action will be focused, and who will be the TV-chosen spokesperson.

Television, then, is now an intimate part of foreign policy-making, for better or for worse. Television can use remote-sensing satellites to uncover hidden military and economic formations. It can pretend to cover the world when in reality the number of correspondents and bureaus is shrinking. The fact that the three U.S. television networks do not have a single correspondent anywhere in Black Africa is bound to limit dangerously the understanding of Americans about that important continent and its diverse peoples. To make up for loss of resources on the ground, television may project a single available personality or an isolated event as representative of a far larger but perhaps quite distorted whole. Once a story is in the library, no matter how incomplete, it becomes established "truth." The network news computers can search thousands of feet of computerized film, tape and press clippings. These can be quickly called up to add background and credibility to a new story. The tendency is to regard file stories as accurate, no matter the quality of the original coverage. Yet the viewer of the new event will assume he has seen the true story. After all, pictures don't lie.

Or do they? Some viewers may wonder.

In November 1981, 40.2 percent of U.S. television viewers, about 32.7 million people, watched the nightly news programs on all three networks. That percentage has dropped steadily. In November 1988, about 8 percent fewer Americans were watching network news shows: some 29 million viewers or 32.4 percent of the possible audience.[23] Many viewers turn instead to the Cable News Network (CNN) for a more comprehensive, if still unsophisticated, continuous reporting of events. CNN covers foreign affairs with daily feeds from Europe, Africa, Asia and Latin America—an offering never approached by network news departments. Perhaps most prescient, CNN's "World Report" carries three-minute reports from Third World countries. These telecasts convey the perspectives of the local reporters, not that of an American interviewer or reporter. This is the forerunner of far wider networking of global viewpoints in the Age of ISDN. What C-SPAN does for public service coverage, and CNN provides for foreign affairs viewing, the Discovery Channel (TDC) does for other forms of "lifetime learning": history, nature, science and technology. All of these presently alternative channels provide increasing competition for the large networks. And the new technologies provide far cheaper instruments of coverage, enabling small producers and local TV channels to provide new competition to the giants.

Television in all its aspects enjoys greatly increased power, rivaling state power. But television does not always demonstrate the restraint which most officials display under television's own spotlight. Yet aren't we expecting the impossible of television some of the time? We anticipate instant analysis of complex events and insight into the unknowable. Instead, we should plead for documentary-style programs which provide better-researched, broader views and deeper background and format for perspective on today's events.

Journalism, however, is a difficult business. It is difficult for many of the reasons we have set forth. Yet it is inescapably a business. It faces all the problems of any other business, especially that of competition for profit in a market economy. But journalism differs from all other businesses. It is also a calling. It has duties other businesses do not have. Journalism is more crucial to the entire nation's running fairly, efficiently, productively, and safely than all the airlines, steel plants, health services, and police forces combined, because journalism is the link that binds all aspects of the society to one another.

Thomas Carlyle put it well exactly one hundred years ago. He wrote:

> Literature is our Parliament too. Printing, which comes necessarily

out of Writing, I say often, is equivalent to Democracy: Invent Writing, Democracy is inevitable.[24]

For Carlyle, printing was democratizing. For us, print plus broadcasting is democratizing. In Carlyle's age as in ours, however, trends toward state or corporate monopoly, self-interest, biased practices, or crude commercialism at the expense of truth have marred the full realization of the media's potentials. But most often statist controls and influences have far more widely hampered the full use of ideas for the parliaments of men and women.

Yet the prospect for the Age of ISDN is emancipatory. To paraphrase Carlyle: Invent ISDN. Democracy is inevitable.

We have a third chance coming. "Print technology created the problem"—individuals reading according to personal choices, says Marshall McLuhan. "Electric technology created the mass"—individuals abandoning their individuality to a fragmented "audience" created by the broadcast technology. With the ISDN Age—a network of networks—individual selection will again be central. All manner of subjects, in different formats, will be callable at the touch of a keyboard.

Invent ISDN. Democracy is inevitable.

Provided, of course, there is the political will, and the sharing of information power. Jacques Chirac, before he became Prime Minister of France, said privately, "The best press law contains just one sentence: 'The press shall be free.'" Then he added, "You should say that publicly while you are out of power, to commit yourself. Otherwise, the temptation to act differently is too great, once you come to power." He never did say this publicly. Like every politician, once in power he didn't resist the temptation to ignore his own sound advice.[25]

2.
Revolution Without Losers

World community is a fact because instantaneous international communication is a fact...So, while nations may cling to national values and ideas and ambitions and prerogatives, science has created a functional international society, whether we like it or not. And that society, like any other, must be organized.
—Secretary of State Dean Rusk, 10 January 1964.[1]

A QUARTER-CENTURY ago Secretary of State Dean Rusk foresaw an electronic world community forming. Its full realization may still be forty years off. But its foundations are being actively laid today. That world community will not arise everywhere at once. Some countries, such as the United States, will be affected dramatically almost immediately. Other countries, still in early stages of economic development will be fully affected years later. But every country will inevitably see lifestyles change. Ahead, then, on varying timetables is personal, national and international change of revolutionary proportions.

In the past, revolutions always produced apparent winners and losers. Losers usually paid with their lives, or at least their fortunes, for living at a time of revolutionary change.

Not so with the information revolution. It will, indeed, alter the way everyone lives, learns, engages in work, and participates in public affairs. All change generates some tension, and the information revolution will not be an exception. But in most countries it will follow years of social and political change which has already accustomed citizens to such uncertainties. The information revolution will produce visible gains to soften the blow of uncertainty. It will provide exhilarating potentials for new employment, and personal participation in the life of countries such as most citizens have never experienced: a revolution without losers.

* * *

THE COMMUNICATIONS SYSTEMS which Secretary Rusk observed were relatively primitive by today's standards, but he had the prescience to understand the extraordinary opportunities which science and technology could create for men and women everywhere. These innovations would alter all forms of communication, and therefore all manner of business transactions and human relationships. Perhaps most revolutionary, the nascent electronic world system could reorder the fundamental conception of the nation, and of international civility.

Since the city-states of ancient Greece, sovereign power has been the essence of the state, whether it is organized as democratic, Communist, Fascist, Socialist, monarchial, African-socialist, tribal, traditionalist, Islamic, Judaic or another. In the inherently different social and international system evolving, information power is replacing police and military power as the functional instrument of international relations. International law, an estimable objective, is presently unenforceable. "Order," therefore, is linked not to a universal system of justice, with penalties for violators, but to the acknowledgment of latent force. This asocial, cynical analysis would be more fearful had not the power-based diplomacy of the two superpowers produced nearly forty-five years of peace between them. That, however, has not prevented scores of "small" wars from killing millions of people since World War II ended.

In the places where state power has been most obtrusive, it has failed to achieve *secure* sovereignties and better lives for their peoples. This realization is generating fundamental political and social reassessments in the Soviet Union, the People's Republic of China, much of Eastern Europe, and in many countries of the Third World. Reluctantly, they acknowledge that their socio-political systems are inadequate; moreover, that the information age promises decisive advantages—but at a considerable price: The nations' leaders must sufficiently trust their own citizens with the instruments of information which are needed to create a new society, and an international community.

As in no time past, the tension between state power and information power is decisive. Information in all its forms—news and commentary, data flows, retrieval of past records, and myriad informational services, including instantaneous monetary, banking and market reports—can directly influence the power of nation-states, those who rule them, and all who inhabit them. Not the least affected will be the press in all its forms: print, broadcast, cable, and scores of information services which deserve inclusion as amplifiers of news and information media.

Since the first Sumerian tablets were carved, rulers have sought to control writing. Before there were censors, the nobility and the church retained scribes and only later proscribed certain writings. Yet, in the tradition of the Old Testament prophets, there were also the defiers of authority, those who spoke their mind to kings (ethical challenges often couched as the words of God). The prophets were the first "watchdogs" of the governors, a role now assumed by the press when functioning responsibly in democratic countries.

The prophets set the ethical agenda for their rulers, and often for the people. The power of modern, independent news media rests not solely on their stating what is "true," but—no less important—in framing the context in which the report appears. By picturing the background against which a new event or personality moves, the journalist delimits future discussion of the event. The form of the news report, as well as its content, establishes the relevance of further reports. In effect, the news media create the universe.

All other players, including governments, act on that stage. The power to set the stage, therefore, is a considerable boon. Governments by their nature resent this assumption of power by the independent media. Most governments—three-quarters of all in the world today—rein in independent journalists, thereby distort the free flow of news and information, and control center stage.

But the larger question—whether the world will ever again return to secret diplomacy or secretive government—is settled. Inquiring, informative, investigative journalism is here to stay. Governors must now conduct public business with one eye on the domestic constituencies and another on world opinion. But Dag Hammarskjold, frustrated in his negotiations for the United Nations, wrote of the dangers in untimely revelations:

> The best results of negotiations cannot be achieved in international life any more than in our private world in the full glare of publicity with current debate of all moves, unavoidable misunderstandings, inescapable freezing of positions due to considerations of prestige and the temptation to utilize public opinion as an element integrated into the negotiation itself.

Commenting on this, Abba Eban, the former Israeli foreign minister, declared:

> The American media have celebrated great triumphs of exposure in such issues as Vietnam, Cambodia and Watergate. These experiences have given birth to the assumption that secrecy is intrinsically sinful, while publicity is inherently virtuous. The effect has been to create a fallacious identity between privacy and conspiracy...But the politician or diplomat can reply that there are some higher values than the immediate satisfaction of public vigilance and curiosity...It exists primarily in two fields: in areas where publication could endanger human life; and in areas where exposure could prevent peace settlements and even result in war.[2]

Eban recalls that Henry Kissinger lied about his absences from Washington while en route to prepare the U.S. opening to China. Asks Eban,

"Is not this alleviation of international tension a superior interest to a journalistic scoop?" He cites the *New York Times* in 1956 withholding publication of a proposal for ending the Suez-Sinai war:

> If these proposals had been published before I had had time to secure Ben-Gurion's agreement, there would probably have been a rejection in Jerusalem and a negative reaction from Cairo. Without this compromise, war would have been renewed.

He acknowledges the right of the journalist to dig tenaciously for the truth, but suggests that "the efforts of statesmen and diplomats to preserve reticence deserves a measure of respect, and should not be condemned out of hand."

Both the official and the journalist are poised on the same slippery slope. The official, by withholding information for less than valid purposes, undermines the democratic contract. The journalist can also damage the social process, even in the name of "the right to know." Neither the official nor the journalist need slide toward the point of danger. Both have the responsibility to achieve consensus on critical information; and, if necessary on the part of the journalist, voluntary abstinence at least for a time. But such temporary restraint by the journalist places an added burden on the official. He should tell the public in good time what has been withheld, and why.

These issues arise repeatedly in all societies. Open societies struggle with this dilemma and sometimes suffer the pains of openness. Closed societies justify their restrictive information policy by pointing to the problems which journalists create wherever they are free. In between, countries fearful of freedom, yet embarrassed by its alternative, choose "guided journalism."

We shall examine later these journalistic variations, the dangers inherent for journalists, and the distortions of information suffered by the publics where journalism is merely a handmaiden of rulers who decide what is news.

Deciding what is "news" troubles one editor in China. "As a news editor, I am often irritated by receiving 'news stories' that are not really news," complains Ma Jianbing of *Guangming Daily*. He cites one story: "A good party secretary who doesn't take bribes." Does one make a "'good party secretary' simply by failing to accept a bribe?" the editor asks. He continues, "Things are valued because of their rarity," and moralizes that "since certain things that should be natural and common become newsworthy, it means they are too rare." In the United States, that is known as the journalism of exception. Anywhere, it means that if bribery is common, an honest official is "news." But nowhere is that likely to be reported. And, everywhere, officials won't like to hear the whistleblower.

In Mexico, where the press is almost entirely under government domination, one newspaper, *El Norte,* shocks readers because it is both independent and professional. It reports corruption though thirty-five Mexican journalists have been killed the past fifteen years. In 1988, thirty-eight journalists in nineteen countries were murdered because they were engaged in reporting or commenting on the news. Another twenty-eight were shot. And fourteen journalists were kidnapped, 225 arrested, 428 harassed, and twenty-four expelled from the country. Some forty newspapers and radio stations were shut down. Death threats were sent another forty-three journalists, and forty were clubbed or beaten. Eight newspaper or broadcast plants were bombed or set afire. All in 1988. [See Appendix A.]

These are the "journalism morbidity" statistics maintained for some years by Freedom House. Such terrorism directed specifically at the carriers of news and information is bound to have extensive effects. In most cases, but not all, governments either are directly or indirectly responsible for these assaults. In some places, insurgent movements or drug traffickers are the attackers. Whoever the assaulters, the victim is not only the journalist, but the public itself. Acts of violence or demonstrations of state power aimed at journalists dampen the ardor (to put it mildly) of all journalists in the vicinity. In certain exemplary cases, however, repeated murders of journalists intensified the commitment of those surviving. Sixteen newsmen were murdered in the Philippines in 1985, yet Filipino broadcasters stayed at their microphones and continued to oppose the benighted Marcos regime. That, however, is unusual. The drug-traffickers' assaults on Colombian journalists have sent many fleeing the country. Still, three journalists were killed in Colombia in 1988 for persisting in covering the drug story.

The hope for developing a peaceful domestic polity, as well as an international society (Dean Rusk's term), is an explosion of news and information and, paradoxically, the fragmentation of information power. The diverse information delivery systems now forming should make any one reporter or transmitter far less subject to physical reprisal. Indeed, the networks-to-be will carry incredible volumes of information of all kinds available instantaneously. That accessibility to the public can translate into power significantly challenging to state power. Yet there is not likely to be a central source of information power either inside or outside a country's borders, at least not if the Age of ISDN is constructed as a democratizing rather than a monopolizing or neocolonial force.

ISDN, Integrated Services of Digital Networks, is the ultimate networking. It is global in scope. It is virtually all-inclusive. On line would be news from everywhere, including a great volume of "development news" not now carried by the major global news services—an exclusion deplored by more than 100 countries and Unesco. Also included would be instantaneous monetary, banking and business news from around the world; data flows carrying new messages as well as archival material; myriad encyclopedic information; and libraries of information on the natural sciences, social sciences, literature, history, and "practical" guides.

The technologies to produce such revolutionary storehouses of information are largely available today, either in use on a limited scale, or as prototypes. The decisions to form the ISDN linkages—the network of networks—are mainly political, not technological. One of the earliest stages will be passed in 1992 when the European Economic Council (EEC) begins linking the telecommunications systems of the twelve Western European member countries. The EEC has already rationalized some of the technologies that must be meshed, and is conferring on the pricing and trade barriers that must be reduced.

Canada and the United States in 1988 concluded an historic agreement that will permit free trade, including the transfer of electronics and other communications equipment on a preferential basis. It remains, however, for the U.S. and the EEC to mesh their telecommunications systems so that the U.S. does not fall victim to European trade protectionism, or find significant communications instruments incompatible in either the North American or European systems. This would be a serious loss for the early achievement of a universal ISDN. It would also set American manufacturers at odds with European and Japanese communications firms on the world market. Not the least of the likely markets in the years ahead are the more than 100 developing countries. How they regard the U.S. position in international communications, and how the U.S. sees the Third World, may be a decisive factor in twenty-first century U.S. global relationships.

The threat of "commercial saturation"

Since the first non-face-to-face communication began, there have been only a few communicators who told their story and many receivers on the periphery who could only accept the view from the center. The former wielded word power; the latter felt its effects. There must always have been critics of the earliest parchments; later, books; still later, photographs; then telephone and radio; next, television; finally, satellites—critics who said the communications systems were unfairly loaded in favor of the few who wrote, photographed or spoke. All the while, with each new invention of communication technology—mass printing to mass broadcasting to mass telecommunications—the variety of messages and available viewpoints has exponentially increased. Still, some writers foresaw massive retrogression, possibly for all time. Among these latter-day critics was

George Orwell who predicted that despite the promise of greater individual access to new communications, there would be Big Brother's control of all the systems, and a lashing of the human personality to the word machines.

Some social scientists today share Orwell's forecast. They base their analyses on present *private* ownership of the means of communication in the national and international spheres. Their concerns deserve careful attention. Indeed, whether or not their contemporary analyses and long-term projections are entirely valid, their warnings are certainly in order. For this massive communications revolution under way could be short-circuited by monopolistic controls before the limitless diversity of information channels is full networked. Monopolies are most likely to be secured by statist controls, however, nationally and internationally, as Orwell envisioned.

Nearly all of the formal academic research and analysis of communications systems, however, focus on present-day information channels which are largely in the hands of corporate managers. The leading researchers in the field generally apply Marxist analysis to what they regard as the commercial domination of communication systems. So pervasive have communications become that several academics presume that if Marx were writing today he would base his analyses on "Communications" instead of "Capital."

"Information production" has always been controlled "and has led to social stratification based on unequal access," Herbert I. Schiller told a conference on microelectronics in May 1983. Schiller, a communications scholar at the University of California, San Diego, generally opposes the "information age" analyses of the late Ithiel de Sola Pool and Daniel Bell. They regard the new techniques as liberating. Bell saw *development* as a democratizing force for former colonial regions. Pool particularly viewed *communications* as a democratizer. Schiller agrees that information today is central in all spheres of material production and throughout the economy, "affecting the organization of the overall social system." There would be little quarrel here with the less radical analysts. Schiller sets himself apart, however, by regarding the commercial and privatizing aspects of the new technologies as purposely dominating instruments of political as well as economic power. He sees this power used primarily to magnify investment profits for transnational and national—mostly American—corporations. These may be communications firms or noncommunications affiliates. These corporate drives, says Schiller, have extensive impact on developing societies the world over. He regards the integration of communications systems, and ISDN particularly, as a loss of free and equal access to information, not only in the Third World but to citizens in the developed countries as well. (He would be hard-pressed, however, to discover free access to existing communications today for citizens of most developing countries. Their leaders reject the concept.)

Nevertheless, these and related critiques of present and future communications-power arrangements should be addressed. The principal charges of the critics may be summarized:

1. "Mass media now saturate the life-space of all Americans with a ritual serving the industrial establishment, and presenting images of society in the world to which there is no equivalent challenge."[3]

2. "The world of television" is made to "uniform specifications of institutional service and sales."[4]

3. "There is an undisguised, optimistic faith, which appears not to require substantiation that there are readily available technological/communication solutions for social problems."[5]

4. "It is all-around changes that are required"—in the political and social systems, not just the communications systems.[6]

MASS MEDIA DO, indeed, strongly influence the lives of Americans, and increasingly of all the world's citizens. That, in itself, is not necessarily evil or deplorable. Content is the key to the value of the dominant aspects of a culture. Communications should provide a readily accessible mix of information and viewpoints to help each person fulfill his or her need and potential. Disease born of ignorance and the absence of information; performance and representational art constrained by limited implements and facilities; lives shortened and lifestyles hampered—these are the evils of the pre-information age which, for centuries, and still today in some places, are "saturating" and "deforming" the lives of millions of people.

Technology can strongly influence, but by itself cannot alter, a social system. A communications system can so inform citizens, however, that political and industrial changes are seen as necessary and, in the interest of domestic tranquility, imminent. It is far better, one suspects, to have ideas and images dominant through telecommunications and human interaction rather than dormant. New forms of communication through the centuries have accompanied improved conditions of human life, but only where communications were sufficiently diverse in describing the society, providing news and analysis of reality, and offering some implicit or explicit choices for citizens. The effectiveness of such communications systems is proved not primarily by whether they are privately operated and commercial, but whether they fulfill the real needs of people. In three-quarters of the countries where governments, not private entrepreneurs, own or largely influence all the communications media, the diversity of ideas and certainly human choice is harshly restricted.

The "world of [U.S.] television" is, indeed, largely made to produce sales of its own time and its advertisers' products and services. Sale of TV time is the fuel which drives the private engines of communication. Before television, radical critics complained that newspapers were driven largely by the dollars and politics of advertisers. Now some of the same critics favor newspapers over television. They regard the print press as more diverse, particularly since the newspapers require a literate public. But this is an elitist view. The vast number of illiterates in the world deserve a medium they can understand. Television, it is further charged, is a leveller. It has "abolished the old provincialism and parochialism." Residents of hovels and penthouses now see the same TV shows. This pervasiveness is somehow feared, when it should instead be regarded as another, not the only, sign of democratization.

Call it an "undisguised optimism." Such analysis is supported by the entire flow of human history. Since the Dark Ages—with some exceptions still today—the general human trend is toward a more participatory, more open and a freer society. Technology does not guarantee human improvement, per se. But technology specifically constructed to raise the living standard and enhance the variety and the fulfillment of real choices of people, can do so dramatically.

Yes, "all around changes are required"—in the political will and the fairer distribution of power. Those who see only danger in the networking of diverse communications forms and systems may be substantially refuted in the years directly ahead.

Futurists often err by making straight-line projections from the present systems. That, I believe, is a failure of those who forecast television "saturation" by reference mainly to today's TV schedules. ISDN will use television screens and computer keyboards, but TV as we know it today will be a small part of the visual products.

There will be far more diversified news, analysis, raw information of the past and present, classical literature, art, music and poetry from all lands and cultures, basic mathematical and physical theories and processes, architectural and technological theories and models, developments in many physical and social sciences, and countless other data flows. There will be room for the avant garde dramatist or artist to test the most innovative ideas and techniques, for the most dissident political or social experimenter to air views that run counter to the mainstream. For in fact, in such a massive network of networks there will no longer be a mainstream. The flow will come from all directions to all directions; from continent to continent; from the periphery overseas and at home, as well as from the center. This will be made possible by private, corporate entrepreneurs long before some still-oppressive governments permit such diversity.

The "undisguised optimists"—including the present author—urge the unreconstructed communications radicals—the neo-conservatives of the information age—to focus at least as persuasively on the major subverters of the democratization of communications: the wielders of governmental power in most countries. Changes are indeed required "all around"—all around the world.

As ISDN is developed, even an undisguised optimist should proceed with caution. Governmental as well as commercial power-seekers may want to manage the systems in order to predetermine the personal choices of the users. Technology may be used for surveillance, to restrict access, or censor content. But—optimism again!—the penalties will fall hereafter on those who contemplate restriction. Open communications for all citizens will be increasingly necessary if a nation is to attain and retain economic and political power through technological inventiveness and development. The Soviet Union has become a second-class superpower largely because its communications have been closed to all but a small elite. The United States has fallen behind Japan in scientific and technological innovation because its communications systems—at low and high levels—are geared to short-term commercial profit, replacing open-ended competition for new ideas and diversity, while its information channels are saturated with momentarily saleable, old-line products. Both the USSR and the U.S. are coming to recognize the penalties paid for less than maximum diversity in information, and inhibitions, whether planned or not, to the free flow of ideas.

One may ask the radical critics for their preferred alternative system. Would they have governments and intergovernmental agencies police the flow of information domestically and worldwide? The millennial history of official news and information has thoroughly deformed human communication. Is the technology at fault: does networking inevitably lead to centralized, monopolistic controls whether in the hands of benign officials, community-minded entrepreneurs, or socially-oriented cooperative leaders? How, except by the networking of networks, can widely diverse channels of information be made readily accessible to a worker in Zimbabwe and a researcher in the Library of Congress?

The answer is, indeed, change "all over." New generations of citizens everywhere should be educated to participate in an interdependent, information-oriented world. They should learn to use the opportunities denied their forebears but open to them. Low-cost, "small" information technologies can be as near to them as the telephone. Telephony will provide the facility to broaden minds and enhance beauty, as well as create structures and products. Market economics, functioning under broadly accepted social rules, can be both inspirer and provider. There *can* be a revolution without losers.

The historic tension

Governments cope with known adversaries but fear the unknown. News media are the most threatening "unknowns." They can circulate news and information which, if not immediately damaging to officials or policies, may raise embarrassing questions about both.

The Soviet Union's Glavlit, formed in 1922, employed some 70,000 censors at one time. For decades, Glavlit determined forbidden topics and discredited leaders. Glasnost, the new policy promoting wider discussion, has lifted some prohibitions in order to devise a new conformity. The United Kingdom is slightly liberalizing its Official Secrets Act, even while restricting journalists administratively. South Africa threatens to license journalists. India frames a tough defamation act to hamstring the press, but then withdraws. Sudan temporarily bars Western journalists from covering the 1988 flood because they hint at official maladministration of relief. China pulls the plug on the news satellites in May 1989.

Governments congenitally dislike inquisitive reporters, especially when they come from abroad, armed with the latest communication gear. (Their own journalists, however, they can restrain in countless ways.) The international news business is barely a century and a half old. Yet it both challenged and yielded to state power at the beginning. Indeed, the first international system, Agence Havas, created in France in 1835, actively sought to mollify the government. Havas wanted to concentrate on business news and sell the same information to as many newspapers as possible. In one fourteen-year period there were 520 court actions against French journalists. While Havas struggled to remain independent, the agency accepted government financial support in return for providing overseas services. Havas also received preferential treatment from the postal and telegraph services. Havas's successor, Agence France-Presse, still calls itself independent. The French government purchases subscriptions to AFP for government bureaus at home and abroad. These subscriptions have helped balance AFP's budget for several years.

Such arrangements have apparently been satisfactory for a long time —and not only for AFP. In the Franco-Prussian war of 1870, Jonathan Fenby documents,[7] the French government's own intelligence was so bad it signed a contract with Havas to supply war news. The British Ministry of Information learned of the Nazi advance in 1940 from United Press. Shortly before he died, Stan Swinton, then director of the Associated Press world news service, told me of his visit with Ho Chi Minh in 1945. Off in another room a subordinate was transcribing the AP's news report for personal perusal by Ho. Indeed, still today, particularly in developing countries, the four global news services are purchased as a stand-in for intelligence operations which these countries cannot afford.

That suggests the four services, Reuters, AFP, UPI and AP, enjoy a certain credibility for accuracy and efficiency, and deservedly so. But one might not suspect that, to hear the litany of charges made against them since 1972 by Third World and Soviet-bloc countries. These four services are accused of serving principally commercial interests, ignoring social and economic development which is said to be the principal concern of four billion people, and fashioning news to serve the interests of Western governments. The news services reject all these charges, especially the alleged ties to governments.

The debate is confused by the split-level history of the issue. In the nineteenth century, the British, French and German news services had intimate ties to their respective governments. Americans had already gone through decades of the party press and penny press. Supposedly independent U.S. newspapers had close ties to political parties and popular political leaders with aspirations to public office. The international field was further complicated by the 1871 agreement between the major newsgathering organizations of the world. Under this cartel arrangement, Reuters would cover Great Britain and her colonies, and Egypt, Turkey, China and other places within Britain's sphere of influence. Havas would have exclusive reportorial rights in France, Switzerland, Italy, Spain, Portugal and Central and South America. The German Wolff Agency could cover Germany, Scandinavia, Holland, Russia, Austria-Hungary and the Balkans. That left only the United States for the Associated Press. News of the U.S. covered by the AP was distributed abroad by the foreign agencies.

The Americans disliked the cartel not only as business competition, but (said the AP) because the European agencies "saw news as something that might be bent or twisted as necessary to serve diplomatic or imperial interests." Precisely the same charge is made today by Third World critics of the American, British and French global news services.

The AP avoided the cartel bind in 1917. The State Department offered to help AP distribute news in Latin America, where the Germans had secured rights. AP was given preferential telegraph rates from New York to Buenos Aires. Kent Cooper, then AP's assistant general manager, next asked the U.S. government whether the 1919 Versailles Treaty ending World War I would include some provisions for press freedom and a free exchange of international news. President Woodrow Wilson related the freedom to exchange information to making the world "safe for democracy." This was the beginning of the end of the European news cartel.

An AP director went through the Far East in 1919 and reported that Reuters's control of news channels was hampering U.S. interests in the western Pacific. Congress subsequently directed the U.S. Navy to permit AP the use of radio circuits at low rates to enable AP to compete with the British and Japanese. These Navy transmission services were later expanded. They were defended by the Navy as "internationally necessary for the national prosperity of the United States in its inevitable economic conflicts with rival business interests of other nations."[8] AP argued that this was not a subsidy to the wire service, but to the scores of newspapers using AP news. The papers owned the agency as a cooperative. The AP had sought and accepted political help and financial assistance from the United States government. Such relationships have long since been broken, and UPI and AP are as alert as Washington in avoiding even the appearance of a financial or editorial relationship. There was no indication during the early decades of this century that U.S. government interests dictated AP's coverage. But it is likely that much overseas coverage reflected a commonly held view of what were the major economic interests of America, and the chief military concerns of Washington. To address these concerns in news coverage was probably automatic, and little examined either at AP in New York or in the government at Washington.

Significantly, no one seriously questioned in public the governmental assistance given AP. This was not seen as a threat to the separation of power guaranteed in the First Amendment. No cry was heard predicting that peacetime use of naval communications by profit-making pressmen would start American journalism down the slippery slope of government intervention in the content of news reporting. AP paid lower tariffs, competed more successfully, and left editorializing, for or against the U.S. government, to its individual newspaper clients.

In 1934, AP broke the 1871 cartel agreement with the European news agencies, and assumed the right to send news everywhere. Before World War II ended, the AP urged Secretary of State Cordell Hull "to incorporate in the international agreements to be made after the war stipulations regarding freedom of the press throughout the world, and equal access by all to the news and the means of transmitting it." This fell short of seeking preferential treatment. It asked a "free flow" from everyone to everywhere. To be sure, this meant finally destroying the British and French news monopolies. By then Reuters and Havas (taken over by the Nazis) were severely crippled. AP sought an open field for everyone. AP turned to government because it was the only channel with access to all the other players. Such an appeal did not compromise the independence of the news media. Indeed, AP insisted that "The peace treaties should insure purification and extension of international news exchange by private interests." AP added, "It also seems desirable that the vanquished should accept the principle of press freedom."[9] The American Society of Newspaper Editors (ASNE) called for a "world guarantee of freedom of the press." Both Houses of Congress in September 1944 adopted a resolution supporting a "worldwide right of exchange of news...protected by international compact."

The U.S. campaign for "free flow"

As World War II ended, the American press and government began a concerted campaign to press for the free flow of information. But the campaign had antecedents in the Paris peace talks in 1919 when President Wilson's communications expert, Walter S. Rogers, stated that "the ultimate basis of peace is common knowledge and understanding between the masses of the world." The U.S. press agreed. Today, however, they almost unanimously resist efforts at intergovernmental meetings to link "peace" and "information." The press usually contends now that such linkage implies an objective for journalists beyond covering the news. They take this position because Soviet and Third World spokesmen extensively expand this basically hortatory statement linking information to peace by adding a litany of commandments for journalists. The 1978 Mass Media Declaration at Unesco is entitled: Declaration on Fundamental Principles Concerning the Contribution of the Mass Media to Strengthening Peace and International Understanding, to the Promotion of Human Rights and to Countering Racialism, Apartheid and Incitement to War.

The press and some Western governments today have forgotten the history of the free-flow debates, and the fact that sides have reversed almost 180 degrees. The U.S. press about seventy years ago actively engaged the assistance of the U.S. government to have the free-press guarantee included in the Treaty of Versailles, and later in the League of Nations. To enlist government support, however, was to poise on the slippery slope of governmental participation in the affairs of independent journalism. Yet for the record, the American press did then what Third World governments are doing today: They invite governments and intergovernmental organizations to do what the press itself cannot manage, that is, to generate state power to assure greater freedom of access and dissemination for the independent information systems.

At a meeting called by the League of Nations in 1927 the idea of foreign correspondents carrying international press cards was approved. Today, that is regarded as tantamount to the government licensing of journalists, and is strenuously resisted by Western journalists. In 1933, an effort was made to restrict the activities of American and other foreign correspondents in Europe by "forcing them to send news 'compatible with the interests' of the country in which they are stationed." U.S. newsmen regarded this demand—as they do similar proposals today by Third World countries under the rubric of their "right of communication"—as government retribution for objective coverage by independent journalists. The governments annoyed in the 1930s were the Western Europeans whose press services covered events most of the time without embarrassing the respective Foreign Offices.

When World War II ended, however, the European economies were decimated. The AP's advocacy of press freedom was viewed cynically by *The Economist* of London. It wrote that Kent Cooper's "ode to liberty" implied that the "huge financial resources of the American agencies might enable them to dominate the world." Cooper responded: "Democracy does not necessarily mean making the whole world safe for the AP."[10]

The U.S. tried at the organizational meeting for the United Nations at San Francisco in 1945 to include a freedom of information guarantee in the U.N. Charter. But both the British and the Soviets objected, and the proposal was dropped. Press freedom formally came to a U.N. body in May 1946 when the Human Rights Commission was formed. Eleanor Roosevelt, the late president's ubiquitous wife, was chairman of the commission (and, incidentally, honorary chair of Freedom House). Five separate proposals on press freedom, from ASNE, AP, UP and delegates of Cuba and Panama, were on the commission's agenda. With other major issues before the commission, particularly the draft of the historic Universal Declaration of Human Rights, press freedom questions were shunted to a new subcommission to deal with free-press matters.

The initial debate revealed the stark differences between the Americans and most of the rest of the world on questions of press freedom. Power-sharing by the press and U.S. government, mandated by the First Amendment, was duplicated nowhere else. The U.N. debates revealed this quickly. Mrs. Roosevelt expressed the two different philosophies of journalism. "Some people believe freedom of information implies that all kinds of information should be available, and that the public can be relied on to sift the true from the false." This was straight from Milton's *Areopagitica* 300 years earlier. Mrs. Roosevelt summarized the second viewpoint saying, "Some kinds of information are deliberately falsified and slanted to give the public an incorrect impression of the facts" and most people are unable to distinguish the lie from the truth. Consequently, in this second view, she said, "freedom of information implies some kind of control over propaganda for protection of those who cannot recognize it."

That paternalistic argument persists to this day, and is the rationale for government control of the news and information media. The Americans, then and now, were concerned by the subcommission's consideration of the "rights, obligations and practices [that] should be included in the concept of freedom of information." The Americans countered, as they have at Unesco and the U.N. ever since 1945: If such issues were to be examined, the U.S. would insist that the subcommission study such obstacles to the free exchange of information as: "censorship of press and radio, control of correspondence, discriminatory cable rates, powers (beneficial and otherwise) of press agencies, etc."

Through the years, the U.N. and its specialized agencies would spend much time debating the "right of correction" (1962), freedom of information (1958-61), principles on the use of satellite broadcasting for the free flow of information (1972), principles for the use of satellites for direct television broadcasting (1972), support for "a new world information and communication order" through the U.N. General Assembly and its Committee on Information (1985 to the present), and other related questions.

The great accomplishment in all these years was the inclusion of Article 19 in the Universal Declaration of Human Rights completed in 1948. The article states: "Everyone has the right to freedom of opinion and expression; this right includes freedom to hold opinions without interference and to seek, receive and impart information and ideas through any media and regardless of frontiers."

As early as 1945, however, the intensive effort by Americans to garner universal support for press freedom was focused on the United Nations Educational, Scientific and Cultural Organization (Unesco). While the U.N. was contemplating this issue, Unesco's constitution committed the organization to "collaborating in the work of advancing the mutual knowledge and understanding of peoples, through all means of mass communication and to that end recommended such international agreements as may be necessary to promote the free flow of ideas by word and image."

The head of the U.S. delegation was William Benton, then an assistant secretary of state, formerly a successful advertising executive. Benton set the stage for Unesco's communications program for years to come. He also set Archibald MacLeish, distinguished poet and Librarian of Congress, the task of writing many of the stirring, almost poetic, words in Unesco's charter. "Since wars begin in the minds of men, it is in the minds of men that the defenses of peace must be constructed"—the most stirring words of Unesco's constitution were written by MacLeish. Americans thus played a major role in assigning Unesco specific authority to devise programs in international communications. When some of those programs in the 1970s addressed the *content* of the flow of news and information, Americans (and some other Westerners) claimed the organization was going beyond its mandate, and had become "politicized."

I asked Archibald MacLeish in 1981, several years before he died, how Unesco's founders regarded "communications." A Unesco official had just implied that Benton and MacLeish set the stage for present governmental efforts to change the content of domestic and international journalism. Mr. MacLeish replied 30 November 1981 in a handwritten note to me:

> Of course the London Conference which founded Unesco was dedicated to a free press. One of the most despicable aspects of fascism was its control of the press to suppress the truth (the murder

of the Jews) and its distortion of the truth. This was taken for granted at London and in the first meeting at Paris: Unesco would always support a free press. Recent developments have therefore shocked and embarrassed me and I am grateful to you for an opportunity to set the record straight...

> faithfully
> Archibald MacLeish

In the very beginning, however, some American journalists introduced proposals for international treatment of the news media which would be thoroughly condemned by the U.S. press in 1988. Benton had asked Edward W. Barrett, then editorial director of *Newsweek*, to consult with U.S. journalists on proposals for Unesco. Among the responses: proposed international accreditation for newspersons and a code of professional behavior for journalists, both considered onerous today.

At the draft convention on freedom of information in 1948, the U.S. proposed a "right of reply." Offended nations would send responses to the government of the news service, and the government would relay the reply to the medium. The French offered an international "right" of correction. Both "rights" are heartily condemned today by American journalists and officials. Their views have been influenced by the prevalence of press controls worldwide.

The international free-press question was somewhat clouded by the Commission on Freedom of the Press headed by Robert Hutchins, then president of the University of Chicago. The Hutchins Commission, a privately funded group, strongly criticized the commercialization of American newspapers, and called for a stronger display of "social responsibility" by U.S. journalists. Then and since, many governments have defended their control or censorship of the news media by claiming that journalists do not act with "responsibility." Yet the Hutchins report was a carefully nuanced statement. If written today it would probably have made clearer the distinction between "the press" acting responsibly, and the individual journalist conducting himself/herself so as to consider the implications of sensitive reports which have socially harmful consequences—regardless of the truth of the reporting. That is not to argue for censoring the unpalatable; but if important to society, finding balanced, uninflammatory ways to present the issue. At the time, the Hutchins report was perceived to support critics of American journalism. Hector McNeil, delegate from the United Kingdom, agreed that foreign news agencies sometimes carry false and damaging reports about small countries. The solution, he said, was deepening journalists' sense of moral obligations rather than subjecting them to government-imposed standards.

In 1947 Benton described his efforts at the U.N. and Unesco, but admitted he had not "progressed very far toward our goal." He said "the world is in worse shape now with respect to freedom of information than it was in 1919." He estimated that 75 percent of the world's population was "living today under some degree of censorship." He added, "in some important areas this censorship, and the deliberately fostered distortions that accompany it, are more virulent than ever before." Interestingly, in 1988 the Freedom House assessment showed 76 percent of the nations with less than free broadcast systems; 66 percent with less than free print media. Our criteria today include more variations of control and are probably tougher than Benton's subjective standards four decades earlier.

The early press-freedom debates at Unesco set the tone for years to come. Many countries felt in 1948 that American pressure for freedom of information did not confront the question of abuses of the press. The Americans, for their part, resented calls to follow any standards except those created and monitored by journalists themselves. The result was a standoff, destined to become even more bitter in the mid-1970s and 1980s.

After several years of debate, an international freedom-of-information document was framed which severely clashed with American press-freedom objectives. The issue was passed from one U.N. committee to another. Clearly, the U.N. was finding press freedom impossible to handle. Slowly, the delegates considered moving the press freedom issue to Unesco. The State Department's Samuel DePalma, however, saw "one minor drawback." He noted "the unwillingness of Unesco to deal with anything of a political nature." That, he felt, would reduce Unesco's ability to deal with freedom of information issues. That was in 1949. Nearly forty years later, DePalma, as a member of the executive of the U.S. National Commission for Unesco, with this author, grappled there with U.S. charges that Unesco had "politicized" communications and other programs.[11] The U.S. withdrew from Unesco in January 1985.

The events leading up to that withdrawal are significant as a demonstration of the power of information in conflict with the power of State. In reviewing the history of American charges against Unesco in the mid-1980s it is useful to recall the words of Carroll Binder, editorial page editor of the Minneapolis *Tribune,* who served as a U.S. delegate to the Special Committee on the Draft Committee on Freedom of Information (1951). "We" (the United States), said Carroll, "are godparents of the international approach to problems of freedom of information. We may not like the way in which the child has grown...Some may wish we had never sponsored it. We cannot deny it is our baby."

As the U.N. passed the press-freedom baton to Unesco in 1951, Isador Lubin, the American representative on the ad hoc committee of the Economic and Social Council, rejected "the idea that Unesco's activities in the field of freedom of information and of the press should be confined to the purely technical aspects." He underscored this position: The U.S., said Lubin, "could not accept the idea that an organization set up to disseminate information should not act when political obstacles hinder such dissemination. Unesco was quite as much concerned with the nature of the information to be disseminated as with the practical means of disseminating it." Clearly, that meant that Unesco should "combat the effects of any harmful propaganda and attempt to stamp out any form of discrimination liable to hinder man's cultural development."

Such views were to be repeated twenty years later by Soviet and Third World delegates pressing for a new world information and communication order, and were vehemently rejected by American and other Western journalists and governments who called this interference in the free flow of information. The seven years of American efforts to move press freedom commitments through the United Nations' program had failed. But the seeds had been sown in Unesco for still more bitter controversies to come.

The Unesco fiasco

Westerners, particularly Americans, who moved the press-freedom action into Unesco in the early 1970s, still saw "the free flow of information" as an idealistic objective. The Soviet Union had long since attacked free flow as American "cultural imperialism." In Marxist analysis, "the importance attached to 'free flow' then, as well as today, arises from its indispensability to the successful operation of a worldwide business system that requires, along with a free flow of information, the free flow of capital, labor, and commerce. These are inseparable elements of a privately administered world economy."[12]

The Soviets, in fact, introduced anti-free-press resolutions at the Unesco general conferences in 1970 and 1972. The toughest came in December 1975 at Paris after the U.S. and other Western delegates walked out of the planning session over an action not to seat Israel.

The centerpiece formulated at the Nonaligned Movement in 1975 was the plea for a new world information order, linked to the already mooted new international economic order. The "order" was the brainchild of Mustapha Masmoudi, then secretary of state for information of Tunisia. Masmoudi was a young, energetic spokesman. He understood the inherent power of information in developing as well as developed countries. We met for the first time in his office in Tunis shortly before he went to Paris in 1978 as one of sixteen members of the MacBride Commission —Unesco's International Commission for the Study of Communication Problems. Since there is no one official version of the "new order," Masmoudi's outline may be regarded as definitive—at least as an expression of

Power, the Press and the Technology of Freedom

the most vocal, activist member of Group 77 (the nonaligned countries, now numbering more than 100).

Masmoudi's premise was clear: The global news flow, as other injustices, cries out for "revolutionary equity after centuries of imbalance and injustice." The effort to provide objective and balanced news has been "obstructed," particularly by "the monopoly which has been exercised ipso facto in some of the developed countries by powerful news media." Masmoudi charged, "The news-trusts erroneously construe their role and adopt vis-à-vis a given country a different or even diametrically opposed attitude to that country's own [attitude], or that of other nations or the international community." He denied that the "new order" was meant "to set the poor against the rich, the unaligned and third-world countries against the countries with powerful news media."[13]

Among the contemplated actions: support the Nonaligned Countries' News Agencies Pool (initiated in 1976) assist every developing country to "exercise full sovereignty over information"; emphasize the "tendentious" use of the "free flow of information" by the big news transnationals "who extend their pernicious activity devoid of any international control or regulation"; and examine the possibility of drawing up "common legislation to govern the activities of the big news transnationals in the nonaligned countries." (These are extracts from the final report of the Nonaligned Symposium held in March 1976 in Tunis.)

Masmoudi turned the free flow concept to his own (and other governments') use. If it is imperative for an individual to get objective news, he argued, "one cannot contest the right of states and groups of people to speak, nor refuse them the use of every means of communication." Yet Masmoudi insisted that he had no "intention to supplant the big information systems in the East and West, and even less 'go to war' against them," but rather to create a new formula for the exchange of information. That new formula would be regarded in the West as an insurgency if not a war.

I debated Masmoudi in Washington, October 1978, before the Carnegie Endowment for International Peace. That was just days before the Unesco general conference in Paris would approve the long-contested Mass Media Declaration. Masmoudi and I debated his "new order" but he obviously softened both the tone and the objective for his media-dominated audience in Washington. He conceded, however, that "the significance of the new order to which the developing countries aspire is fundamentally political in nature." He also said the information order is "inseparable" from the new international economic order.

The linkage is an undeniable statement of fact: international banking, monetary systems, and commerce generally are today unthinkable without international communications. Masmoudi, however, suggested more than this. He seemed to say that information power, so essential to state power, is a political, a geopolitical fact. And the developing world wanted its share of that power. The threat to "free-flow" arises from the methods used by many governments demanding greater information power. They already control or censor domestic news and information, and want to make that the standard and practice in the international flow of news and information.

The opposition of the Reagan administration to Unesco's several geopolitically-oriented programs—communications was the most crucial—led to a bruising, three-year showdown.

Not only Unesco's programs in communications, but its faulted conception and communicating of its own communications programs, more than any other factor paved the way for the United States decision to leave Unesco on 31 December 1984. Since its inception, Unesco had conducted or commissioned research in many aspects of interpersonal and mass communication. It called for wider sharing by developing countries of new communications technologies and substantial change in the nature and content of the flow of news, data, and diverse cultural expressions. Nearly 75 per cent of Unesco's member states are "developing." Some of their polities may be called free; others, partly free or not free. All recognize that the nascent information revolution is steadily gathering momentum. Expanding the Third World's communications capabilities is, therefore, crucial. No less important is persuading developed-nation communicators, including those independent of governments, that they *should* improve their own coverage of Third World developments and should do so by permitting more words and views of Third World people to pass through the major world news and information systems.

That is an intellectually persuasive appeal. It could be still more persuasive if most developing countries opened their own news and information systems to diverse views and granted greater access to information by domestic as well as foreign communicators. Such an appeal could also be more effective if Unesco itself had encouraged Third World journalists, rather than mainly information ministers and officials, to participate in communications debates, and had examined the diversification of domestic as well as global messages.

As persuasive as the Third World appeal might have been, it did not persuade the nongovernmental news, information, data-processing, and cultural disseminators in the industrialized countries. A natural inclination to defend the status quo was not the sole reason for their unresponsiveness. The promotion of communications sharing under the rubric of an undefined NWICO frightened the independent journalists. They saw governments as censors, not friendly sharers of the information flows.

That, in turn, motivated Western governments to resist most change in international communications. Change came to be regarded as threatening to the freedom of individuals, corporate and cooperative news services, and Western governments. All invoked the free flow guarantees of the Unesco charter and the Universal Declaration of Human Rights. Fears that the NWICO was a prescription for governmental control of the media of communications—domestic and transnational, independent media as well as government-owned—were amply fueled by Unesco debates and projected programs. At Unesco general conferences in 1970, 1972, 1974, 1976 and 1978, there were increasingly bitter discussions of proposed universal standards to influence the *content* of transnational communications.

Despite reports by usually trustworthy news media, they failed to state that not one of these debates resulted in Unesco's approval of a single resolution that supports censorship, the licensing of journalists, or other government controls of independent news or information media.[14] Tensions and bitterness preceding votes on drafts perceived as censorious, however, remained prominent in Western memories; the defeat of all such proposals was forgotten. There remained an inchoate belief, particularly among U.S. opinion makers, that "Unesco"—the organization itself, not only delegates of some member states—has an agenda, sometimes hidden, that will ultimately fashion censorious norms.

Marxists and rightist authoritarians (and opponents of both)—each in their own ways—regarded the NWICO initiatives at Unesco as seeking to establish a particular universal norm for global communications. That norm would replace present systems with one that took directions from some governmental center. The Marxists and others favoring central control systems (whether "rightist" or "leftist") were activists; their opposition, mainly in the industrialized West, were generally reactionists. The most authoritative Marxist, Kaarle Nordenstreng, describes the decade-long debates that preceded the 1978 Mass Media Declaration:

> ...the Mass Media Declaration was introduced on Unesco's agenda in 1972. In fact, the Declaration came to serve as a symbol and catalyst for conflict between the forces of the new order and its adversaries....*in the world arena as well....pressing the West into positions of retreat and even despair* (emphasis added).[15]

Article 12 of the Soviet's "Nairobi draft" epitomized the Marxist position: "[S]tates are responsible for the activities in the international sphere of all mass media under their jurisdiction." Yet that view was rejected in 1976 by most developing and industrialized countries. And subsequent versions of the revised Soviet draft were similarly rejected in favor of the Mass Media Declaration of 1978. The Soviets were shocked at the sudden gutting of their text and a substitution approved by acclamation. Yet, as late as 1984, an official U.S. briefing paper used an adversarial reference to the bland 1978 declaration in order to support the U.S. decision to withdraw from Unesco. The paper referred to the Mass Media Declaration as "Soviet-initiated."[16] It is as accurate to say that the formation of NATO was Soviet-initiated because the actions of the USSR motivated Western Europeans to form a defensive alliance. Thus, the extremes meet, and the bland declaration is regarded by both poles as a step toward centralized controls.

While Nordenstreng would support such a straight-line development, he accepts that

> Unesco served as a central forum *for the articulation and mobilization of a new policy-oriented generation of communication research. But it would be misleading to understand the new perspectives as a creation of Unesco; all that Unesco did was to facilitate the expression of certain political and intellectual tendencies of the day* (emphasis in original).[17]

Unesco did "facilitate" the expression of both censorious and noncensorious views, but it supported by vote only opposition to censorship.

Western press reports, however, saw Unesco differently, and that version dominates the data banks of even the best U.S. newspapers. There, the NWICO is generally defined as Unesco's "plan" to control the press, though no such objective has ever been approved. Thus, when the Soviet Union introduced a resolution in November 1983 that would authorize Unesco to monitor the press, the *New York Times*'s[18] spread the story over six columns, under a large headline. After that Soviet draft was killed, the fact was not reported in the *Times'* short concluding report of the negotiations (27 November 1983, p. 3). Thus, Unesco's name remained blackened over communications issues.

Unesco debates were, indeed, exploited by regimes that already owned or controlled their domestic media and limited access to information sources by foreign correspondents. Ideologues of the Soviet bloc and Third World nationalists readily supported politically right as well as left developing countries which would establish a universal norm: statist journalism under central control. When the leftist military in Peru nationalized all that country's newspapers, a nonexistent "mandate" from Unesco was cited as justification. At the Nonaligned Information Ministers Conference in Jakarta in January 1984, President Suharto called for stronger control of the news media in developing countries to counter "domination" by Western news services. He reflected countless declarations heard at Nonaligned Movement meetings and "recalled" at Unesco.

The valid critiques of Western journalism and the Third World's natural yearning for better coverage and infrastructure were lost in the years of Western reporting of these acrimonious debates. Emphasized instead were the objectives of those who seek to change the content of news and information by defining journalistic "responsibility" and who would control staffing of news media through licensing devices linked to "protecting" journalists.

Unesco, meanwhile, failed to communicate to its global audience the distinctly differing objectives of those participating in the NWICO debates. Still more self-defeating, Unesco tried repeatedly to forge consensual statements out of principles that were regarded as non-negotiable by many participating delegates and the citizens they represented.

The report of the MacBride Commission is an example. Every sentence, paragraph and chapter is balanced to reflect the equalizing of two completely disparate approaches to news or information: one centralized and controlled by government, and the other diversified and independent of government. No single standard can ever satisfactorily encompass these fundamentally different systems. Before the MacBride Commission met, I was asked by the commission to recommend an approach it might follow. I urged the commission to provide "two distinct sets of standards or objectives: one for government-controlled journalism, and another for nongovernmental journalism."[19] In similar fashion, I added, Unesco's 1978 General Conference "would be well advised to reconsider adopting a single draft declaration" on the mass media. In seeking for a decade to define the NWICO as a single, universal norm, Unesco fell victim to all the ideologizers on the political right and left.

For the attacks on Unesco's communications programs set the stage for the United States' withdrawal. The debates over licensing journalists, assigning them "responsibilities," and monitoring their output blackened Unesco in the eyes of the influential liberal wing of the American press. U.S. conservatives had long opposed many aspects of the U.N. system. The Unesco communications issues fit neatly into their anti-U.N. argument. American liberals and conservatives therefore joined in attacking Unesco.

The U.S. withdrawal notice also charged budgetary excesses, politicization of other programs, and general mismanagement. There is, indeed, much to reform; everyone, including the director-general, agreed. But the American ultimatum probably could not have been as readily produced if American opinion molders had not been persuaded over a decade that the NWICO was aimed at *them* and the content of their media products, and that change was to be government-generated. The crisis might not have been as easily sparked, moreover, if Unesco itself had not failed to differentiate between analyses of communications flaws and some still-undefined intergovernmental actions to eliminate those flaws. It should have been possible to examine the needs of communications-poor countries without seeming to threaten the independence of communications-rich media (generally now free of governmental control). It was ambiguous for Unesco officials to disavow proposals to censor or license journalists, while Unesco programs discussed aspects of such objectives. It was no less ambiguous for American officials to single out Unesco for withdrawal when the same complaints can be directed to most agencies in the U.N. system. Despite official denials, it appeared that the United States was using Unesco to alter or withdraw from the wider U.N. system. Such an objective deserved open and full debate.

Unesco must address the implications and opportunities inherent in the information revolution. It should explore the limitless applications of communications technologies to personal and national development and do so by advancing diversity of news and systems of delivery, not by debating restrictions intended to achieve critically defined objectives that are set forth as universal norms.

And Western communicators should understand that they *do* dominate international communications and are therefore subject to sober analysis and appeals for constructive change.

After the Unesco fiasco

With the withdrawal of the United States and the United Kingdom from Unesco, the organization suffered a severe financial loss, followed by a struggle to name the successor to Director-General Amadou Mahtar M'Bow. After a year of campaigning by the incumbent and others, Federico Mayor, a Spanish biochemist, was elected. He took office in November 1987, and quietly enunciated a communications program *without* a new world information order. Instead of a free and balanced flow of information he substituted "free and uninhibited."

Mayor told Unesco's board in October 1988 he would remove ambiguities and misunderstandings from the controversial communications programs. He rejected the licensing of journalists and reaffirmed Unesco's constitutional commitment to a free flow of information. He said the departure of the U.S. and the UK was mainly over communications programs he will now revise. Later, Soviet Foreign Minister Edward Shevardnadze met with Mayor and said the Soviet Union bore a share of the responsibility for the U.S./UK departure. "Sometimes we were acting on the principle of a 'tooth for a tooth' under the pressure of confrontation," he said, "trying to oppose the attempts to impose on us ideas which we could not accept." He said there had been "excessive introduction of ideological dimensions" in Unesco debates.

The debates were moving from ideology to pragmatic support for Third

World communications infrastructure. Victories of sorts had been achieved both by the Nonaligned and the Westerners. Since the Bandung Conference of Asian-African countries in April 1955, the Third World had pleaded for its share of economic assistance. That cry produced favorable resolutions in the U.N., some sizeable grants-in-aid, but no real change in the economic power structure. The information-order challenge to the West was intended to pry loose communications technologies for the development of Third World economies. Only after the severe ideological attacks on the Western news media at Nairobi in 1976 did the U.S. formulate a trade-off: reduce the ideological assault and we will provide significant communications technology training and transfer.

During 1976-78, when the Mass Media Declaration was debated, the bitter attacks flared up repeatedly. The bland declaration, however, encouraged the West, and led to tech sharing through the newly created Unesco program, International Program for the Development of Communications (IPDC). The U.S., while conceiving IPDC, never contributed to its fund to support Third World communications. The U.S. did give grants "in trust" to bilateral projects on the IPDC menu.

The 1983 Unesco general conference also produced a further tamping down of communications controversies, partly to ward off the American/ British withdrawal, and partly because a satisfactory compromise had been negotiated. A "new world information and communication order" (NWICO) was hereafter to be seen as an "evolving and continuous process." That format implied that no preconceived "information order" was contemplated, or would be suddenly imposed. As a member of the U.S. delegation, I negotiated those communications issues at the conference. Shortly afterward, despite the generally favorable climate, the U.S. announced its withdrawal.

With the U.S. and UK gone, Unesco entered a crisis. Staff morale dropped. Funding for programs was severely cut. Recriminations among delegations, and between them and the staff, abounded. Although previously scheduled communications programs rumbled along, no controversial project was completed. An inflammable communications issue had flared just as the U.S. was debating withdrawing. In mid-1984, a Latin American left-of-center press group planned a meeting the following March on the controversial subject of "protection of journalists." In the past this had been tied to governmental identification of who is a journalist, and whether one was meeting standards of practice. Unesco had provided several thousand dollars to help the planners of this meeting. Western journalist associations immediately protested. I regarded the incident as a serious barrier to the U.S. remaining in Unesco. For, while I had been the most active critic of Unesco communications programs since 1976, I believed the U.S. should remain in the organization and fight for American beliefs and interests. I also felt that many valid questions about global communications should be reasonably discussed. And Third World communications should be helped to secure better facilities.

I publicly expressed these views as vice-chairman of the U.S. National Commission for Unesco during the years when the withdrawal decision was being made. The commission was congressionally mandated to advise the secretary of state. The president accepted instead the position of Assistant Secretary Gregory Newell who had been dispatched from the White House staff to manage the withdrawal from Unesco and run an aggressive challenge to the United Nations and other specialized agencies. In my several debates with Secretary Newell he confirmed that as the administration came into office in 1981 it had temporarily eliminated Unesco from the first budget the administration could submit. There had been no "careful study" of the Unesco question as one Newell assembled, with questionable accuracy, several years later. The 1982 funds were restored, but the decision to leave Unesco remained with the few officials who managed the pullout.

Withdrawing—reforms notwithstanding—would satisfy those Americans who have little or no interest in Unesco's value for Americans and others. They see that organization as merely a chip in the larger game called the United Nations system. And trashing Unesco, for them, was a shot across the bow of the entire U.N. system. The U.N. needed to be reformed. But that was another and larger issue which was yet to be examined frankly, openly, on its own merit.

I regarded the contemplated Latin American "protection" meeting as a crucial test of Unesco's official insistence that it rejected governmental controls over the press. The consortium of press-freedom organizations expressed their concern, but that appeal seemed to arouse Director-General M'Bow's ire. I asked him privately to cancel Unesco's role in the protection meeting. [Our exchange of correspondence appears in Appendix E.] M'Bow made several useful clarifications. The "protection" meeting was not held as scheduled.

After M'Bow left the scene, the IPDC, in 1988, was finishing a large World Communications Report. This contained nearly 200 pages of tables on print and broadcast personnel and facilities the world over. The report was also to include a major essay on developments in communications. An early draft contained many of the highly controversial terms and concepts that brought Unesco to the brink just a short time earlier. The material objectionable to the Western critics was later removed.

The basic debate on information vs. state power was transferred in 1985 from Unesco to the General Assembly of the United Nations. There, at the Committee on Information, a new cast of characters—the political delegates—would learn anew the litany of the Unesco debates. At each new session, in 1985, 1986, 1987, 1988, they would argue again the NWICO theme, using the Unesco formula, but not always with the same results. The U.S. ran its U.N. information battle on a tight rein from Washington. In 1986 and 1987, the U.S. stood alone against final drafts of an information resolution—usually because several words ("establishment" of a NWICO) suggested approval for a formula *less* acceptable than the new NWICO litany. Ironically, the U.S., having recently withdrawn from Unesco, was now calling for the Unesco formula in the U.N. information debates.

These latest U.N. debates based on geopolitical approaches to global information—sheer state power over communications—underscored the MacBride Commission's questionable conclusion in 1980. It said the American and French revolutions established a first generation of individual political and civil rights—the rights of man. The Russian revolution created the second generation of economic and social rights—food, housing, education. The third generation was to be the collective—States' rights. This was exemplified by the governments' right to communication.

We may foresee a fourth generation: the Age of ISDN, in which all these rights—particularly those of individuals—have equal access to information, and to the transmission of information through limitless electronic linkages. And the *receiver* will decide what he/she wants to retrieve.

Meanwhile, says Colleen Roach, a critic of U.S. policy, every "NWICO-related issue or subject ('social responsibility of the press,' 'protection of journalists,' 'right to communicate,' etc.) was reduced to the slogan of 'government control of the media.'"[20] The reason for this strategy, says the critic, "is not merely the U.S. predilection for oversimplification of complex issues, or even the historical commitment to the First Amendment, although these factors are not to be neglected. The emphasis on the government control argument reflects, above all, the need to ensure that the NWICO would not reinforce government-run or public sector communications media at the expense of the private sector." For, in this analysis, "the implication for U.S. transnational telecommunications business interests is concretely reflected in the 'deregulatory fever' and the move toward the privatization of the public sector. This coupling of the deregulatory and privatization movements necessarily undermines a strong public sector of the economy whether on a national or international basis."

Apart from the serious issue of governmental control over information, this argument ignores the failures of public-sector services. Public communications in many countries have histories of inefficiency, at best, corruption and high rates for services at worst. Privatization and deregulation are sweeping Western Europe, the Pacific area and Latin America, following the lead of the United States. Pluralism in carriers, communicators and manufacturers provides choices for individuals and groups. Choice is a basic democratic factor. Competition in the communications field has brought steadily decreasing prices with ever-expanding technologies. One can support privatization and deregulation—when they prove effective—and still fault American policy for having been shortsighted and inept in playing the game at Unesco in recent years. One can also plead for greater public discussion of communications issues, including the question of journalistic integrity and professionalism—perhaps a better term than "social responsibility"—without aiding the avowed press controllers.

Writing between the lines

It is time to turn the attention of the international forums on communications to those places where news and information are most in need of "correction."

In Vietnam in 1974 I asked a journalist why a short item on the front page of a daily newspaper appeared there. The story told of a child in a small southern town dying after eating cactus. The journalist replied: "You must read between the lines. The girl would not have resorted to eating cactus if she and her family had not already suffered hunger. The paper could not publish that." I have since read between the lines in South Africa, Paraguay, China, the Soviet Union and many other places. Such writing or speaking requires an eloquence and intelligence beyond usual journalism. When successful, writing between the lines also inspires an unusual sense of awareness in the reader. Both the writer and reader are drawn together by the knowledge that they share a common key to language and, more important, an understanding of reality which the government denies or forbids to have examined. Such writing also defeats the censor by rejecting self-censorship. Writing between the lines is the courageous way of challenging the censor and his chief, without permitting the issue to reach the surface. Neither the censor nor his superior can afford to admit what both know: the writer and the reader are aware of reality, not the authoritarian government's version of reality.

The most eloquent description of between-the-lines writing has been set down by Miklos Haraszti in the *Velvet Prison*.[21] He describes his native Hungary:

> Communication between the lines already dominates our culture. This technique is not the speciality of the artist only. Bureaucrats, too, speak between the lines: they, too, apply self-censorship. Even the most loyal subject must wear bifocals to read between the lines: this is in fact the only way to decipher the real structure of our culture.....Our messages between the lines are suggestions sent to the same state and the same public that our official lines continue to serve. Debates between the lines are an acceptable launching ground for trial balloons, a laboratory of consensus, a chamber for the expres-

sion of manageable new interests, an archive of weather reports. The opinions expressed there are not alien to the state but are perhaps simply premature. This is the true function of this space: it is the repository of loyal digressions that, for one reason or another, cannot now be openly expressed.

Writing between the lines can continue only after a certain degree of information-power sharing has already been acceded by officials. They must have realized that totalitarian journalism is suspect in every aspect—even in weather reports and street maps. Consequently, some credibility must be secured for official news if communication is not to be solely administered through the barrel of a gun. *Moscow News,* for example, noted that U.S. commentators were not entirely favorable to President Bush in covering his 1989 inauguration. "The ability to critically treat words spoken from the loftiest of rostrums represents an enviable trait of the American way of life," wrote the Soviet reporter. But this could also be read as a "between-the-lines" commentary on the *Soviet* way of life.

Increasingly, new technologies of communications will add many new "lines" for writing "between." It will be impossible for officials to monitor them all. Typewriters have been licensed in Romania, copier machines forbidden for unofficial purposes in the Soviet Union, and direct-dial telephoning denied citizens in many countries where censors bug the calls. But these restrictions are being swept away by the urgent need of inefficient, backward authoritarianism to climb aboard the technotronic carrier before it leaves the twentieth century.

There may, finally, be much between-the-lines writing in the USSR. An historic, almost unnoticed, step was taken by the Soviets in September 1988. They announced a joint venture with San Francisco businessmen to enable Soviet citizens to purchase computers, software, audio-visual products and printed materials. This is one of the first steps to allow the average Soviet citizen to engage in copying and printing. Since central surveillance of idea-flows will not have ended, Soviet citizens will now be able to write and read between the lines. The USSR has exercised tight control over access to information machines. They are all owned by the State. Lack of access to such services has been an obstacle to the many unofficial newspapers and journals that have sprung up under glasnost. A spokesman for the U.S. Information Agency called the new venture "extremely radical because even most Soviet government bureaucrats do not have access to copying machines." The joint venture was an outgrowth of the Gorbachev-Reagan Moscow summit, said Soviet Deputy Consul General Gennady Zolotov in San Francisco. The project will include eighteen consumer outlets in Moscow and a central printing plant.

Driving such projects is the economic crisis in the USSR. But the most long-range, radical effect of this information-power sharing by the State may be the experimentation with new ideas by average citizens. They are still inhibited, however, by an April 1989 ruling that "anti-Soviet" uses of reproducing machines may bring a seven-year prison term.

Small liberating technologies, already used in some crucial situations, replace the need for writing between the lines. During the anti-Noriega demonstrations in Panama in 1988, expatriates broke the government's censorship by using the telefax to send newspaper clippings from Washington to Panama. The clippings, reports from Panama to American papers, told Panamanians what their own newspapers could not report. The fax reports went to churches, labor unions, and schools, which then reproduced the news on small copier machines. It may be recalled that the Shah of Iran was overthrown with the help of audio cassettes smuggled into the country from France. Leo Tinofeyev, the editor of *Referendum*, to whom we spoke in Moscow, prints his unofficial magazine on a desktop computer (also "unofficial"). Nonviolent dissenters in Nicaragua and Chile use miniprinting devices to disseminate their views and announce meetings.

In the early days of automation, the mechanization of ideas was both hailed and feared—feared mainly because it was assumed that grave centralization of information power would follow. George Orwell's *1984* dramatically forecast "newspeak." Not only would all messages be centrally controlled, but the language would be altered to convey only official words and thought. In many countries today, crude versions of *1984* have been in place for decades. But there is a creeping liberalism in many "closed" societies. Writing between the lines is one indication. Tentative discussions of democracy is another. The Beijing newspaper *Bulletin of Theoretical Study* carried a plea in 1988 for "political modernization" in China. Said Bao Xingjian, "The Socialist State power organs should be public servants, but part of the functionaries now have become the lords over the people, by taking advantage of their power." He said capitalist democracy is not the threat to "socialist democracy." Feudal autocracy is the devil. Both "democracies" oppose it, said Bao.

The most hopeful sign is the arrival in closed places of the machines of enlightenment. They may produce the Age of ISDN—if political will permits, and it may be forced to, in many places now deemed unlikely.

Guided journalism: four models

In the decades before a full networking of new communications technologies can link all countries—developing as well as developed—power-wielders in most countries will resist yielding control over centralized and censorious information systems. The news and information controllers will use all the technical and political devices they can muster. Some of their rationalizations for retaining information power will sound eminently plausible.

Since the overthrow of the Allende regime in Chile in 1973, the Pinochet dictatorship has thoroughly restricted newspapers, magazines, radios and television. It has "guided" those which continue to operate. Journalists charged with violations under new laws have been repeatedly freed by the civil courts; some must also face harsher military courts. Such actions have a chilling effect. And more than 100 journalists have been left unemployed. Pinochet rationalizes the closure of newspapers and harassing of reporters and editors by claiming he is resisting a return of Communist-leaning journalism, and moving, even glacially, toward the reinstatement of democracy. Chileans still have great pride in their long democratic history. The Organization of American States (OAS), after extensive study in 1985, repeated the finding made by its Inter-American Commission on Human Rights (IACHR) in an investigation eleven years earlier. The OAS found Chile still violating "the human rights of its citizens by denying their journalists the power to report freely." The IACHR stated in clear terms the classic power struggle between the state—fearing a dissenting idea—and freedom of opinion and expression:

> Whatever the consequences of actions based on a particular ideology, in any event, and whatever the value judgment merited by that kind of thinking, it is clear that ideologies cannot be eliminated the way an epidemic disease or a serious social vice is eliminated, if the basic principles of a representative democratic system of government are to survive.
>
> This deviation from the recognition of freedom of opinion is undoubtedly the result of temporary political circumstances and emotional factors. It is to be hoped that once they have both been overcome, the upholding or disseminating of particular ideas will cease to be punished as a crime. However, it must be noted for now that this has been classified as a criminal act in Chile, for the avowed purpose of "eradicating" a particular conception of society and of the causes of historical change. Undoubtedly we can disagree with that concept, but the only way of eliminating it without paying too high a price is by the appeal to reason and persuasion.
>
> It is inadmissible that, because of the mere fact of upholding and disseminating a certain ideology, a man becomes a kind of "untouchable," whom it is considered legitimate to deprive of the possibility of working, deny him the free expression of his thought, and even send him to jail.[22]

Perhaps the classic exposition of state-guided journalism has been made by Prime Minister Mahathir Mohamad of Malaysia. His view deserves extensive reporting. He has provided a philosophical, political and pragmatic defense of government guidance for all the mass news media.

When I visited Malaysia in 1985, the prime minister had begun to warn journalists that they were threatening the stability of the country. Malaysia has a large ethnic Chinese minority. For a decade ending a quarter-century earlier, ethnic Chinese communists terrorized the Malay majority. But a political settlement was reached, and the two communities share the geography, if not the political power. Fear was just below the surface. I sensed this in speaking privately to Malaysian journalists. One said, "I come to work every day ready to go to prison." Yet the prime minister and several members of the cabinet spent many hours September 1985 explaining to more than 100 ASEAN journalists how free journalism is in Malaysia.

Dr. Mahathir Mohamad began by asserting that "there never was this individual man, born free, living completely unfettered in isolated splendor." Thus, "a code had to be developed and imposed by common consent... that could not but restrict individual freedom." Then, an "enforcement authority" was needed. The news media, too, are actors within a human community. They have become "so powerful a force in fact that kings and presidents bow and scrape" before them. There are, he said, four basic models regarding the concept of freedom and the role of the press: the authoritarian, Communist, libertarian, and social responsibility models.[23]

Each system, he said, has its own assumptions and "none are completely without virtue, not even the communist model." He added, "none are without flaws...not even the libertarian model that so many in the Third World, unable to break the shackles of psychological and intellectual neocolonialism, sometimes aspire to with such wide-eyed enthusiasm." He said he had "no negative assessments about the curbing of press freedom in Britain and the United States through the introduction of censorship during the First and Second World Wars." It should be "plain to the inventors of the doctrine of 'clear and present danger,' that [many societies today] have no choice but to do what needs to be done." He declared himself "a firm believer in the greatest freedom consonant with the vital interest of society." Therefore, he added, "for most countries most of the time the morally proper choice is the social responsibility model." The prime minister then proceeded to "demolish" the other three forms he had described:

> Both the authoritarian and the communist model believe that the mass media is a servant of the state. Both assert a monopoly of wisdom by those in authority. However, the communist model requires the mass media to be more active, positive tools for the use of government or the party for the achievement of socialist goals. Communist systems demand more than just nonobstruction and

noncriticism and a little help now and then from media practitioners. The media must be constantly active propagandists, agitators, and organizers of public opinion—every day of the year and in every column inch. Secondly, the communist model requires state monopoly of all the means of mass communication. Under the communist model, because there can be only one truth—the truth as defined by the Communist Party—the media must work assiduously to mold opinion to ensure a oneness of perception and thought..."the correct view"—is the ideal...a variety of views is not only unnecessary but immoral.

He criticized the authoritarian and Communist model: "Because it is in the authoritarian and communist state that abuses of authority and power are likely to be greatest, ironically it is essentially in the authoritarian and the communist state that morality demands that the media must be a check, that the media be in a confrontationist mode. The watchdog role of the media is needed most in communist and authoritarian systems—where, of course, it is tolerated least. All wisdom does not spring from a single source, truth from a single mind, even a collective mind made up of a large number of intellectual giants. If nothing is to be published, broadcast or televised unless it has been approved by those in authority, power must always be the determinant of truth."

The prime minister continued with this eloquent attack on Western-style journalism:

There are many things wrong with the libertarian model. First, it must be quite clear that man is as much an irrational animal as a rational one.....Even the wisest of men have often consistently been led up the garden path..

Second, is it right that truth, the whole truth and nothing but the truth, must always be told, at all times?...History is littered with examples where it was justified not to tell the truth, the whole truth and nothing but the truth.

Third, the libertarian model in its unremitting advocacy of the adversarial role may be justified in the case of an authoritarian or Communist or evil government....The basic assumption that government must always be corrupt and evil is also absolute and silly nonsense.

Fourth, if it is assumed that power tends to corrupt and absolute power tends to corrupt absolutely, by what magical formula is the media itself, with all its awesome power, exempt from this inexorable tendency? Is power the only cause of corruption? Freedom, too, can corrupt and absolute freedom can corrupt absolutely.

Fifth, the libertarian assumption of a free marketplace of ideas where there is a multiplicity of voices, where each individual has a chance to have his say, can exist only in the realm of theory....The concentration of media even in the United States, the haven of the libertarian model, has concentrated power in the hands of a select few.

Sixth, the libertarian model is based on the childlike assumption that the media will generally, if not always, adhere to ethical practices and aspire to the public good.

The prime minister made this unusual revelation of instability as part of his argument:

> For a society precariously balanced on the razor's edge, where one false or even true word can lead to calamity, it is criminal irresponsibility to allow for that one word to be uttered. It can be no surprise that it was in the United States itself that the doctrine of "clear and present danger" was formulated. Comparatively few countries in today's world are ultra stable states where full, free and utter license can be allowed to run riot. Even in these ultra stable states such license has not been allowed. There is and has never been such a thing as absolute freedom. It is my view that regardless of circumstance or time, the best model is the social responsibility model.
>
> Its basic assertions are simple. The individual has rights. So too does society. Whereas the authoritarian and the communist will boldly say that the rights of society must take precedence over the rights of the individual, and the libertarian will take the equally rigid view that the rights of the individual must override those of society, I believe that it is a question of qualitatively and quantitatively balancing the two rights.

Then, the key question: *"Who is to decide on the balancing of the two rights? In a democratic state with a democratically elected government, it is the task of the democratically elected government....*To put it another way, so long as the press is conscious of itself being a potential threat to democracy and conscientiously limits the exercise of its rights, it should be allowed to function without government interference. But when the press obviously abuses its rights, then democratic governments have a duty to put it to right..." Almost slyly, the prime minister ended: "Now let us see how this little speech of mine is treated by the media."

The press treatment of this pronouncement was itself instructive. The major Malaysian daily, which has significant financial support from individuals in the ruling party, carried four-inch high, front-page headlines over a report that began: "Datuk Seri Dr. Mahathir Mohamad said today the media must be given freedom but stressed that this freedom must be exercised with responsibility." The headline read: "Media must act without prejudice and malice—Freedom with responsibility."

The less party-dominated paper under a five-inch banner headline—"Limits of Press Freedom"—began the story, "Datuk Seri Dr. Mahathir Mohamad assured the press today that it will be allowed to function freely if it conscientiously limits the exercise of its rights and is conscious that it is a potential threat to democracy."

The subtle differences in the headlines and reporting were themselves an indication of the partial freedom under which the Malaysian press operates. Both reports were accurate summaries of the prime minister's talk. One report stressed press freedom, the other press responsibility. The full text was published in the *New Straits Times*, which is strongly influenced by the party.

But there was no editorial comment on the speech—and I was told there would be none.

I took the liberty the next day of responding to the prime minister. My remarks were not published in the *Times*, but did appear in the *Star*, the less government-dominated paper. I welcomed the prime minister's discussion of the classic four models of the press and said that, "I have long favored the social responsibility model—but with a difference: Social responsibility, by definition, invokes the responsibility of the journalist to *society*, not the government. The government is no less in need of watching than other organs of society. The watchers over the press, to answer the prime minister, are a more diversified press and an informed citizenry—both of which will monitor press infractions. To expect government to monitor the press—any government, even the most democratic—is to tip the scale inevitably in favor of government overpowering the press. For only government—not the press—has the power of the police, and the threat of a call in the night.

"The openness of a free society promises not everlasting truth, but the freedom to pursue it; not absolute freedom, but a balancing of power, particularly brain power. The canons of professional press conduct—based on a social contract with all of society, not just government—is the surest way to strengthen both democratic government and social stability."

Some months later, the prime minister pressed a bill through parliament which toughened the Official Secrets Act. Prison terms were provided for passing "state secrets." They were only loosely defined. Soon afterward, major newspapers were temporarily shut down, and journalists arrested. The *Asian Wall Street Journal* was temporarily banned, and its correspondent expelled. A 1988 amendment to the Printing Press and Publications Act (1984) gave the prime minister "absolute discretion" to ban or restrict the publishing or importation of any publication deemed "likely to alarm public opinion." The minister's decision shall be final, and not called into question by any court on any ground whatsoever.

This is the strong arm of guided journalism. The "guide," however, should be expected to act according to (1) the ethic of reasonableness, (2) the equity of advance warning of the rules of information freedom, and (3) the rule of law, with appeal to an independent judiciary.

These protections will reduce the fear of descending the slippery slope when restrictions are promulgated in a crisis. Prime Minister Mahathir posed an imminent crisis as his defense for preemptive action: shutting down the newspapers, imprisoning opponents and generally abrogating the existing laws. But his professed fears sounded to Malaysians like the effort of a weakened politician to retain power by crying wolf (or, in this case, divisive racism). The Malaysian journalists' union widely circulated posters headlined, "Don't Jail Journalists!"

We shall state our own credo of press responsibility later.

The slippery slope

Just as a Malaysian prime minister fears press freedom will lead to license and irresponsibility, so journalists tend to believe that *any* restriction (libel or anti-trust laws, or a court requiring journalists to comply with rules as any other citizen) must eventually lead to thoroughgoing oppression, even in a democratic state. This is the fear of the slippery slope.

Such fear is far more justified for journalists than for governors. Governments more regularly and more viciously victimize journalists. Governments, after all, have police and prison power, not to mention the myriad forms of harassment described in Chapter VI. Journalists have only the power of the press. It may be persuasive, even influential. But it works only through mediators—the public and public institutions. Journalists have neither subpoena power nor the military to back up a news report or an editorial. To be sure, an irresponsible press—the collective "press," not just a single editor or one newspaper—can poison the well of understanding and do harm to a democratic state. In a free market society, however, the readership can eliminate an irresponsible publication. In the late nineteenth century, in Marble Hill, Indiana, a newspaper called the *Era* carved this motto on its masthead (Leo Bogart of the National Advertising Bureau reminds us): "Not for love, honor, or fame, but for cash." The *Era* soon disappeared, says Bogart, "proving that newspapers must generate love, honor and fame through journalistic excellence in order to produce a respectable cash flow."

Public reaction does not always assure the quick victory of good journalism over bad or even adequate attention paid to socially uplifting or useful reportage. When NBC televised the first hour-long interview with Mikhail Gorbachev in December 1987, U.S. viewers were more interested in comedy programs and a special featuring a movie stuntman. Only 13 percent of the Americans viewing television watched an unusually revealing conversation with the leader of the Soviet Union. There may be more breaking away from serious news programming. The deregulation of television by the Federal Communications Commission (FCC) has freed TV news from many of its public-service requirements. Entertainment is replacing news shows. Worse, some news is treated as entertainment or as fodder for sleazy shout-fests. Local TV stations are taking back from the three large networks a major share of the news time formerly used by network news. The local stations do not have the capability of covering world news, so overseas coverage slips again.

Some of the slack is taken up by cable news. CNN provides around-the-clock coverage which often surpasses all three old-line network news departments (ABC, CBS, NBC). In radio, too, deregulation has produced immediate cuts in news broadcasts. With cutbacks in overseas news staff, the temptation is to send reporters and TV teams to places where reporting is easiest—not complicated by press control, complex insurgencies, or wars that keep journalists out (as Afghanistan, Iran, Iraq).

Then there are vast areas of the world deemed not newsworthy. In a time of high expense (it costs about $200,000 to keep an American journalist abroad for a year), and presumed decreasing audience interest in foreign affairs, the news organizations are readily inclined to omit regular coverage of those countries not a substantial trading partner or outside the global power balance. Ideally, nowhere on earth is unnewsworthy. Some places are more so than others, at one time or another. Yet, as Mary Anne Fitzgerald tells us, Africa is no longer politically fashionable. The hopeful spotlight on the grand post-independence experiments in economic and social liberty has dimmed along with national aspirations. Instead, says this writer on Africa, "the continent presents a repetitive litany of coups, corruption and famine."[24] That may bore American editors, but that *is* the world, and Americans should know how people elsewhere live and die. Yet when the faltering few Western journalists make their way to back countries of Africa, says Fitzgerald, "The foreign press, with our mixture of cynicism, dedication, idealism, and disillusionment, bear the brunt of the fourth estate's responsibility to focus distortions and coax change."

All too often, the local rulers make it extraordinarily difficult for foreign journalists (as well as domestic newspersons) to describe the country as it really is. As we show later, there are exceptions to the press-control rule in Africa: Botswana, Mauritius, Senegal, the Sudan; and courageous journalists such as Hilary N'gweno in Kenya.

In most of the world, however, the slippery slope has long since been traversed, and Third World journalists especially are near the bottom of the hill. In the democratic states, too, fear of the slope can be enervating. It may also be counterproductive. Adequate safeguards can prevent the subverting of a democratic system, even if the absolute separation of state and information power is occasionally breached for sound and defensible purposes. The creation of the Public Broadcasting System was, after all, a potential First Amendment violation. Yet it is often an oasis in the commercial "wasteland."

The promise of the democratization of communications—the ISDN networking of networks—is the proliferating of many views of reality. All will become accessible through the dynamics of the new technologies. The sooner they are put in place, the more rapid will be the loosening of political bonds on domestic communications. Technologies already here include optical fibers, multi-use telephones, electronic memories, supercomputers, transistors, direct broadcast satellites, TV cable and dishes, expanded databanks, remote-sensing, videotex and more.

These technologies will present a vast diversity of news, information and opinion to be delivered domestically and worldwide, directly into homes, schools and businesses. Such access will also permit a receiver to become a sender. The two-way flow, so heatedly demanded at Unesco, can become a reality, not only for governments, but for individuals everywhere. For this to develop, the monopolizers, mainly governments, but commercial operators as well, must regard the democratization of communications as a revolution without losers.

3.
The Too-Easy Equivalence of Liberty

"ALL LEADERS MAKE mistakes."

"All politicians lie."

"All newspapers misrepresent."

And, most dangerous, "all systems of governance are, or may easily become, oppressive."

These premises are voiced by frustrated citizens and misguided commentators on the left and the right of the political spectrum in democratic countries. There is a modicum of truth in each statement. Yet each projects a gross distortion of reality.

Yes, all leaders make mistakes, but some engineer their "mistakes." Often these result in a major disaster. Stalin's murderous protection of his cult resulted in the deaths of millions of Soviet citizens, and lifelong pain and cruelty for countless others. Some "mistakes"!

The Soviet Union today is a metaphor for all centrally controlled societies grappling with the communication revolution—particularly the coming Age of ISDN. There is, indeed, a great distance from Stalin to Gorbachev. Stalin's black deeds are now formally rejected by the Communist Party of the Soviet Union. The new Soviet policies that make glasnost attractive to many citizens in the USSR, and especially to Westerners, are the promises of still greater openness in the Soviet Union.

The first example of this openness has been widely and properly acclaimed and encouraged in the West. Television "bridges" between Soviet citizens and Americans, especially those involving journalists of both countries, are primarily symbols. These bridges symbolize not only a new openness, but a falsely projected similarity of information systems, and uses of information power. And therein lies a danger.

There is—and will remain into the foreseeable future—great dissimilarity in the uses of information in the Soviet Union and the West. To suggest otherwise is to confuse reality with wishes (in the case of the idealist or the naive) or to bend realism to ideology (for those committed to advancing statist information elsewhere).

The Soviets want to stress similarity. Western and Soviet citizens are alike in crucial ways; they are alike in all aspects of the rights they should enjoy as human beings—human rights. In this, all share the expectation that every person shall enjoy the extensive menu of human rights described in the universal declarations and conventions to which the respective nations are signators. This expectation has been largely unfulfilled in the Soviet Union even now under glasnost. In no field is this more apparent than in the highly restricted deployment of information technology, and the severe limitations still placed on Soviet communications to transmit ideas and on Soviet citizens to have access to information—domestic or foreign—and project their own ideas into the large communications channels.

It is misleading, then, to suggest that "we are alike" even though we have different political systems, as Vladimir Posner, the Soviet television personality, frequently claims. His "different" political system, in fact, demands that there be the most glaring difference in information power between the USSR and the West. The Kremlin is the central power station for all basic information flows inside the USSR, and those largely moving into or out of the country. The incremental details of the flow—the selection of illustrative content for the big issues placed on the media agenda at the highest levels—do not begin to approach the freedom promised either in the utopian oratory of the founding father, or by the present rulers of the Soviet Union.

Yet such hortatory promises find ready acceptance by some Western relayers in the news media, the academy, the political forums, and intellectuals of varied associations.

It was true even in the dark days of Stalin. On 16 December 1941—nine days after the United States found itself at war with Germany—a twelve-year-old girl in New York wrote the following letter of praise to Joseph Stalin:

Dear Mr. Stalin:
I wish to congratulate you on the recent events in Russia. I have been very happy to hear such good news and I hope you will continue to do as fine fighting as you have been doing. I wish you the best of luck.
                    Sincerely yours

By then, Stalin had murdered, starved or imprisoned millions of Soviet citizens. The twelve-year-old reflected the prevailing American attitude. She was misled by a sophisticated propaganda campaign which exploited Soviet "front" organizations under Moscow's direction—all in the name of drumming up support for the new Soviet "allies." Her letter, thus, is a symbol of the failure, on a massive scale, to understand the horrors of a system founded not only on terror at home but information totalitarianism as well.

A rereading of George Orwell's preface to his masterpiece, *Animal Farm*, is, therefore, instructive. For British and American intellectuals accepted Joseph Stalin as a friendly cooperator. The signs of glasnost—then or now—should, indeed, be welcomed when they are truly moving toward frankness and information diversity, not merely Soviet domestic or foreign policy objectives. Information diversity requires the real sharing of information power at all levels in Soviet society. There is as yet little sign of such diversity.

There is an understandable receptivity in the West to take Soviet claims of diversity and information power-sharing at face value. George Orwell faced this dilemma in Stalin's time. *Animal Farm* was written between 1943 and 1944 when the Allied military alliance with the Soviet Union was holding but strained. Orwell, on the British left, had gone to Spain in support of the socialist-Communist forces fighting the Fascists under Franco. Orwell soon became disillusioned with the cruel undermining by far-left ideologues of the very people the Communists said they were assisting. Orwell began thinking of *Animal Farm* as early as 1937. It would be a satirical critique of totalitarianism. For Orwell, that meant communism. In recent years, Soviet spokesmen have suggested that Orwell had written a polemic, set in barnyard terms, against the Fascists of his day and this. But Orwell's preface to *Animal Farm*, long buried, was published for the first time in 1972.[1] That preface makes unmistakably clear that Orwell intended his now-legendary parable to be an attack on the system of government in the Soviet Union, and on the Western intellectuals who seemed to welcome Communist theory.

Orwell also reveals the difficulty he had in finding a publisher for *Animal Farm*, a fact not unrelated to the intellectuals' affinity for the Soviet system. "The sinister fact about literary censorship in England," he writes, "is that it is largely voluntary." He continues, "Unpopular ideas can be silenced and inconvenient facts kept dark, without the need for any official ban. Anyone who has lived long in a foreign country will know of instances of sensational items of news—things which on their own merits would get the big headlines—being kept right out of the British press, not because the government intervened, but because of a general tacit agreement that 'it wouldn't do' to mention that particular fact."

One who challenges the "prevailing orthodoxy," says Orwell, is silenced with surprising effectiveness. What was demanded in his day by the prevailing orthodoxy was "an uncritical admiration of Soviet Russia." Everyone knows this, says Orwell, and "nearly everyone acts on it." To reveal undisclosed facts, is to remain unprintable. One could then quite safely attack Winston Churchill, but not Josef Stalin. Orwell notes occasions when this occurred in earlier years. The most egregious models, he adds, are liberal writers and journalists who have no need to falsify their opinions. "Criticism of the Soviet regime from the *left* could obtain a hearing only with difficulty," remarks Orwell. There was much anti-Russian literature from the far right, but "manifestly dishonest, out of date, and actuated by sordid motives."

There was also, Orwell reminds us, an "equally huge and almost equally dishonest stream of pro-Russian propaganda." One published responsible criticism of the Soviet Union under threat of boycott, and worse. Events in that country were judged by a separate standard. The great executions of 1936-38 were "applauded by lifelong opponents of capital punishment, and it was considered equally proper to publicize famines when they happened in India and to conceal them when they happened in the Ukraine," writes Orwell.

So, it was said by the British liberals, Orwell's *Animal Farm* should not have been published.

Times have, indeed, changed since *Animal Farm* appeared. Stalin, month by month now, is known to a new generation for the great horrors he unleashed. But revealing them post facto neither assures this generation —in or out of the USSR—that institutional changes preclude a second coming, or that present Soviet leaders are forever committed to peaceful reform. Indeed, given the present state of Soviet mass communications, and the obvious command of public relations techniques in the Kremlin, it will take highly sophisticated observers of the Moscow scene to separate the latest fact from contrived myth. Orwell's warning in 1945 may be still more appropriate forty-four years later:

"If liberty means anything at all, it means the right to tell people what they do not want to hear. The common people still vaguely subscribe to that doctrine and act on it...It is the liberals who fear liberty and the intellectuals who want to do dirt on the intellect."

* * *

WHAT, THEN, IS the chance for the liberalization of the Soviet Union; or, in specific terms, for the communications revolution to serve as the counter-Communist revolution?

This is as much a concern of Americans and other Westerners as it is of Russians, Lithuanians, Armenians and the Muslims in the USSR. For the Soviet revolution did, indeed, alter world history for this century. But a new century is coming, and the portents do not favor repression, but liberty—if only the developed nations of the West will not fear the vast shift in the correlation of power and ideology. That shift will mean the universal acknowledgement, finally, that the "idealism," utopianism and political investment in Marxism-Leninism has been an unmitigated fraud, a delusion that cost millions of lives and generations of stunted personalities across the vast Euroasian continent, and in outposts in Africa and the Caribbean where that ideology took root.

In America and Western Europe, too, as Orwell reminds us, the cost extracted by that faulted ideology in intellectual and moral perversions has also been great. Liberalization of the Soviet Union may proceed, one century late, if the communications revolution is permitted to develop freedoms of thought, choice and then action—political action, above all—and if Americans and other Westerners do not again allow their own wishes for a liberalized Russia to cloud reality. For while the communications revolution has the power to transform even the Soviet society, as later chapters suggest, a retrogressive crackdown is also possible. And Americans should not again misread reality, fear the possibilities of liberty, or rationalize the delay, anywhere in the world, of deploying the networks of freedom.

4.
Fear of Information Power in "Socialist" Countries

We would continue to document specific acts of oppression; we continued writing our articles and editorials, know that we were, in effect, writing primarily for the censor. But in a larger sense, we were also writing for our collective conscience and for a people who had no voice.
—Jaime Chamorro Cardenal, an editor of *La Prensa*, Nicaragua

THE BUSINESS OF any government—governance—is the management of power. The purpose of informing is to share intelligence. Intelligence is a major key to power. Governments define themselves by the way they treat their citizens and permit them access to information. Power-sharing, then, is inherent in providing access to information. Without it, freedom of expression is a sham, and self-fulfillment, discovering truth and advancing knowledge are hampered.

Freedom of expression thus enables members of society to participate knowledgeably in decision-making. And, no less important, freedom of information and freedom of expression help frame a workable consensus on public issues, and avoid destructive polarization. Limitations on public expression are necessary when society-rending crises threaten, or in order to avoid libel or invasion of a nonpublic person's privacy. In a free society, other restrictions are rarely sought, and still less readily applied.

The information system built by Vladimir Lenin is still used by the Soviet Union, and with indigenous variations by the countries of Eastern Europe, and the People's Republic of China. Socialist dictum demands "common control of the means of production." This includes the production and distribution of information and ideas.

Marxist critics charge that journalism in capitalist countries is largely governed by profitability, by what information will "sell" in the marketplace. While profit margins are a factor in market-economy journalism, the actual news and commentary are not treated as a "commodity" nearly to the extent they are in "socialist" states, where absolute control of the information product by central authority—even under glasnost in the USSR—is fundamental. "Common control"—especially in the realm of ideas—has always meant state control.

"Socialists" often use the bland term "bureaucracy" to describe the controllers. That term is used by the chief U.S. correspondent for Soviet television to describe his main "enemy," the bureaucracy. He softens the description, though, by saying that American journalists face the same "enemy," bureaucracy. As Vladimir Dounaev spells out his problem, however, it looks quite different from American practice: "Some investigative reporters [in the USSR] are put into prison; for a short time, fortunately. It's a fight, believe me, a real fight because these bureaucrats there are fighting for their lives. He [the bureaucrat] is fighting mainly against investigative reporting," Dounaev said recently in the United States.[1]

Then, in the current Soviet effort to equate Western freedoms with Soviet glasnost—"people like us"—Dounaev added that "our press...like [America's]...is part of the establishment." He singled out Tass, the official news agency. "For them," he said, "the most important thing is to mention the bureaucrats." Dounaev likened U.S. to Soviet bureaucrats because he had been at first prevented from speaking in Minneapolis because the State Department had put the city off limits to Soviet citizens for five years in retaliation for travel restrictions on U.S. journalists in the Soviet Union. "We are all part of the establishment, and depend on the people we are criticizing," says Dounaev.

Bureaucrats are not the same in both countries, nor are journalists. American journalists may depend on bureaucrats and top officials for information, but there are ways to circumvent even the most intransigent source: by challenging him in print or broadcast, by finding other corroborative sources, or by using the freedom of information machinery to smoke out a recalcitrant official.

The American press sometimes does seem to be a part of the establishment, however that is defined. But the U.S. establishment is not solely the government. Indeed, the press spends much time acting as adversary to government at every level. Sometimes the news media accept an official position without sufficient questioning. Sometimes, too, the press moves ahead of government and the establishment, and forces a new policy, as in the withdrawal from Vietnam. But sometimes the press, by its own inactivity, delays the full disclosure of events and their complex backgrounds. By doing so, the press is not handmaiden to the establishment. When the press fails to tackle highly controversial issues promptly it lends credence to the charge that it is covering up for government's inaction. In fact, both the news media and officials may choose for quite different reasons to avoid difficult assignments.

A case in point: In 1988, the *New York Times* published a half-page-long article, starting on page one, giving the views of Jews who emigrated from Arab lands to Israel, mostly after the 1948 Arab-Israel war. The article was incisive and important. It told of the sharp differences between the Ashkenazi (East European) and Sephardic (Arabic) Jews inside Israel, the implications for the country's domestic politics, and the possibilities for peace between Israel and some Arab countries if Sephardic Jews come to power in Israel. That article could have been written at any time during the past forty years. Apparently the political climate today in the United States and in Israel makes such an article seem timely, and less subject to the charge of anti-Israelism or anti-Semitism. Since the Israeli incursion into Lebanon, and the 1987 uprising in the occupied territories, American journalism has sought to describe the Arab-Israel conflicts in less simplistic terms.

This has not been the result of policy dictates from the government bureaucracy, or the establishment, which has been little altered over the years. The Reagan administration was no less supportive of Israel than it or its predecessors have been since 1948. The change may be due in main to new and younger actors taking second and third looks at the old premises about the Middle East, and many other areas and issues. The consequence is fresh insight, as in the *Times* piece, and new factors inserted in old equations.

That is possible only because the information system is sufficiently open to accept new premises and new conclusions. The media can move ahead of government policy. This is not so, however, in countries where the "bureaucracy" controls the information flow.

One should not lightly dismiss the more-than-accidental likenesses in centralized control of information, and all else, under all of the following: National *Socialism* of Adolph Hitler; the first fascism (an outgrowth of Italian socialism) under Benito Mussolini; Stalin's murderous cult of personality; the lesser destructive cults in Eastern Europe; and Maoist socialism which inevitably led to the horrendous decade of the Cultural Revolution in China.

In each, the propaganda machinery of the state was directed first at "expropriating the personality" of its own citizens (as Vasily Selyunin, a Soviet economist, argued in *Novy Mir*[2]). Secondarily, Soviet information power was turned to foreign targets in the developed, "socialist," and Third Worlds.

"Information sovereignty," a term heard increasingly in international debates on communication flows, is defined by its users to mean the sovereign capability of states, not individuals, to have access to information. This is not so in American governance, which more than any other system, including all other democracies, regards the individual citizen as the ultimate sovereignty.

Governments, all governments, are suspect. U.S. checks and balances—ingeniously, three distinct arms of national government, and fifty states offsetting federal power dilute central authority—are based on worst-case possibilities. The First Amendment further prevents unforeseen encroachment on individual sovereignty. This elaborate screen protects the ultimate sovereignty: the individual. Some say he/she is too protected, to the extent of stultifying real leadership even by duly elected officials. By clever blockages at infinite entries to the system, it is possible to delay or forestall executive, legislative and even judicial action. And, too, by extraordinary openness in the systems of information run by the government, and the permissiveness granted journalists, it is possible to distort and hopelessly delay government processes. Yet these disadvantages, great as they seem, far surpass the seventy years of harsh disappointment visible in the diametrically opposite systems in the "socialist" world. And these systems, it must be remembered, have been "marketed" since their beginning under the banner of high idealism, and the promise of utopian gifts for all citizens. Those gifts were to be consumerism which would outproduce America and the West, and also ensure the broad development of human dignity. The USSR and China have only recently realized that both countries were moving steadily away from both goals.

The program of perestroika, using glasnost as a developmental tool, an effort to change direction, is, therefore, an historic milestone in the Soviet Union. The experiment of perestroika cum glasnost has worldwide implications, and is examined here as the test of a new information power mechanism. It may be tried increasingly wherever socialist systems have failed to deliver most of what they have promised. The failures may be found in every country—right or left—which has introduced centralized controls of information systems for whatever announced objective. The worst experiences, and the longest lasting, have been those countries which called themselves communist; closely following, are those nations which called themselves fascist. Spain and Portugal broke the chain of dependency, the cults of personality, major state ownership, and worst of all the "expropriation of [citizen] personality" when they created democratic systems of government.

There is no sign that the Soviet Union or the People's Republic of China is moving toward democratic governance; indeed, rulers of both countries actively oppose "bourgeois liberalization," capitalism, or "Western democracy," terms by which other than state control is rejected in both countries.

What, then, are the new information systems in the Soviet Union and China?

The best clue to the Soviet system is the Russian definition of glasnost. It is described in the 1987 edition of the Soviet Political Dictionary as "access by the public to information about the work of government and its activities." Glasnost, says the official definition, "is the supervision by the broadest masses of the population over the organs of government, especially local organizations, and of the struggle against bureaucratism. The broadest channels for glasnost are the mass information media, oral publicity and visual aids such as displays, graphic media, etc. Information containing state military, scientific and technical, industrial and criminal investigatory, medical, and other secrets is not subject to glasnost."

The official Soviet definition carries the cross-reference: "see Revolutionary Vigilance." It is officially described as "unremitting attention" to the forces opposing the consolidation of the advanced social structure" of the USSR. Says the text, "When it becomes clear to even the most aggressive forces that it is impossible to decide the historical struggle between socialism and capitalism by military means, it is inevitable that the attempts of reactionary circles in the West to weaken socialism and to provoke its internal erosion will be intensified. To this end, these circles utilize intelligence operations and methods of 'psychological warfare,' and promote ideological campaigns intended to undermine confidence in the socialist structure, defame the socialist way of life, and try to speculate on nationalistic feelings." Revolutionary vigilance is intended to "render harmless" forces hostile to socialism, without generating "a morbid obsession with spying, groundless mistrust and suspicions, which lead to excessive tension in the social atmosphere."[3]

This official definition is as important for what is excluded from glasnost, as for what is included, and under-valued. The top "organs of government" are far less subject to glasnost, in this definition, than "local" bureaucracies. Yet the major decisions will still be made at the top; and, indeed, the state, not individuals even as "masses" will control the machinery of information—incoming and outgoing—whether "visual aids" or "oral publicity," presumably radio and television. Left out of glasnost, officially, are "state secrets"—a very broad segment of information, particularly in a highly centralized state such as the USSR.

What, then, is glasnost? It is limited freedom of speech and still more limited freedom to write, using only the officially designated channels of communication controlled by the state. Mikhail Gorbachev provides an authoritative description of the need for perestroika/glasnost, and the uses to which they are put. Using the metaphor of power, state power, he comes to the translation of that into information power. He begins by describing recent Soviet history:

Something strange was taking place: the huge fly-wheel of a powerful machine was revolving, while either transmission from it to work places was skidding, or driver belts were too loose...A country that was quickly closing on the world's advanced nations began to lose one position after another...A gradual erosion of the ideological and moral values of our people began...(A) breach had formed between word and deed, which bred public passivity and disbelief in the slogans being proclaimed. It was only natural that this resulted in a credibility gap: everything that was proclaimed from the rostrums and printed in newspapers and textbooks was put in question...On the whole, society was becoming increasingly unmanageable...We can no longer tolerate stagnation...journalist and politician—everyone has something to review in his style and method of work...We will now firmly stick to the line that only through the consistent development of democratic forms inherent in socialism can we make progress.

Nowhere, however, does Gorbachev discuss *his* definition of democracy except to say it is not the Western conception. Nor does he describe the significant modifier of democracy in his prescription: socialism, old or new.[4] When *Glasnost*, the independent magazine, was created in 1987, its editor, Sergei Grigoryants, was repeatedly harassed, arrested three times, and his equipment and archives confiscated by the state. Even in official publications, debates on the "limits of glasnost" have prevailed for several years. There are no conclusive answers. But everyone recognizes that there are limits, so self-censorship has largely taken the place of official censorship of every word printed. But that does not mean that glasnost is a step toward democracy. On the contrary, glasnost is a substitute for democracy, just as perestroika may substitute for a free market system. Both alter the rules so that criticism of the system can take place at officially specified levels, and with targets other than the system itself, or the elite controlling it. Both are managerial tools to make the machinery run better. Supporters of the security arm, the KGB, say it was that organization that launched the campaign for glasnost. The KGB receives the best information about developments inside the Soviet Union and the world outside.[5]

In the case of glasnost, the information system is targeted for improvement. Complaints about many aspects of Soviet life may now be aired. Commercial advertising appears on some Soviet television. One Soviet newspaper rebukes another. A dissident historian is praised in the press. Authors long banned may be read. Plays once buried are produced. A minor KGB scandal is revealed. History revised is now re-revised. Some emigration is permitted, but still highly controlled. Natural disasters and human accidents formerly buried are now revealed almost as they occur.

The meltdown at Chernobyl in 1986, however, revealed glasnost reversed. For three days, Soviet news channels were silent or distortive. Finally, under pressure from Western news systems, the Soviets told more of the massive accident. The Soviet nuclear specialist, who first flew over

Chernobyl and realized the extent of the damage, committed suicide in 1988. And the umbrella of glasnost has protected notorious anti-Semitic meetings of Pamyat (Memory), an organization ostensibly devoted to protecting the national heritage. Some ultra-conservatives who would undo glasnost and perestroika use thinly veiled anti-Semitism.

Izvestia opened up the long-forbidden subject of Siberian labor camps, and revealed that the traditional system of forced labor still operates. And Estonians belonging to the new, independent Popular Front demanded local control over the press and television. The demand was carried in the official press.

Glasnost and its uncertainties can be nervewracking. A Ukrainian writer, Vladimir Drozd, described in *Literaturnaya Gazeta* his fears as he wrote a speech supporting more reforms:

> And I thought, "Why does it have to be me?" You can understand my alarm, and even my fear. Fear for myself, fear for my family. Fear for all of us who can feel their hot breaths on our necks, can see their narrowed wolf-eyes—the eyes of those who do not want the perestroika...
>
> They get together every evening, sit late into the nights, cursing everything new that has come to our lives. Our every mistake or miscalculation gives them unbelievable pleasure—"See what your precious reform leads to." I shall go further. They watch all of us who speak out today in favor of renewing our society, those of us whose souls still live and who feel pain. And they make lists of our names, for when their day comes round again.[6]

No one can yet say how long glasnost and perestroika will last as official policies. Lenin, in his "letters from afar" published before the revolution never once described the kind of state he would head, except to say it would be formed by the "proletarian militia." The Gorbachev party is not being much more specific now.

Perhaps, then, one important revelation of glasnost is this: Soviet citizens now seem to have known all along the broad outline of what was formally withheld from them. They are now told that for fifty years Soviet map-makers had faked the national geography. But then any citizen knew the map of Moscow did not reflect reality. They did not know statistics of those murdered under Stalin, or how many Crimean Tatars were forced to emigrate to the east, or Jews prevented from leaving the country, or how crippled was the state of the economy. But they knew relatives and friends were killed, or disappeared; that lines at food stores grew longer, and consumer goods became less available; health services deteriorated; alcoholism was rampant; and the state of women debased. Blue-sky propaganda for domestic consumption further reduced the credibility of the entire information system. Just a small opening in the window of truth gave greater believability to the whole system.

More than that, the "masses" had a good idea of how the *nomenklatura* was treating itself. Privileges of the elite were examined in a public opinion poll published by *Moscow News* in July 1988.[7] The results showed what must be long-simmering objections to many of the nomenklatura's privileged ways. The sociologists who polled 548 of all age groups in Moscow were "quite sure that most people take a negative view of privileges, but we wanted to know why." Some 61 percent dispelled the fear they supported "barrack-room equality"—egalitarianism. They said it is legitimate for some people in a socialist society to have "very high incomes." (Some 25 percent, however, objected, but they are described as "people over sixty, retired, and those with a poor education.") From 60 to 84 percent think these privileges enjoyed by government executives are "unfair":

Food packages, goods from exclusive shops (84 percent)
Free availability of seats in theaters, movies, etc. (80 percent)
Flats in superior housing (67 percent)
State-owned dachas (65 percent)
Exclusive health care centers (60 percent)

The respondents also graded the recipients of privileges by the degree they deserved privileges. Those whose privileges are "largely undeserved" or all "undeserved" are the Communist party apparatus, ministerial and departmental executives, trade union executives and Komsomol executives.

Another poll[8] asked 400 Muscovites "of all ages and socio-professional groups" how they felt about the press and glasnost. Asked about exposing the "blank spots" in Soviet history, 65 percent approved and 21 percent "deplored" such publications, saying "They sap belief in our ideals." Six months earlier, only 49 percent favored glasnost and 34 percent said it should be restricted, according to the poll. In the later poll the "more supportive" people were nineteen-year-olds and under (78 percent), twenty- to twenty-nine-years (75 percent), students (88 percent), researchers and teachers (69 percent), and workers (67 percent). The less supportive were people over sixty (56 percent) and economic managers (50 percent). The subjects the respondents felt were least adequately covered even under glasnost were Soviet foreign policy, law enforcement and army life, and low-income problems.

Another poll tested the views of 100 "Soviet experts" on foreign affairs. Only 14 percent agreed that contradictions between capitalist and socialist countries precluded cooperation between them. Yet 73 percent said the "American threat" exists in the military sphere. Fifty-two percent said the most effective means for ensuring the USSR's security is political arrangements. The "correlation" of U.S./USSR military forces should be 50/50, they replied. They felt (49 percent) that "ideological contradictions" hampered the normalization of U.S./USSR relations "to a great extent." Yet 85 percent said humanitarian contacts should be developed "regardless of the political climate." They added (81 percent) that the observance of human rights in the USSR and the U.S. is not "strictly their personal matter." To reinforce that, they responded (92 percent) that human rights cannot be a "zone outside criticism." This was a far cry from the routine Soviet responses at Helsinki review conferences that domestic human rights issues were solely a matter of internal Soviet affairs.

These are the views of knowledgeable urbanites, now a major part of the population of the Soviet Union. They have a closer link to the intelligentsia and the elite than to the rural population which Stalin and Brezhnev represented. These are the people who read newspapers and write letters to editors, letters which in their tens of thousands form the best continuing "poll" of what Soviet citizens think. These letters are the "masses" speaking, but only the literate, mainly urban masses. They have become more outspoken, more critical, and probably more polarized. Eventually, glasnost will have to deal with the problems which even limited freedom creates. The irreversibility of glasnost, therefore, is a question Mikhail Gorbachev discussed openly at the Party Congress in 1988. He denied, of course, that it will or can be turned back. He is right. One cannot say that formerly distorted truth, now revised, can ever again be said to have been correct in its obviously deformed state. But harsh military or police crackdowns could once again silence the expression of what Soviets truly know. Andrei Sakharov, the Nobel physicist/human rights activist, warned early in 1989 that "conservatives" may muzzle or replace Gorbachev.

An unspoken aspect of glasnost may be its most important feature. It may mean more openness for the nomenklatura, for the elite to know the players and their positions. In Lenin's time, the work of the political bureau of the secretariat of the Soviet Communist Party (the rulers of the country) was far better known than in the years since. Party members and others outside the party were kept informed of the diverse positions of the members of the Politburo. Communists in Lenin's time could read the minutes of the Politburo's sessions. Differences were spelled out, and not revealed, as later, in elliptical phrases. This was the classic "writing between the lines." The advantage even of such limited "openness" is clear: citizens can know what are the real issues, and perhaps no less important, who stands behind them.

The debate over television in July 1988 at the party conference provided just such an inkling of the divisions. Some 100 million viewers were estimated to have watched. This was an important contribution of glasnost to the system of governance and to the sharing of information among the "masses." Gorbachev made use of the historic conference to unleash the media and his supporters in order to cut down the power of the Old Guard. He was not able to remove many of those remaining from the Brezhnev period, but he used public television to indicate that opposing perestroika would be unsafe.

Yet the critics of glasnost were also using the public information channels. The Yeltsin-Ligachev controversy concluded with the put-down of Yeltsin, the proponent of more permissive glasnost. The message seemed to be that glasnost would encourage criticism to advance economic reforms, yet would nevertheless restrict political liberalization that might seriously alter the positions of the elites.

I had discussed this five years earlier with a central planner in China. He made it clear that when excess capital (profit) reaches a critical point, Beijing will raise production quotas (the equivalent of taxes) to siphon off economic power before it converts to political power; so, too, information power can be controlled. In the USSR, glasnost is a tool, still being honed in daily use, to permit Gorbachev and his supporters to strengthen their power within the party, and the state apparatus as well.

One observer said Gorbachev used the media the way Mao Zedung used the Red Guards. Both sought to destroy the power of the bureaucracy. Since the Soviet press is an important part of the bureaucracy, many "reformist" speakers at the crucial conference in 1988 criticized the Soviet press for being imbalanced, inaccurate and excessively negative (criticisms often voiced against the Western press by Third World spokesmen). Other conference speakers attacked the press, principally the "liberal" press such as *Ogonyok* magazine, for failing to emphasize positive aspects of Soviet society.

During the 1 July 1988 debate, the seemingly ultimate exchanges developed. First, President Andrei A. Gromyko, the veteran Politburo member, was publicly attacked, made to step down, and "answer for everything and do so personally." Then, a member turned to Gorbachev, still on television, and, with his permission, asked the leader a question. "Would you rather have no mistakes, no one upset, and lower the role of the press? Or am I right in thinking you want to raise the role of the press, even if it means some mistakes?"

The question answered itself. The press would be "raised," even if some mistakes were made. And the Soviet audience gave the televised conference rave reviews. "That such a thing could be on television," one viewer commented. "We've waited patiently for a long time," said another.

The most novel aspect of Gorbachev's struggle with the bureaucracy is the presence of television and the satellite to bring both pictures and printed word of at least some of the proceedings quickly to the vast Soviet public. These were not available when Khrushchev tried to reform the system he inherited from Stalin. Gorbachev has the intelligentsia on his side but it does not dominate the administrative power structure. One Muscovite told me in July 1987 that Gorbachev could achieve the drastic restructuring he seeks if he could use Draconian totalitarian power to impose the changes he seeks. Since he cannot do so, I was told, slow, incremental changes can produce chaos and the defeat of Gorbachev. It is clearly too early to make such a prediction. But the analysis has merit. It leads one observer to write "Gorbachev's best chance of fostering democracy will probably involve nondemocratic means." But, then, Gorbachev's definition of "democracy" is "nondemocratic" in Western terms.

During his visit to Washington late in 1988, barely a year after Gorbachev released him from internal exile in Gorky, Andrei Sakharov chided the Soviet president for the recently drafted changes in the USSR's political structure. The changes, said Sakharov, would centralize too much power in the hands of Gorbachev, and might even encourage a coup. Sakharov urged that the process of political change be slowed.

American press coverage from Moscow throughout 1987 and 1988 revealed mainly details of change which officials wanted publicized: particularly cultural and historical openings. It was somewhat less difficult than in preglasnost days for a foreign correspondent to interview the key actors and get substantive response on where the society and party are going.

Lest glasnost be misunderstood, at home or abroad, *Moscow News*[9] carried Gorbachev's words: "No one is above control in our country...This applies to the mass media. The Soviet press is not a private shop. Let us recall again Lenin's premise that literature is part of the common cause of the Party...Editors should have a sense of responsibility. I don't want to give names. We are talking in a comradely way. But it should be remembered that a magazine, a publishing house, or a newspaper are not someone's private concern but a concern of the entire Party, of the whole people."

The Soviet press will remain a state press, and glasnost is a tool to manage better the responsibility of the press to party dicta. One can discover this in examining the prizes awarded to journalists in 1988 by the presidium of the board of the USSR Union of Journalists.[10] Of the twenty-six honorees for domestic journalism, twenty were named for their work on perestroika, "democratic" processes, improving management, fighting "negative phenomena," preserving national property, intensifying labor, criticizing "existing disorders in industry," and related subjects. Four writers received prizes in the field of international reporting.

But new truth, that is another matter. Can new distortions, misinformation, disinformation be floated? Of course. The Soviet news and information media are increasingly better placed to conduct domestic and foreign propaganda than before glasnost replaced crude censorship, and encouraged critiquing and the appearance of full openness. Only now it will be harder to detect or dissect self-serving statist information, certainly from outside the country. There will now be the hardest kind of propaganda to examine: that based on partial truth, maybe even large-part truth, but still statist and not wholly accurate. For Americans—tired of the cold war, high-cost arms, and frightening rhetoric and geopolitical contests—the inclination will be to give the benefit of the doubt to Soviet statements and actions, particularly those accompanied by sounds and acts of "people-to-people friendship." Certainly television "bridges" from Moscow to New York are interesting, even colorful. But the easy equating of "people like us" as though they share our freedoms, particularly information freedoms, is dangerous.

When Vladimir Posner, the articulate Soviet television commentator, appears on the Phil Donahue show (as he has many times), he seems "just like us." He is introduced as a Soviet journalist. He is not. He is a civil servant doing the state's business, and doing it well on international satellite linkages on which he is equated with American journalists. These Americans, ironically, are also adversaries of the United States government some of the time.

On one telecast bridge in February 1987 Posner in Moscow waved an American book which, he said, analyzed objectionable references to the Soviet Union. As the camera zoomed in on the cover of the book, Posner intoned authoritatively what he said were "the words that are used most often in the American press to describe the Soviet Union and Russians...Here are three major American news magazines, *Time, Newsweek* and *U.S. News and World Report.*" It must have seemed to the viewer that the words to be mentioned are the terms the three U.S. magazines actually used to describe the Soviet citizen. Posner continues with great authority. "And the words most often used to characterize the Russians are—and I quote—'savages, dupes, adventurers, despots, barbarians.' (The Soviets') methods of behavior are—again I quote—'brutal, treacherous, conniving, unmanly (strangely enough), aggressive, animalistic.'"

I obtained a copy of the book, *Faces of the Enemy*, by Sam Keen, an American psychologist. The book revealed Posner's distortion upon a distortion. First, Keen did not perform the original research. He referred to another's work. Second, even Keen did *not* write that Farrel Corcoran, the initial researcher, found the words "savages, dupes," etc. in *any* of the three American magazines. Third, Corcoran did not find those words in the magazines. They were his own characterization of the magazines' content based upon *his* premises and analysis.

One must then examine Corcoran's full thesis. He analyzed the coverage in the three U.S. magazines during the funerals of Soviet leaders from Stalin to Brezhnev. He described that coverage as "myth, ideology and victimage ritual." Neither Keen nor Corcoran cited any uses of the "enemy" words in the U.S. magazines. The closest Corcoran comes to using the terms Posner broadcast to millions of viewers was to say the coverage produced *"metaphorical* characterizations of Russians as savages, etc." One may excuse Posner for not going into the academic detail provided by Corcoran, but then one may conclude that Posner as propagandist has few qualms about delivering misinformation with verve and gusto. His show-biz technique was highly effective.

Of the millions of viewers, who knows that Posner pulled a rather clever deception? If he were an American, it is likely some investigative journalist would have gone after the facts, and later revealed them. Posner is slick, and effective. And Americans must become used to the fact that Soviet "journalists" are quite different from the American variety. As Posner himself is fond of saying, "I'm a Commie."

One may readily understand how an average American viewer can be taken in by "bridges" that pander to state-selected Soviet "people like us," and falsely equate the two societies. But American academics should be expected to avoid such ideological "equations." They sometimes succumb to it. For example, in twenty-four pages of data and analysis, John D.H. Downing[11] describes Soviet coverage of the invasion of Afghanistan. He begins by noting that Western coverage of foreign news, especially Third World events, is incoherent and subject to politico-economic pressures. Downing turns to Soviet press coverage of the war in Afghanistan. He shows how the themes varied over eight years, as thousands of Soviet deaths wore down the invading army and especially the morale of people back home. Except for a one-sentence, statistic-free, bland reference, the author does not record the utter absence of reporting in the Soviet press of the horrendous losses of the Afghan people. He does not mention one million dead, five to six million forced to flee to other countries.

Downing, however, points out Third World complaints of inadequate coverage of their news in the Western media. These are of a far different character: complaints of sensationalism, not of intentionally blacking out massive military campaigns, as the Soviets did for eight years (with no substantive analysis from Downing). He covers it all by quoting "truth is the first casualty of war." True enough. But what about this war? How was truth most frequently, cruelly a "casualty?"

While American reporters were describing the bloodiest evils of the Salvador death squads, Soviet reporters were hiding the grim murders of Afghans by Soviet troops. Said Mikhail Kozhukov, who had reported from inside Afghanistan the longest of any Soviet writer, "All the truth about this war will never be written or told. There are some things which I know that I will not even tell my son." He added, "I consider that war so bad that those people who are lucky enough not to have seen it must not know the details." He said that some day, "when glasnost gets to this topic fully, when all the details are known about the decision of 1979, about all the faults of the [Afghan] revolution, and the party and of those Soviet officials involved...when all of us have a clear picture of those nine years—then maybe we'll know."[12]

In his last six pages, Downing compares American press reporting in El Salvador with the Soviet propaganda in Afghanistan. He finds "intriguing parallels." Downing concludes the U.S. reporters were "more professionally skillful" and better able to subvert informational clarity than their opposite numbers in the USSR. Praise, indeed! Downing finally allows that there was "no distance" from Soviet policy visible in the Soviet media. But he tells us that American press coverage in El Salvador (like the Soviet propaganda in Afghanistan) undermined the informing of U.S. citizens, and increased "the menace" of "global nuclear holocaust, nuclear arms negotiation notwithstanding." A rather extreme conclusion for a false equation.

If comparison is to be made between the American and Soviet press systems, it may be said that in the Soviet Union, government is the chief *editor* of all publications, telecasts and broadcasts. In the U.S., government is the chief *reader* or *listener*. That is not to say that democratic governments never try to manage the news; they do. But they are also open to strong attacks from the media when suspected, or caught.

President Johnson was such a "reader" of the press. He reversed Vietnam war policy when he read the *New York Times* and *Washington Post*, and saw nightly television, and believed the mass media more than his intelligence sources.[12] Intelligence may have been more accurate, but the press was influencing the public, it was deserting the president, and Johnson made his decision accordingly. The press did not play that role in the Soviet Union during the Afghan war. Other considerations—a faltering Soviet economy, Third World dissatisfaction with the Soviet invasion, unease at home over visible caskets and crippled veterans, and the need for a better relationship with the U.S.—all these pointed to a Soviet withdrawal, not based upon any coverage in the Soviet press or broadcasting.

A word, finally, about the structure of the Soviet press in this age of glasnost. The daily *Pravda* (Truth), the creation of Lenin even before the state was established, has been called the most important newspaper in the world. It was never the best written, or most representative, but it has always been important. For seventy years *Pravda* set the tone for the rest of the Soviet press. In a country of large-circulation papers, many magazines and smaller papers, *Pravda* is regarded, at home and abroad, as the voice of the Kremlin.

Pravda's unique role persists, despite occasional glasnostian "firsts" published in *Ogonyok* and *Moscow News*. Some of the most surprising revelations of the Gorbachev period have come from these latter two publications, but *Pravda* has improved its format, shortened some of its articles, and examined—particularly in its letters column—subjects which could not have been printed in *Pravda* or anywhere else in the USSR for its first sixty-seven years.

When I interviewed the editor of the English edition of *Moscow News* in 1987, he told me proudly, "I read *Pravda* to get the line, but *Pravda* reads *Moscow News* to see how far they can go." That remarkable statement probably is accurate.

Pravda, read regularly in Russian and English at Freedom House, is still generally stodgy and boring despite elementary efforts being made to liven the copy and format. But serious foreign students of the Soviet Union must still read it. Party functionaries must assimilate it to know where the Soviet is going and who is taking it there. *Pravda* is the Partyman's guide to policy and personal promotion, and now perestroika.

Pravda is the case par excellence of the hazards of a state-embattled press. Before the Bolshevik revolution of November 1917, *Pravda* was frequently closed down by the Tsar, and reemerged. The day after the revolution, even before Lenin's Press Decree took effect, the Bolsheviks closed down the independent ("bourgeois") newspapers, and expropriated their plants. All printed copies of the independent press were burned in the streets. "Even tsarism had never practiced such a massive settling of accounts with the press," wrote Nicholai Sakharov, a Menshevik leader.[14]

The new information order was established in the USSR. Its first act was to silence the opposition newspapers. Within six months Glavlit, the official censor's office, was created to monitor all publications. The last note of controversy disappeared from *Pravda* by the end of 1923, and did not reappear in the subtly controlled sense of glasnost until 1986.

Pravda has wielded immense power. One of Khrushchev's speeches was altered by the party newspaper. Despite Gorbachev's insistence on greater truthfulness, the large birthmark visible on his forehead on television, disappeared in *Pravda's* retouched photographs. In eliminating it, *Pravda* apparently still considers itself—until told otherwise—the mobilizer of the masses in fulfillment of the party objectives.

Pravda objectives were spelled out by Viktor Afanasyev, its editor-in-chief, on the newspaper's seventieth birthday:

> We inform the masses of the decisions of the Party and government, propagandize these decisions, mobilize and organize the Soviet

people to carry them out, accumulate and mold public opinion, and concentrate people's efforts on solving precisely those tasks which are most important and most necessary for the country and for the Party...We help the Party to bring up the new man...the greatness of his affairs, his successes, needs and interests, the difficulties he encounters, the problems he solves; we criticize shortcomings, mistakes and omissions, and we try to find ways of overcoming them. We raise the Soviet peoples, striving to make real patriots and internationalists of them, ready to defend their country and the gains of world socialism.[15]

This is a large commitment for a government, let alone a newspaper—but then *Pravda* is an arm of government. It is much more than "the press." It is the model for the rest of the Soviet press—even when glasnost encourages several other papers to carry a sensational story occasionally.

For the most part, though, the Soviet system of governance remains sacrosanct. Low-level "mistakes" can be cited. A young guide in Leningrad said of Stalin's mass murders, "all rulers make mistakes." Still, the main thrust of Soviet reporting is an exhortation and the use of models—coal miners, farmers, industrial workers—meeting or beating their quotas, or, these days, encountering foot-dragging middle-level bureaucrats who slow production. With glasnost, Soviet life as reported in *Pravda* and other publications has become more believable.

The state of Soviet journalism—seen through the eyes of Glavlit, the state censor—was revealed 30 October 1988, in *Moscow News*. Five years earlier, the article began, Glavlit's censor removed a phrase before publication because "it might make the reader think that censorship exists in this country." Today, Mikhail Gorbachev answers a press question by saying "definitely" that censorship exists now to prevent divulging state secrets and illegal propaganda for war and violence. Revealing that there is, indeed, censorship is regarded as a step toward liberalization of the information media.

For nearly seventy years, Glavlit not only precensored state publications but appointed editors and changed editorial boards. Now, presumably, that has ended. And *Moscow News* takes the same position on censorship as Washington correspondents when they are blamed for publishing official leaks. Says *Moscow News*, "Secrets aren't born in newspaper offices. They can only reach editors through the negligence of officials...Wouldn't it be more sensible to guard such secrets at the places where they are supposed to be?" The Washington response in recent years has been to crack down on officials. But that, in turn, has been criticized by the U.S. press—a major distinction which ends further thought of equivalence between Moscow and Washington. Soviet journalists are still in the earlier stage of welcoming the relaxation of the censors.

Next, they call for a reduction in the list of secrets which, says the *News*, "has grown to unnatural proportions." Again, a familiar cry in Washington. The paper also calls for censors to "openly publish the lists of state secrets (without disclosing them)." There are, after all, says the *Moscow News*, "always a responsible editor, author and the criminal code." The article concludes with an optimistic assessment: "The very development of the media is forcing preliminary censorship to gradually give up its hold. There are more 'live' programs on radio and TV, and less and less subjects forbidden to the press." And now, writes *Moscow News* finally, perhaps with fingers crossed, "there's this article, too. If, of course, they let it through..."

The editor of *Moscow News*, Yegor Yakovlev, told a Prague paper in May 1988 that the problem of self-censorship remains. He had printed the first report of a young German pilot's landing a small plane in Red Square in 1987. "No one prevented any other paper from doing the same thing," said Yakovlev. Others probably waited to see what would happen to *Moscow News*. Self-censorship. "The problem is," he added, "we are professionally unprepared to use all the opportunities open to us." He admitted that even *his* "head" may be "spoiled" after more than 30 years in his profession. Other journalists, presumably, are more "spoiled" or more cautious, recalling that earlier glasnosts, like earlier detentes, were reversible.

Through 1988, the government drafted and secretly debated a new press law, presumably to enshrine the openness of glasnost for Soviet journalists. The first revealed draft, however, severely restricted the development of a more open media system. Andrei Sakharov called it "a great step backwards." The draft enumerated many subjects the media could not discuss, and created obstacles to the movement toward an independent media system. The draft also would fine or imprison authors or publishers of "unofficial publications" such as Sergei Grigoryants' magazine *Glasnost*.

* * *

IN EASTERN EUROPE, glasnost has been either a welcome importation or a perceived threat, depending on the different conditions in each country, and the degree to which each manipulates the mass media and the public. *Pravda* is seldom found these days on newsstands in East Germany or Czechoslovakia—the top-of-the-line Soviet newspaper is too "radical" for some communist satellites. Yet West German television has become so popular in East Germany that some workers leave their jobs in the Dresden area because Western TV signals cannot be received there. Consequently, the authorities have allowed West German television into the Dresden area over the government's cable line.

In Hungary, the process of greater openness has been evolving for some years—at first, with some trepidation at how Hungarian reforms would be regarded in Moscow. There was no sharp break with the past in Budapest as there was in Moscow, no spectacular change to set analysts to work in the East and the West. When Hungary's economic reforms seemed to be going well in the mid-1980s, relaxation of press controls appeared to be part of a reformist plan. The economy no longer sustains optimism, but some greater openness in the media persists because the still-reformist tendencies seem to require glasnost. The word from Moscow, moreover, favors it—indeed, regards Budapest as an advanced test of perestroika and glasnost.

Glasnost came to Poland well before Gorbachev rose to prominence. The independent trade union Solidarity, in the early 1980s, joined the intellectuals in demanding radical changes in the mass media. They wanted access to the channels, and some right to manage the media in the name of social participation. People should "speak with their own voice," was the cry. The crackdown on Solidarity coincided with the ending of a brief testing of reformist media.

There are two varieties of glasnost in Poland: the official relaxation of the formerly strong censorship, and the unofficial samizdat which has produced an underground publishing industry unequalled anywhere. The government uses with great care its power to rein in the dissident press. Now harsh, now somewhat more lenient, the Jaruzelski regime permits a degree of glasnost not found in the German Democratic Republic or Czechoslovakia. The presence of an active Roman Catholic church in Poland, and the submerged Solidarity union, remind the regime that holding a longer rein on the press can avoid a rupture, and a new crisis. The regime apparently feels sufficiently secure to permit public criticism of past injustices in order to reform the government structure under Party/military control. By seeming to share more information with the public, the regime defuses some opposition and maintains power over the major information flow.

This is not the case in the German Democratic Republic. The Soviet changes, and glasnost in particular, are all but ignored in the GDR press. Reforms are said to be an internal affair of the USSR, not for export. Gorbachev seems not to want to press the GDR.

No real changes are apparent in Rumania, though some mention is made of glasnost. The process has not changed press coverage or attitudes in Rumania very much. Indeed, some restrictions have been added in the name of national security. Glasnost is projected as unnecessary because Rumania—in the terms of its own leader—was doing so well. Infractions of minor officials are publicized, but that is deemphasized by the self-adulation of the Ceausescu regime.

Of all the states in Eastern Europe, Bulgaria regards itself as having the greatest affinity for the USSR. The Russians acquired the Cyrillic alphabet from the Bulgarians. And Sofia readily follows political patterns set in Moscow, but, so far, not glasnost, at least not wholeheartedly. The press criticizes some less worthy bureaucrats, and a few small scandals have been reported, but usually well after the facts have been widely known. Glasnost is used most widely to serve an immediate governmental purpose. In Bulgaria, the incremental use of glasnost is controlled; in the USSR, the larger process of glasnost is centrally controlled, but the issues and incidents for "open" treatment are selected mainly by the editors.

The uses of glasnost in Czechoslovakia are complex. There is still the haunting memory of the Prague Spring (1968) when Soviet tanks, appearing overnight in Prague streets, destroyed the reformers' hope of creating a "communism with a human face." Today, some consider that spring the first coming of glasnost. Publications abounded in new freedom and diversity. Then came the crackdown. Today, a limited form of glasnost is permitted. But not enough to bring in a new crackdown—as discredited as the earlier one now is. New reforms are discussed, and malpractices are spotlighted. But the news media are not permitted to consider a major restructuring of the society. There is, then, greater openness, but it is likely to serve as little more than a safety valve, certainly not as a stimulus for serious social or political reforms.

The East European Communist countries are using glasnost as a means of retaining power—information power—in an era of change and uncertainty. They recognize, in different ways, that telling their people what they already know is not likely to undermine the power of the elite. Nor is removing some of the rigidities in bureaucratic management and censorship of information. But permitting diversity in the sources of information—official and unofficial; and, especially, allowing diverse views and then dissent—that is not what glasnost is meant to do. And nowhere in Eastern Europe, as in the Soviet Union, does it come close to doing that. Glasnost may, however, nurture a deeper feeling for freedom of expression simply by virtue of the incremental changes now visible. Then Eastern Europe, perhaps well ahead of the Soviet Union, may determine whether glasnost, once unbottled, can be forever regulated.

Though statist monopolization of mass media prevails across Eastern Europe, scholars and journalists, particularly in Hungary and Poland, discuss alternative systems. They recognize the failures of the present information channels as a consequence of the failed political and economic systems. Jadwiga Pastecka, the Polish scholar,[16] makes the ultimate correlation between information and political power. She enlarges the terms of the Universal Declaration of Human Rights (Article 19) to include the right to "co-create" as well as seek, receive and impart information. Co-create, she says, "means not only participation in the communication process as such, but also in decision-making, in governing the country, the work place and the like, wherever important decisions and solutions are being created, which are of no less significance than news or messages." This radical enlargement of the demand for information power is precisely the objective most feared by communist and rightist authoritarian regimes; it is reluctantly welcomed, though sometimes subtly resisted by democratic states as well.

* * *

CHINA SPRING IS a melancholy reference to Prague Spring—the ending of the short period of democratic trends in Czechoslovakia in 1968. China's "Spring" seems to come and go like the seasons. Chinese theorists have begun to "reevaluate"—criticize—a Soviet book that has influenced Chinese Communists for decades.[17] Stalin himself had written some chapters in the "History of the Communist Party of the Soviet Union (Bolshevik)." The book reached China in 1939, and all Chinese party officials were told they "must" read it. Mao drew heavily from it. Today, however, the Chinese press says the book "contains many fallacies." They are saying this book impelled some countries to move too swiftly into communism. China is now in "the primary stage of socialism, which will probably last 100 years," *China Daily* reported, and socialism must precede communism. It said that Stalin acted "rashly when he eliminated the private economy" when the country's productivity was low.

Yet China's development is no straight-line movement. In 1957, after the presumably liberalizing "one hundred flowers" campaign, Chinese intellectuals were encouraged to criticize the Party and government, and when they did they were imprisoned or persecuted, some for twenty years. After the inferno of the Cultural Revolution (1966-76), there came the relaxation of information control. A wall in Xidan in downtown Peking became "democracy wall." People were allowed to put up posters in the fall of 1978. They became increasingly critical of the party and the government. Peking closed down the wall in late 1979, ending the brief interlude of a free and relaxed atmosphere which came to be known as Peking Spring. Wei Jingsheng, China's most famous dissident writer, was sentenced to fifteen years in prison where in 1987 he was reported (without confirmation) to have died. Beijing also closed most of the scores of new independent publications that had started in the "Spring."

An association of the remaining unofficial magazines emerged in 1980. A representative conference of the association was then held. The report of that meeting, published in *Zeren* (Duty), 20 September 1980, was prepared by He Qiu, later arrested for his role. He wrote of "the second spring":

Winter winds give way to spring breezes, and the land is reborn, clean and fresh. Privately, new publications, rather than shriveling and dying in the wind and frost, sprang back from their ordeal with renewed vigor. New sprouts pushed up from the once-frozen ground. While still dwarfed by the mighty trees of the forest, they are the symbol of this age and the hope of China. They will continue to grow despite all obstacles.

About 100 unofficial Chinese magazines were listed by Freedom House. These were the "new sprouts" dwarfed by the "mighty" establishment publications. One sprout was a "virulently anti-Marxist" pamphlet reported in February 1981 by Agence France-Presse. The magazine, *Today's China* (1981), included a poem accusing the Beijing regime of being "constantly in error and fear of criticism." The poem called for a society permitting competition of all ideas. The man who gave the journal to a European traveler allegedly said, "Our next edition will be better printed, if we are still alive." The Central Committee of the Communist party's Document No. 9, 1 March 1981, ordered a crackdown on all unofficial publications and organizations. More than twenty-five editors were arrested and some summoned to "thought reeducation."

For several years, the clampdown held. During the subsequent shuffling of political forces in Beijing, there was intermittent relaxing and toughening of information controls. By the beginning of 1987, the campaign against "bourgeois liberalization" attacked those who had moved too far, too fast toward diversity. Many leading writers and teachers were fired.

The National People's Congress in April 1987, however, seemed to halt the conservative crackdown as necessary to encouraging radical economic reform, and another "opening" to the world beyond China. The campaign against bourgeois liberalization was quietly dropped. Some expelled writers and academics were brought out of seclusion, and a new "spring" dawned.

In that atmosphere the magazine *China Spring* entered China. It had been created in 1982 in New York City by a group of Chinese students. The magazine was quietly carried into China and shared with students in the People's Republic. Out of that readership has come the Chinese Alliance for Democracy (CAD). It has members not only in North America and Southeast Asia, but in Beijing, Shanghai, Tranjin and Canton. Despite *China Spring*'s significant accomplishment in circulating inside China, it has not yet provided a rallying point for intellectual discussion in the PRC. *China Spring* believes the conservatives still lurk in the background, ready to turn any new spring into fall, if not winter.

Meanwhile, the struggle for the control of China's mass media takes many forms. Young journalists from the print and broadcast media and

Xinhua, the Chinese news agency, are sent each year to journalism colleges in the United States. I have interviewed groups who completed their studies during 1985, 1986 and 1987. The first year, the dozen young men and women were reluctant to speak, even after spending a year in the U.S. They seemed anxious to return home, and say as little as possible about their experience abroad. That first year, I told the Chinese journalists this story:

When I was in China in 1983, *China Daily* (May 19), carried a story with a five-column headline, "McCarthy era—two kinds of Americans." The story described America's "dark era" through the eyes of a Chinese actress, Wang Ying, well known in the 1930s. She lived in the U.S. in the '40s and '50s, but her autobiographical novel was recently published.

The young couple in the novel was detained and "maltreated" at Ellis Island, New York and their return to China prevented for some time after "liberation." There were, however, "hospitable, helpful, trustworthy and sympathetic Americans who did all they could to help the young couple," the news report said. But there were also the demagogic Senator Joseph McCarthy and his followers. Perhaps the most important part of the story for Chinese, however, was the dramatic conclusion. I paused a moment before reading it to my Chinese journalist guests: "During the turbulent years of the 'cultural revolution' [having returned to China, the author Wang Ying] was accused of being a spy and put behind bars by the Gang of Four. Her health deteriorated rapidly and she died in prison in 1974."

Reading between the lines, I told my visitors, this is a story about two kinds of *Chinese*. There was some shuffling of feet, but no verbal reactions from my guests. Indeed, the two-hour session produced little dialogue. These journalists were the children of the Cultural Revolution, whether sons and daughters of the victims, or Red Guards themselves, I never knew. But for them, safety still lay in silence.

Not so by the time the 1987 group arrived. We engaged in a two-hour discussion of press freedom, American style, and the implications for China of a far more diverse press. There was much notetaking, cross discussion and frank disagreement with me and others around the table. This serious group included writers for newspapers, radio and Xinhua, and a professor of journalism. They were intent on adapting new ideas to their professional assignments.

As a regular reader of *China Daily* I am sometimes encouraged by the "spring"-like rhythms. These include professional discussions such as that about the newsprint shortage, and the advice of journalists to "the state": "Invest more and set up paper mills and timber-producing bases."[18] And this account by Di Yi, in *Guangming Daily*, cuts closer to the key issue: Leaders, says the writer, "encourage criticism and the reporting of bad news only on the surface, but get frightened when such reporting and criticism actually appear in the press...It's certainly an improvement for the leaders to admit that they do have some 'dirty linen to wash.' But to prove their sincerity in encouraging public criticism and objective reporting, they should not only put up with this, but also accept the supervision of the public including the press."

That simple observation probably describes accurately the central authority's fear of exposure to examination, let alone criticism, and the limited freedom therefore permitted to write analytically. Di Yi's comment, after all, is highly generalized and thus not a threat to anyone or any bureau. That much is permissible in today's China. Yet even that is an advance over yesterday.

So that tomorrow's journalism may be still more "open and bold," some 472 middle level Party members and others were asked in April 1988 to evaluate the Chinese news media. Most members of the National Committee of the Chinese People's Political Consultative Conference strongly criticized the Chinese media.[19] Under the three-column headline: "Media urged to be more open," *China Daily* (7/11/88) reported: Asked "Are you satisfied with the openness of the news coverage?" 63.2 percent said "No." A member cited an example. Last summer, quite a few months before the Thirteenth National Congress of the Party, foreign newspapers he read began to report the Party's forthcoming personnel changes at the highest level and, as it turned out later, these were all true. But at the same time, Chinese readers were told nothing by their own news media.

Asked whether they are satisfied with the critical reports of the news media, 57.9 percent said "No," while 23.3 percent said "Yes." The remaining gave no replies. Reasons for dissatisfaction were "news media criticism is usually concerned with low-ranking officials instead of high-ranking ones," 81.1 percent; and 66.5 percent, "it evades crucial problems." A member put it bluntly, "The news media dare not criticize high-ranking officials by name."

Unreliable reports also drew criticism. A member listed such things as "reporting good news only, statistics that are not compiled in a scientific way, and biased criticism or praise." Roughly 68 percent said they "basically believe what the news media reported," but only 1.3 percent said they "fully believe" it. About 20 percent said the media are unbelievable or unbelievable in general. The remaining 10 percent said they were not quite sure.

Those surveyed also gave opinions on a variety of issues. About 77 percent disputed the claim that openness in news coverage would affect the stability of society. Eighty-nine percent said they agreed on reporting of "big cases of official corruption." Those who call for the major restraints of the Party and state leaders made up 74 percent. More than 80 percent agreed that the news coverage should include the difficulties in reforms and the open policy, and the negative side for society in the course of reforms. About 91 percent said the news media should become the forum for the common people to discuss state affairs. About 94 percent said the principle that all are equal before the law should be applied to news media criticism. Asked whether citizens should be allowed to start their own newspapers, about 56 percent said yes, and 31 percent said no. The remaining 13 percent did not answer.

This is China Spring—of sorts. The interviewers are asking the right questions, and directing the finger of public opinion toward "party and state leaders." They are also reporting the amazing fact of disbelief—that only 1.3 percent of China's knowledgeable actors questioned "fully believe" the Chinese press and radio. The creators of this survey, obviously reformers themselves, seem to want to make their society more open even if it takes boldness and perhaps courage to create a constituency for change. But must they not be aware that among the present state leaders are conservatives who may soon mount a new campaign against "bourgeois liberalization," and bring on the fall? Fearing that, one question takes on particular significance. Should not equality before the law, the pollsters ask, apply to criticism of all, equally, in the news media? Shades of Watergate!

* * *

CHINA IS ABOUT a decade ahead of the Soviet Union in attempting to reform "traditional" communism. The PRC's analysis of Gorbachev's reforms is, therefore, significant. The authoritative weekly *Beijing Review* in January 1988 cited problems that have plagued China's efforts to overhaul its socialist economy, and predicted that the USSR faces severe difficulties in managing economic reform. The hurdles mentioned by the Chinese:

1. Grassroots resistance from traditional ideologues, and those clinging to egalitarian expectations; interference from both radical and conservative factions; a struggle between the young and the old (a division borne out in the *Moscow News* poll cited above).

2. Opposition from cadres and officials who are afraid of losing their posts in elections, and corrupt officials fearing exposure.

3. Ordinary people who fear change, particularly those who accepted the decades of Soviet propaganda that described the country as an earthly paradise, where there is no need to worry about food, clothing or shelter.

The article seemed to reflect China's disillusionment over such Soviet propaganda, as well as Chinese experience with a leadership that could not deliver efficient services at the grassroots level. Specifically, *Beijing Review* said "deep-seated egalitarianism in the [Soviet] wage system must be corrected or it will be impossible to accelerate social and economic development." Soviet policies are "complicated and full of contradictions." The next two or three years "will be critical if the program is to be a success," the Chinese predicted. In this critique, China seems to be engaging in cross-border glasnost.

At home, Chinese academics are drafting a press law expected to be completed in 1989 or 1990. An early version would permit private individuals to set up newspapers, magazines and even radio and television stations. Journalists could attend open meetings of the government and the party, censorship would be eliminated, and the right of reply for false reports would be assured. Journalists would be licensed, however, to maintain state control. The draft would provide penalties for journalists who break the law, and citizens who obstruct the work of the press.[20] Such innovation, if allowed to proceed from law to practice, could institutionalize the Chinese Spring.

* * *

IF THE SOVIET Union is a metaphor for all closed societies struggling to exploit new communications technologies, but fearing their openness, a satirical view from *Moscow News*[21] expresses the nervousness which glasnost generates. Some excerpts from "Glasnost Blues":

> I consider glasnost a noble cause, provided people don't overdo it. But if there are no brakes on it at all, it may prove to be too much for your psyche.
> I had a field day when we turned from sugary lies to the triumph of truth. Early in the morning, I would go to buy newspapers. Everyone shouts about what they have read recently in the papers. In the evening you're glued to the TV set, as if to oxygen.
> In short, I liked glasnost until it began to give me the blues. A day would hardly pass by without some accident. Accidents, of course, happened before, but we were kept out of the picture so that we wouldn't be upset. We were kept posted only about the accidents that occurred abroad, since it was quite clear why they had accidents out there: profit-minded capitalists did not give a damn about industrial safety. But, nowadays, our nerves are on edge day in, day out. You only have to look at the sorrowful face of the anchorman and your heart sinks: here we go again.
> Not even the Black Sea makes you feel happy these days. Last summer I went to Sochi only to find warning signs on its beaches forbidding bathing on account of the high level of pollution.

For good measure, the television people have come up with an idea to widen my horizon. The other day they showed a Swiss food store on TV. The Soviet reporter asked the store's manager about queues, but the manager couldn't figure out what he was talking about. I'm not against other countries' methods. But if no one is going to adopt them, what's the point of bothering me? It would be better to stick to the good old ways, showing their homeless people, their down-and-outs digging in the garbage cans, who we will sympathize with, taking heart in our employment-for-everyone policy. A bitter pill should be sweetened. For example, when you show rotten potatoes, in the next breath you say something like "this year we had a bumper crop, hence the problems with transportation and storage." Today, nobody tries to console me or set my mind at ease. On the contrary, I am told: "Look that's what we have come to! Wake up," they say, "we cannot go on like this". But I don't feel like waking up. I envy the ostrich. Smart bird!

5.

Power and Press in the Developing Countries

THIRD WORLD LEADERS are not like the movie primitives cowering at the sight of a lighted matchstick in the dream sequence of "A Connecticut Yankee in King Arthur's Court." Yet, Third World officials say they are sufficiently aware of the power of modern communication to fear the uncontrolled flow of information from abroad. But some fear more, it seems, the unrestricted access by their own people to domestic news and information. Many developing-country governments, therefore, either own the mass media or "guide" the news and information flow.

Some leaders argue that a state-run press can be free. Others, conscious that foreign perception of a state-run press costs them some credibility abroad, seek to justify their stringent control by changing the world's standard. They call for a "new information order" that is intended to legitimatize state control for those who want it, and, on behalf of less censorious governments, create a universal code of journalistic conduct to influence the independent media. Under this order, government-designated objectives would replace the present amorphous global news concept based on the free flow of information enshrined in intergovernmental declarations and covenants.

To provide the kind of news and information they desire, developing countries have created national and regional news agencies, and a global Third World agency, all run by governments. There is also an independent Third World news agency committed to carrying developing-country viewpoints. These alternative news systems are linked to one another but not operationally to the Western global news networks. The Western services do not yet accept the alternative agencies as credible news sources.

Yet in the coming Age of ISDN, when all networks will be linked, the credibility of a government information agency will matter less than that its voice be heard on the global spectrum which ISDN will provide.

The receiver of the Third World flow then may be not only a journalist but a scholar, businessman or interested First World citizen. The journalist will not be needed as intermediary for other users of a terminal anywhere to tap into the Third World news flow. That will, indeed, be a new information age, if not "order."

* * *

IN A SMALL classroom in Bandung, West Java, in 1987, sat the chief librarian of the University of Guelph, Canada. From an Indonesian island of traditional culture, John Black spent two hours linked to a computer in Canada. This was managed, he says, "using a landline to Jakarta, an Intelsat link to the United States through one of the U.S. international record carriers, then into Tymnet (a U.S. packet-switched network), back into Datapac (a Canadian public packet switched data network), and then to the computer in Guelph. Less than five seconds was required to make the initial link contact, and then two hours of solid use without as much as a character distorted by a transmission error." Comments Black, "We tend to take it all for granted."

We do. And some also tend to regard such linkage as threatening, either to traditional cultures or, inevitably, to highly modernized cultures as well. Some see modernization not only destroying tradition, but also homogenizing the cultures of the industrialized countries which produce electronic wonders. The vast projecting power of the satellite-linked computer, coupled with the attractiveness of the modern word and image, are assumed to be irresistible. Electronics, plus an idea, can move mountains. Who, then, will provide the idea? And whose mountains will be moved?

Developing countries are fearful that their idea will not be projected, though their cultural mountain may be bulldozed. So far they have expressed the most fears, and most loudly. They have done this in Unesco and other international forums. They say they cannot now speak to one another as sovereign nations unless their messages, public or private, pass through Western telecommunications, data banks, or news services. The Western media are attacked for biased, fragmented and incoherent reporting of foreign news. Middle East news coverage is simplistic, repetitious, and generally one-sided. Latin America, despite its size and diversity, is woefully under-reported. Central America appears only during a military or political crisis. African news is tied mainly to Western trade and geopolitics. Indeed, foreign coverage generally, say critics, reflects Western political, economic or military interests; based too often on sensationalized or exotic events.

The truth is not so simplistic, as studies of news influence in the developing countries have shown. Often, even when Third World editors receive news of distant developing countries, they choose instead to carry reports of developed countries, particularly trade partners, and only nearby Third World nations—just as Western editors do most of the time. Sadly, that is reality.[1]

Everywhere in the world there is a Babel of news and propaganda on the airwaves. Radio broadcasting, particularly in developing countries, is the most effective means of reaching large audiences. The BBC, Voice of America and Radio Moscow are the major shortwave broadcasters, but Radio Egypt is widely heard on African airwaves, as is Taiwan in East Asia, and the Voice of the Andes in Latin America. Third World voices add to the cacophony.

How, then, has the modernization of information flows affected traditional and recently decolonized countries?

Modernization

New technologies influence even the most traditional or decolonized developing societies. The lines begin to blur between communication and education, between essential service and social-driven contact, and between social intercourse and political action. A telephone installed in a village to maintain contact with the central authority soon becomes a social instrument used by others than the chief. And the demand grows for more than a single phone. A common radio receiver in a tribal meeting place conveys more than word from the leader.

Radio is a greater multiplier. It is used as an educator of children, a health adviser, and conveyor of planting and harvesting information. Yet the written word, still a necessary part of cultural transmission, is sorely disadvantaged in the developing countries. In 1981, the consumption of printed materials in kilos per thousand inhabitants in Bolivia was 1,224; in Peru, 1,772; and the U.S., 44,422. Where reading levels are low, can computers be readily introduced? In the U.S., poor children have access to the new technologies, particularly computers, through the public school systems. But in Latin America, argues the Peruvian Center for the Study of Transnational Culture, "It is the private school which more easily acquires the equipment, having the economic strength to do so; therefore, won't access to computers perhaps be a means of marginalizing, which will only widen the gap between social classes in Latin America?"

It would be as wrong to blame that problem on the technology, or the creator of the computer, as it was for citizens of La Gaude, a tiny village in the hills behind Nice, to tell me in the 1970s that their growing air pollution was an "American" import. The unwanted effects were, rather, the consequence of envying U.S. consumerism. Frenchmen were perpetrators as well as victims. The cars and motorcycles (mostly of European manufacture) that cluttered the narrow roads and emitted evil fumes were chariots of choice. The result was a shared global problem induced by the process of modernization. The process has no nationality, no insidious manipulator. The problems come with the perceived advantages: mobility, in the case of the automobile and other locally desired technologies, which compete with indigenous culture.

Modernization inherently favors choice-making. This may undermine traditional patterns. New products, especially new communications, require adaptability as well as continuity. Since new things in developing countries are generally importations, the individual mind is turned outward beyond the village or the nation to the world outside—and certainly beyond the self. Traditional values give way to more universal interests, if not values. A middle-aged Algerian complained to me in Algiers that the passing mini-skirted woman absorbed too many French ideas. That was said ten years before the generational riots in Algeria in 1988.

Modernizing the traditional or decolonized societies focuses on the very development which undercuts tradition. Citizens are now geared to developmental change—more crops, new industry, mass communications —rather than older, habitual occupations. Little-restrained population growth creates immediate demand for more of everything. The simple structure of the old society is replaced by new institutions. These institutions may replace the religious with a secular base. The sacred is made to compete with the profane.

All of these changes, most occurring within one generation, destroy the sense of commonality. At the same time, however, nation building or rebuilding makes essential the use of mass communications in some form. Communications, which helped destroy traditional social patterns, are quickly regarded as the primary builders of a new pattern, a new social structure with communications at its apex. Suddenly, print and broadcast media become government's prime channel to the people, overshadowing the tribe, the priest, even the family—unless, as in Iran, the mosque remains dominant and exploits radio and other technologies to achieve religio-political objectives. Rulers soon realize, however, that mass communications exact a price from traditional societies. Sudden changes in values and living patterns cannot easily sustain the homogeneity of traditional life. Change can generate misunderstanding, a loss of solidarity, even active discontent.

Who is to blame?

"America," as the villagers of La Gaude responded to their own gasification of a traditional landscape? The Western news media, as most Third World countries cry in Unesco? Or some combination of causes which begins with the decision to modernize—and thereby suffer certain crucial consequences in the expectation of ultimate advantages?

The lesson soon learned is that new communications technologies, mainly in the hands of authorities, give government great power to control word and act, and manipulate the public. Citizens, in turn, have little power to resist the central voice. They had more power in the traditional tribe and extended family, where decisions were made by consensus. Modernization of developing societies may readily generate collectivization, bureaucratization and alienation. Sensing these hazards, the ruling elite often decides to use the double-edged sword of communications to inform for developmental purposes, and control for ideological/political objectives.

Such twin use of the media was demonstrated in the Asian-Pacific area in the 1960s. Extensive programs were mounted to train independent journalists to report scientific and technological achievements. Such coverage was designed to advance agricultural and industrial development. Local farmers, workers and managers profited. Soon, however, governments distorted the programs by adding national, political or ideological objectives to "development news reporting." That combination of objectives later entered the Unesco debates. The valid demand for coverage of technical subjects was overshadowed, in the Western perception, by the new purpose to assure governmental control of media content.

Can a state press be free?

A Third World journalist clearly stated this press-control objective a decade ago. Ludovick A. Ngatara, then general secretary of the Journalists Organization of Tanzania, addressed the International Institute of Communications meeting in Dubrovnik in 1978:

"In the two decades since various Third World countries, especially those of Black Africa, started emerging as independent states, the masses in these countries witnessed all media instruments, which were supposed to be their voices, being utilized by one regime or another for swaying them like helpless willows to all corners of the political winds." The emerging regimes, he said, "have opted to use the instruments of mass media as a means or shield to protect them on their unpopular thrones." Ngatara continued, "A regime which opts for dictatorship or totalitarianism in the guise of socialism employs mass media instruments to achieve its end. Other unpopular regimes take different courses to justify their ends." When a new regime takes over, he added, "the media instruments also change hands to usher the masses to yet another ideological cause."

Can a state press be free? Can it provide diverse views, even dissent from government policies? That is the "Third World dilemma," said Hilary Ng'weno, Kenya's most prominent editor-journalist. He wrote in June 1979 while still publishing the *Weekly Review*, a remarkably free and sophisticated magazine. Sadly, newsprint shortages and financial problems brought the magazine under government patronage. "Theoretically," said Ng'weno, perhaps pondering the future of the *Review*, "there is no contradiction in the concept of a state press which is also free." He continued: "Most states profess to exist solely for the purpose of serving the national interest. Thus, a State should find no difficulty whatsoever in owning or controlling a press which has the freedom to serve the national interest through vigorous investigative reporting aimed at minimizing the abuse of power and expanding the rule of law."

In the real developing world, however, there is only one branch of government, the executive, with no check on its power, said Ng'weno. Other institutions—trade unions, religion, universities, and the press—do not have the muscle to rein in the executive. Regarding "all-pervading power of the government," said Ng'weno, "nothing has really changed from the bad old days of colonialism." He said that in colonial days "the independent press was made up of committed anticolonist freedom fighters working in appalling conditions, fraught with the danger of bankruptcy or detention or both. Theirs was a thankless task which pitted them against the full might of imperialism. But after independence, the private press in the Third World was no better off than before independence as far as financial and human resources were concerned. The best minds in the country joined government bureaucracies or private manufacturing companies which could pay handsome salaries."

This suited the new governments. "Before independence, when they had been struggling for power against the colonialists, Third World nationalist leaders were keenly aware of the power of the printed word, the power of propaganda. In fact, in many Third World countries, it was the printed word which brought down the colonial governments; there was no military struggle, for the pen was rightly considered to be mightier than the sword in many colonial countries....It was not by accident that in virtually every Third World country the new rulers who took over from colonial masters made sure that the most potent instrument of propaganda—radio—was invariably controlled by the state. And the key word is control...Governments moved in to control other instruments of communication. Newspapers were taken over and those which were totally opposed to being incorporated into the government propaganda machinery were closed down. In each newspaper taken over by the government, the state (or in some cases the ruling party) ensured control through appointment of trusted men who carried the title of editor but who were in reality no more than extensions of the government apparatus." That, said Ng'weno, "is the atmosphere against which an investigative journalist in a state press in the Third World actually works, if at all."

Investigative reporting usually means seeking corruption in high places, abuse of power and violations of human rights by those in authority. "The trouble is," Ng'weno continues, "people in authority often consider their first order of business as being the exercise of power. That invariably means hanging onto power. Investigative reporting tends to challenge the legitimacy of the power structure, it tends to hold up the bitter reality against the rosy image painted by those in authority, In short, it tends to make the ground underneath the feet of those in authority somewhat unsteady."

Worse than that, this superb journalist reminds us, "in most countries investigative reporting is looked upon as a sign of disloyalty, anti-statism, dissidence or even treason and it is a foolhardy editor, indeed, who will encourage investigative reporting by his reporters if the newspaper of which he is editor is owned, and therefore controlled, by the state."

Can a state press be free? There are moments of seeming freedom in a Third World state press, Ng'weno acknowledges. Even in a one-party state there are power struggles between individuals or factions. Sometimes the state press becomes a pawn in such a struggle. Reporters are let loose to gather incriminating evidence about one or another faction. Invariably, Ng'weno adds, "the curtain comes down as suddenly and as unpredictably as it had risen." Once party or state control is decided, investigative reporting comes to an end. "It could not be otherwise," Ng'weno concludes. "All states tend to perceive the national interest as being merely another name for state interest. A journalist's perception of the national interest is by the very nature of his profession, unless he is a party hack or mere government functionary, broader. More important, it is tentative, undogmatic, not determined in advance of all the evidence, and to that extent it cannot be given meaningful, free expression within the confines of a state press."

No, Hilary Ng'weno finally advises, a state press cannot be free. An independent press in even a poor country can survive—if the prime minister is permissive and the editor, despite official sanction, courageous. One such editor-in-chief is Bona Malwal of the *Sudan Times*. The four-page tabloid carries the line, "modern daily," on its masthead. Malwal has told me of his difficulties in producing a thoughtful newspaper in a country in transition and engaged in civil war.

Prime Minister Sadiq El Mahdi promised to permit a wide range of press freedom and other reforms after he took office in May 1986. The *Sudan Times* was created in 1985 to test the possibility of a free journalism. Malwal, with academic and journalistic experience in the U.S., has tested the political waters courageously. With the war in progress he headlined one article, "Is there dissent within the Sudanese army?" Another headline, in the same issue[2] stated "The OAU has become increasingly irrelevant." Other editions carried the full texts of the U.S. State Department's country reports on the human rights situation in Sudan and neighboring states. These included extensive critical analyses of the freedom permitted the individual citizen and the press. Malwal, in editorials and signed columns, provides sophisticated monitoring of the political freedoms Sudan permits, and the occasional steps by officials at least to consider or threaten to withdraw some of those liberties.

Sudan, for example, has what is probably the only democratic parliamentary system in Africa. It was achieved after a popular uprising ousted a military dictatorship. The legislature's sessions are broadcast for the public, and recently television was permitted to carry the proceedings. In a significant editorial that revealed the high degree of press freedom, as well as the sometimes tentative nature of the use of information power by the news media, the *Sudan Times*[3] editorialized against the government's television station having edited out some coverage of one parliamentary debate. "This made viewers wonder if some kind of censorship is being introduced," the editorial stated. "Let us hope this will not be yet another encroachment on acquired liberties as we are repeatedly threatened by the Minister of Information and Culture who, we understand, is in the process of introducing a revised version of the only recently revised Press Laws."

This editorial was one of three[4] in which the *Times* found it necessary to defend itself against a speech by the prime minister attacking "certain sections of the English press"—obviously the *Sudan Times*. The editorial called this attention "flattering" but assured the prime minister that his concerns are "misplaced." The paper "does not support armed insurgency against a democratic system" if only because the "*Sudan Times* owes its existence to democracy, and therefore cannot have any interest in supporting those who threaten it with force."

The prime minister's attack apparently was based on the *Times* having published commentary from European and American journals that he found distasteful. These publications provided viewpoints which probably could not have been as well sourced and as critically stated by Sudanese writers, even one as amply situated as Malwal. "We publish such articles," said the *Times* editorial, "because we believe the Sudanese people have a right to know what is being written about them abroad." The paper added, "The fate of any state is determined to a greater or lesser extent by foreign powers, be they friendly or unfriendly." The editorial said "It would be more constructive if the government undertook to answer foreign criticism, point by point, using its own not inconsiderable media resources."

This exchange reflects the universal clash between state power and information power, conducted on the terrain of a developing country. It is especially significant because the central authority has granted—and up to this writing permits—sufficient freedom to independent journalists to engage officials in a rational debate of the fundamental press-freedom issues.

The editorial concludes with this clear summary of the Sudanese situation and, indeed, the universal challenge to the central power structure, whether in developing or developed countries:

> The *Sudan Times* realises that it is in a fortunate position. We have, so far, been allowed complete freedom enjoyed by very few countries on the African continent. The government certainly deserves praise for the maintenance of that freedom; it cannot have been easy to practice such restraint during the turbulent times we have recently experienced. But it must be stressed that freedom of speech and freedom of the press are fundamental to a system which claims, as ours does, to be a multi-party democracy; such freedoms are not privileges but rights. Anyone who interferes with these rights, whatever their motives may be, threatens the whole system with serious and permanent damage. If the press is restricted by government we will find ourselves on the slippery slope to authoritarianism.

Later in the year, Malwal took the global news media to task for ignoring the thousands of deaths in the Sudan that year:

> What happened to the rock stars, media personalities and all those can-do aid agencies? And most of all, what happened to the Western television networks who filled so many homes of prime time viewing in 1984 and 1985 with moving pictures of thousands of bone-wracked Ethiopians...Was it simply a scheduling problem? Perhaps the Sudanese famine couldn't be fit in between the Seoul Olympics and the American presidential election? [5]

But Malwal saved his harshest words for Sudan's state radio and television. He charged the government media with never failing to

> annoy, amaze, exasperate and completely fail in their designated task of informing and educating the public on the issues of the day. One of the major weaknesses of democracy...has been the failure of state radio and television to free itself from the dictatorial control of the government...

He chided the media for paying little and only belated attention to important negotiations to end the war in the South.

> This kind of reporting might have won the state media accolades in a banana republic dictatorship where news that does not mesh

with the views of the rulers is simply ignored, but it ill-behooves a government like that of Sadiq El Mahdi's which loudly proclaims its democratic credentials.

Malwal concluded:

> The state media must be freed from the fetters of governments which see no role for it beyond entertainment and slavish obedience to its policies. The democratic claims of this government will always be suspect as long as the state media serves up a steady diet of pap...[6]

In transition

But suppose a state itself is in transition or relatively free, though poor—still too poor to support an independent newspaper or broadcaster, let alone competing news media. Can a state that defines its political system in relatively free terms permit state-owned news media to provide diverse news and commentary? That depends on the state's definition of freedom.

The Chinese pondered this question early in 1988. Yang Baikui, a political scientist, argued in *Theory and Information Gazette* (excerpted in *China Daily*) that some fellow citizens' "over-anxious aspiration for democracy, which is often ambiguous and poorly defined, poses a stumbling block for current reform." He gave his version of "democracy" and "freedom."

He contrasted Rousseau with Montesquieu. The former, he said, provided the model for a totalitarian "people's democracy," in which "the powerful could bully the weak, and the majority could lord it over the minority." He termed China's bitter Cultural Revolution (1966-1976) a "good taste of the so-called 'grand democracy,' which resembled Rousseauist democracy." The democracy championed by Montesquieu, said Baikui, "emphasizes the power structure, legal procedures, and regulations and rules. Under the premise of the people's sovereignty, it stresses the rule of law." Baikui warned that many people, "when they fail to impose their wills on others or on the authorities, turn to violent and radical actions." This, he said, "is fairly common in developing countries, especially those starting to modernize."

The commentator next makes a useful distinction seldom uttered by Third World theorists: "Democracy basically means the right of the majority to participate in decision making, and the obedience of the minority to the will of the majority. Freedom, however, means the basic rights of people as individuals, which in many cases have nothing to do with decision making." Yet, he continues, "democracy, in the sense of popular participation in decision making, constitutes a means to guarantee the rights and freedom of individuals. Freedom can be regarded as an end in itself.

It embodies people's needs as human beings. Without individual freedom of speech, man's humanity is constrained."

For people in Third World countries, says Baikui, "democracy is not only a matter of definition but also of how to bring it about." He fears that "rash introduction of democracy despite the powerful inertia of traditional social and political structures, would only bring social unrest, corruption, violence and tyranny in the guise of democracy." (That fear of too-swift democratization is the usual rationalization for delaying steps toward constructing a democratic system of governance.) Meanwhile, however, as Baikui hints, it should be possible—even in a giant state such as China—to cease constraining "man's humanity" and permit "individual freedom of speech." Slowly, in a pattern of zigs and zags, China has been opening the printed pages to theoretical "new definitions" such as Baikui's. Chinese "glasnost" has been far less dramatic than the Soviet Union's, though Beijing's came first. In both countries, however, the right of the individual—whether to write, speak or act—is still fully determined by the needs and policies of the state. Neither Rousseau nor Montesquieu would approve.

Probably no foreign teacher, past or present, can adequately instruct developing-country leaders in democratization. Each country's history and contesting social and political forces are unique. But other small developing countries with workable democratic systems can be instructive. Of the freer countries in 1988, nine are poor developing nations,[7] with state news media that provide diverse coverage including the political opposition. Countries such as Botswana have developed effective systems to minimize partisan manipulation and corruption in state-owned properties. Democratic institutions function well in Costa Rica, where the limiting of state power—and virtually unlimited news and information power—have become internalized in all citizens, especially the elite.

Where such internalization begins to occur, even if incomplete, as in Nigeria and Turkey, there is recognition of the tension between state power and the free flow of information inherent in a democracy. In both countries the news media display considerable diversity. They criticize government policies but avoid the few major taboos. The press can play a crucial role in limiting abuses of state power, and serving as a public conscience. That is the function of important segments of the Indian press. There, as in Nigeria, the press has fought off authoritarianism by use of the judiciary and direct appeals to the ideal of democracy. Indeed, the Nigerian regime's attempt to control the press helped unseat the ruling military leaders in 1985. Nigeria was fortunate in having a diverse, sophisticated press, and a network of communications schools run by teachers with advanced training. Professionalism must accompany the freedom to practice the profession, if democratization is to succeed.

But that is not the way of life in 63 percent of the developing countries.[8] The love-hate relationship between governments and the press is putting societies on edge in Southeast Asia. Officials in Malaysia, Singapore and Indonesia are turning to press controls as the blunt answer to better educated, younger citizens who question economic and political failure and cannot understand official secrecy.

Newspapers generally support the government in power. In Malaysia, for example, every paper is owned by some member of the ruling coalition, and in Thailand various pressure groups are represented by the newspapers they own. Every top-level journalist in Singapore must be approved by the government. Still, these and other ASEAN regimes regard the news media as necessary evils. All ASEAN governments crack down on the press when coverage becomes uncomfortable.

The curbs on the ASEAN press during 1987-88 shocked journalists. An Indian observer, M.G.G. Pillai writing in the *Statesman*[9] noted that "In one [ASEAN] country, one is not allowed to criticize the government's poor performance or touch on political problems that reflect badly on those in power; in another, the Finance Minister's business connections are out of bounds; in a third, the government, while not particularly cooperative to reporters, especially foreign correspondents, demands its right of reply to even minor errors." All of these restrictions, says Pillai, "are justified as necessary in the early stage of nation-building." He adds, "This view is not peculiar to Southeast Asia."

Indeed it is not. In the Arab world, every state exercises direct control or indirect pressure over all the domestic news media and, in some cases, over Arab publications headquartered abroad. Such pressures stem not solely from the need or desire to govern, or to retain state power for whatever purpose. Islamic fundamentalism, particularly since the Iranian revolution, inspires greater governmental controls over the press as of other aspects of Arab societies. Religious fundamentalism exacerbates the more repressive controls previously projected by Nasserism, or Arab nationalism. The relative freedom of the press in Lebanon was destroyed by the civil war. Some diversity ended in Kuwait early in the 1980s. Journalists in both countries enjoyed some measure of freedom, though the governments were watchful and leery.

In the Arab countries, governmental uncertainty—if not yet instability—dictates tight controls over the news media. Officials seek to transform traditional societies into technotronic nations without inspiring a revolution from below. In addition, the Arab-Israel war has for forty years deepened the Arab countries' sense of defeat and impotence. One may imagine an Arab ruler pondering: if a relatively free press in Israel can create divisive problems for the victor, how much more threatening would be a questioning press in an Arab country?

In Algeria, Iraq, Libya, Syria, South Yemen (and to a lesser extent Egypt and Sudan), the press is used by the state to mobilize the people in support of political and economic development. William Rugh calls this the "mobilization press." He terms Bahrain, Jordan, Qatar, Tunisia, Saudi Arabia and the United Arab Republic the "loyalist press." They support the regime loyally despite the fact that the press is privately owned.[10] The Arab dilemma is made worse by Arab nationalism and Islamic fundamentalism, says Ghassan Tueni of the newspaper *An Nahar* of Beirut. Yet, he adds, "Both tendencies are inevitable responses to the Arab world's failure to find lasting solutions to its problems of national legitimacy and regional integration in the face of a hostile world." Nationalization of the press is the simplest means of control, as in Syria and Iraq. Laws banning contentious issues provide some greater latitude, as in Morocco and Tunisia. Control can be fine-tuned by press licensing as in the Gulf States and Jordan; or, as with the Palestinian press in Israel and the occupied territories, repeated banning, detention and expulsion limit press coverage physically and editorially. The net result everywhere in the Arab world—as, indeed, in three-quarters of all countries—is self-censorship.

Whether in the Arab world, Asia or Latin America, there is a common defense of Third World governments for controlling the flow of information within their countries, and pressing at Unesco and elsewhere for a drastic revision of the international flow of news. This common defense is based upon the formation of a Third World doctrine which began with the Bandung conference in 1955. That doctrine maintains that poverty and underdevelopment began with political domination and economic exploitation. Even after decolonization, unfair trade and monetary systems, and usurious loans are claimed to continue the domination. To maintain their advantages, developed countries are said to slow development in the Third World by withholding information of all kinds—scientific, trade, technological, agricultural, as well as general news—essential for modernization and development. A few developed countries, mainly through private entrepreneurship, control most of the world's news and mass cultural channels. By struggling together, this rationale concludes, the Third World can speed the trend of historic justice, and secure the information, skills and material substance needed to transform poor, dependent countries into informational, economic as well as political sovereignties.

This is a heady doctrine. It is accepted with varying degrees of antagonism toward the developed countries. Marxist-oriented spokesmen blame the market economies for all past and present problems in the Third World. Developing countries having their own market or mixed economies believe the present global system may be structured to produce inequality, though they acknowledge that developed countries may not seek to perpetuate it. Leaders of all Third World countries, however, share the uncertainties of a transitional generation, largely educated in Western ways, which is still influenced by traditional modes. Indeed, most of the populations over whom they preside have accommodated far less than the elites to modern Western forms. Many different kinds of political power are found in the more than one hundred Third World countries. In all, however, a discernible elite is at the center of power. That elite observes a domestic population of far more traditional background and with little internalizing of democratic ideas. Personal freedoms, where permitted, are also based on tradition. But governance generally remains the domain of the elite.

So, too, are the structure and practice of the information systems, even when they are not owned or directly controlled by the government itself. But, then, the same claim is made about the influence of the ruling Western elites over the news media defined as independent and nongovernmental. And, of course, the information systems in the Marxist states are by definition fully controlled by the governing elites.

There are, however, significant differences in the degrees of control, and in the diversity permitted the independent press in developed countries. That diversity includes treatment of indigenous government as an adversary and, perhaps more disruptive, the unmaking as well as making of presidents and policies as soon as they surface. Observing this, Third World leaders say such disarray is not for the weak, but only for the strong and rich. The Third World campaign does not principally seek to overthrow the political institutions of developed countries—many secretly admire them—but intends, rather, to reduce Western financial and information power.

Information parity

The Third World has sought information parity with the developed world for more than a decade, first, by pleading, then by persuading, and finally by threatening. Some 75 percent of Unesco's members are developing countries. Though there are obvious divisions among them—from the most to the least politically free or educationally advanced—even the most radical exponents of "a new world information and communication order" represent an authentic exposition of some objectives shared by all Third World countries. Those objectives are described differently by every spokesman or discussant. Unesco, the forum for a decade of debates, has never explicitly defined the purposes or procedures of "a new information order." Indeed, Western delegates fought hard to prevent the defining of global information systems. They maintain it is not the business of governments or an intergovernmental agency to interfere in the content of information flows. And, Westerners insisted, discussions of the news and information flows were premised on controlling the *content* of flows, despite denials from moderate Third Worlders and from Unesco.

The Third World's demand for information parity began in earnest at the first Bandung conference in 1955. It determined that ten years after decolonization began, the New International Economic Order promised at the United Nations was not transferring Western wealth to the former colonies. Indeed, their debts were growing and the global monetary system seemed to shrink the value of natural and human resources. Therefore, ran the Bandung litany, the developing countries must focus equally on information power as the key to economic development and nation building. This is a valid argument. Information is needed for all change, and skilled hands and trained minds are no less essential. Also needed is the flow of news from abroad; it helps make every country a more intelligent player on the world scene.

Delegates at Bandung spoke in terms of national interests. Most came from traditional or corporate societies in which the role of the individual was mainly determined by central authority. Consequently, the advantages of an improved news flow were mainly regarded as national rather than citizen objectives. Indeed, in the next twenty-five years at Unesco and the United Nations, the context of the "information order" debate was primarily governmental. Rarely were Third World journalists or private citizens either present or consulted.

The demand for parity was translated into an oxymoron, which Webster defines as a "combination of contradictory or incongruous words." The demand was for a "free and balanced flow of information." The implied *forced* balancing deprives the flow of freeness. All but the word "balanced" was by then the traditional Western (principally American) definition of global and domestic flows. Since 1946, U.S. diplomats had pressed at the U.N. and elsewhere for the universal acceptance of the free flow of information. The term was enshrined in the Unesco charter as in other international documents. The addition—"and balanced"—was the Third World's reference to parity. *Balanced* meant equal access to worldwide communications channels, sharing the technologies, training and skills of the information age, and no less important, forcing a change in the content of Western journalism as it touches on Third World countries, personalities and interests. That was a considerable load for one word to carry.

There would have been little debate, certainly not bitter denunciation from the West, if "balanced" had referred solely to communications infrastructure. The sticking point was the balancing of reporting and commentary about the Third World. To the West, this smacked of censorship. No matter how often or how earnestly Third World moderates explained that press control was not their objective, the linkage of the two aspects of balancing could not shake Western beliefs that a crucial realignment of political and economic power was the principal goal of the information order. For, the West asked, who is to monitor and achieve the balancing? Governments? So it seemed to critics of the "order." There followed scores of conferences, seminars, and formal plenary sessions of Unesco and the United Nations General Assembly at which the objectives of a new information order would be heatedly debated. With each passing year, tempers were shorter, suspicions greater, and language more verbose. Every word of information-order drafts took on several-layered meanings. Even "the" was seriously debated, and finally replaced by "a" before "information order." "The" implied the inevitable imposition of a set "order" while "a" did not. The full term, new world information and communication order (NWICO), was formally set in lower case to diminish its official character. ("Communication" had been added to include reference to technologies as well as message content.)

Then, in 1973, at the Unesco General Conference where I helped negotiate the communications issues, the NWICO "order" was transformed by adding the phrase, "seen as an evolving and continuous process." This removed the implication of an "ordered" system, presumably imposed from outside the information flow. Instead, there was acknowledged the obvious fact: changes in the forms and content of the information flow take place every moment of every day. That is an ongoing process, never ending in an "order." As several of us pointed out years earlier, any promulgated order was too reminiscent of Adolph Hitler's.

The evolutionary term stood for five years at Unesco, during which time it was also incorporated in several resolutions approved by the Special Political Committee of the U.N. General Assembly. The 19 November 1987 version at the U.N. included a "balancing" term which caused the U.S. delegation to cast the only negative vote against the resolution (there were 105 yeses, 15 abstentions). The U.S. objected to the resolution calling for the U.N. to cooperate "in the establishment of" a new, evolving information order. "Establishment" seemed to contradict "process." Another oxymoron. Isn't there a certain finality to "establish" which is contrary to an evolving process? Or by now have all other players accommodated to the first law of diplomacy: When nonnegotiable matters are set to paper, a certain ambiguity in resolution-writing suits almost everyone.

After fifteen years, all but the Americans in these debates seemed to accept that ambiguity in varying degrees. The new director-general of Unesco, Federico Mayor Zaragosa, tried early in his six-year term to end the divisive clamor over the information order. In several speeches in April and May 1988 before international journalist groups he pointedly omitted references to any new international information order, while promising assistance to developing countries' communications infrastructures. Dr.

Mayor went further: He added his own spin to the demand for Third World parity in communications. He replaced "free and balanced flow of information" with a pledge to support a free and *uninhibited* flow. He was warmly cheered by the audience of journalists.

The power struggle—who shall control the content of news and information moving within countries and around the world?—cannot be resolved by words in a resolution. Indeed, the main players on all sides know that. Yet Unesco resolutions do have importance in the communications field. The resolutions are symptomatic of the real problems which exist outside the Unesco arena. And these will not disappear either because a resolution has or has not been consensually approved.

The first contribution of Unesco has been to place the crucial impact of transnational communications on the agendas of governments, and later of journalists worldwide. Unesco did this agenda-setting badly. The organization seemed to threaten the very philosophy of independent news media, if not their existence. Much of the time Unesco was used, sometimes abused, by Third World extremists. And most of the time Unesco was accused by Western critics, with far too simplistic awareness of Third World divisions, of supporting the harshest press-control objectives.

A *New York Times* editorial asked what right U.S. officials had to negotiate the Mass Media Declaration of 1978, or any news-media issues. Americans, whether official delegates or journalists participating in international debates, do, indeed, have the right to address the problems of international communications—as Americans. The problems cast up by the new technologies, and the demands for universal sharing in the fruits of new knowledge and word-power, will not disappear because Americans are absent from the forum. Indeed, since Americans are preeminent in most fields of international communications, it behooves them to participate when the issues are debated.

To be sure, such discussions should be structured so that independent news and information media are not in the dock with governments as the judge and jury. Some of the time, Unesco forums seemed to be so structured. The *Times*, then, had a proper but limited point: in order to improve the quality of news reporting or expand the sharing of communication technologies, the power of government should not be massed against that of the press. That way lies censorship. But the basic issues remain: How shall unbalanced reporting be recognized and improved, and how, indeed, can countries, people and issues now underreported be given a fairer share of media exposure? These are crucial, exciting issues for journalists to ponder—not defensively, but as a challenge to their professional skills and their personal integrity. These issues do not deserve to be written off either because of the faulted manner in which they were raised at Unesco, or because American journalists are overly defensive.

Indeed, a lesson should be learned from the two opposite positions the U.S. has taken over four decades in the information-power debates at Unesco and the United Nations. Starting in 1946 and through the '50s, American officials strongly supported the liberal internationalist view that the free flow of information should be installed as a universal code of practice. That meant that Americans would take the lead in introducing communications issues to U.N. and Unesco forums. That free-flow policy was blocked, however, and no alternative was approved in the intergovernmental forums.

In the 1970s and '80s, Third World countries demanded the imposition of certain codes of information content, and also that communications infrastructures be opened to Third World use. The U.S. strongly opposed this package, thus seemingly setting Americans against information-sharing and facilities-building in developing countries. America's withdrawal from Unesco signalled a decade-long policy of hard-nosed neoisolationism over information issues. That policy, too, has failed. The issues have not disappeared. Americans have simply become less relevant to the development of global policies. The time has come for a reassessment of the U.S. relationship to Unesco, and, more important, of the full agenda of international communications issues.

"Development news"

Modernization of information flows in the Third World has universal impact. It should no longer be news that world order (or disorder) is critically influenced by Byzantine politics, economics, and firefights in the Persian Gulf, the Golan Heights, and the citadel of Panama, or over fishing rights in Tuvulu. The question arises: What is the relationship between national development and the expanded use of communication for political purposes?

That question was intelligently discussed by Frank Campbell, former Minister of Information of Guyana, writing in *Intermedia* (March 1984). He described the "developmental approach to communication" in postcolonial Guyana. He was a key administrator in programs which he says "yielded somewhat mixed results." His analysis, however, is valuable not only for other Third World countries but for developed-country observers who often misunderstand or denigrate the relationship of communications to national development.

I may seem to have contributed some confusion. Campbell quotes Narinder Aggarwala, of the U.N. Development Program, who wrote: "There is a general tendency in the Western media to confuse development journalism with developmental journalism. The latter term was coined by Leonard R. Sussman, executive director of Freedom House...One very often finds Western media leaders condemning development journalism when in fact what they have in mind is developmental journalism as defined by Sussman. It is of utmost importance that we distinguish between developmental journalism and development journalism, or what in United Nations circles is generally referred to as Development Support Communication (DSC) programs. The UN term, although a bit more cumbersome than 'development journalism,' is more convenient and descriptive of the use of various media—not just mass, but any media—for promoting economic and social development. Developmental news reporting is only a very minor element of DSC which in recent years has won many converts among Third World planners and leaders."

I regret having coined "developmental journalism." Unless one reads *Mass News Media and the Third World Challenge*,[11] the book in which I first used the term in 1977, one can easily misunderstand my putting down of "developmental journalism." I rejected the politicization of communications under the guise of economic development. I did not attack—indeed, I applauded—straight-forward applications of "development news" and journalism to improve skills, services and infrastructure.

Campbell, in his careful recounting of his experience with communication development in Guyana, acknowledges that "the government's socialist program" as well as the use of the media for development purposes resulted in the politicization of communications. Former independent publishers and broadcasters sold their holdings to the government when socialist policies dried up advertising income. "As part of its campaign to correct the global information imbalance...and the imperialist information deluge, the government established the Guyana News Agency on 4 January 1981. The GNI has a monopoly on the flow of foreign news to the government-owned media." Campbell quotes Guyana's late Prime Minister Forbes Burnham: "The government has the right to own sections of the media and the government has the right, as a final arbiter of things national, (to mobilize) the people of the country for the development of the country." He warns, however, of the damage done to journalism, the country and the government by a "pathologically narrow interpretation and implementation of a development approach to journalism." He adds, "by a narrow implementation of a government-supporting development-oriented media policy, I refer, inter alia, to the withholding or distortion of the truth; and apparent fear that the criticism of inefficiency in even a single governmental agency would cause the agency, if not the entire government, to crumble; and an exclusion of the opposition voice from the media, especially the print media."

The manner in which government controls or manages the media is another important variable, Campbell says. "Strict, day-to-day political control of the media led to the geographical or professional exile of some of the best journalistic brains. Apart from the bias, which was sometimes very obvious, whole editorials and feature stories occasionally disappeared from the newspapers. Some of this was a function of the brain-drain and, in some cases, the mind-drain. Some, however, resulted from a fear of antagonizing the government; or from confusion about what the government's media policy was. In times like these, journalists seek refuge in—speeches! Almost every news story began with a minister's declaration that something was being done, had been done, ought to be done, or would have been done."

The public, says Campbell, "reacted with creative and understandable cynicism. A new culture developed among our people—a culture of disbelief, especially in the traditional or pro-government press. Sometimes, in fact, people longed for underground media, especially when the traditional opposition wasted the opportunity by overplaying its hand."

To his credit, once Campbell was given ministerial responsibility for information, he campaigned within both government and party for a new media policy—"one that would promote the use of the media for development without the kind of abject onesidedness which had made a mockery and a failure of the entire exercise." He argued that to be effective the message must match reality and be tested "in the crucible of debate rather than forced through in the exercise of our virtual media monopoly."

Campbell won his battle but lost the war. The policies he proposed were formally approved. But, he reports, "After more than a decade of the old policy, the professionals [government bureaucrats] refused to accept the new as anything but a politician's rhetoric." For example, "Accustomed to altering photographs to create the illusion of a crowd, some colleagues continued this practice even where the crowd was so large as to make it unnecessary." Much of the "journalistic skill, self-confidence, courage and breadth...had disappeared or had been subdued." Campbell was removed from his cabinet post and made editor-in-chief of Guyana National Newspapers. There, he says, he assisted professional colleagues who considered themselves politically vulnerable. "The irony," Campbell notes, "was that a political appointee was apparently required to depoliticize the newspaper, in the sense of making it less partisan."

When President Burnham visited the newspaper in August 1983 he supported the right of government media to question governmental policy. But, writes Campbell, "the bad news was that even this presidential statement was not sufficient to bring other politicos in line or to induce adequate professional courage." Consequently, "we have gone backwards to the days of public disbelief."

What are the lessons?

1. Planning for economic development should integrate the use of the mass media, says Campbell, but this was distorted in the case of Guyana. The media became "developmental" (politicized). Says Campbell, "I

reject the Eurocentric idea of men like Leonard Sussman, who argue that press freedom as practiced in the North must be transferred to the Third World, complete with its mechanisms and its methodology. However, Third World leaders should resolutely resist the temptation to use the non-transferability of these myths as an excuse for eschewing the ideals of freedom."

Campbell misreads Sussman. I have never argued that it is either desirable or possible to transfer the mechanisms of the "North" to Third World journalism. Yet, no matter how impoverished, no matter how underdeveloped a country, or its mass media, it is possible to separate "news" from "information," development-oriented or otherwise. Of course the primary business of developing countries is development. Every available technology, mass or other, should be engaged. But news is an element apart. It delivers the realities of each day to citizens who need to know what is happening in their village, in their capital, and in the world outside. News, then, is the intelligence of life. Development news may convey word of the process and progress of national change. To be an effective carrier, however, mass news media must be perceived as balanced in reporting, and reliable in reflecting over time the full range of a citizen's interests, and beyond that, providing an adequate context from which citizens may fulfill their civic responsibilities. This, admittedly, is a tall order for a poor information system in a developing country. But no technology is too small to be deployed for these purposes.

A copier in a tiny village can begin to share news of the neighborhood and, indeed, the world—if central authority will allow. A single radio transmitter can further enlarge the circle of knowledge. It should, however, reflect diverse views whether covering nearby or distant events. These are not "myths," but minimal examples of press freedom which are transferrable anywhere. Campbell himself struggled vainly to provide some diversity and critical reporting in Guyana. Yet he labels my similar approach "Eurocentric." We come together at the end, happily, when he warns his colleagues that they should not use any "excuse for eschewing the ideals of freedom."

2. A democratic approach to the media is more compatible with development than is often admitted, is Campbell's second lesson. "Guyana's present negative economic situation is partly a function of the heavy hand with which we have controlled the media, thus sheltering incompetence and saboteurs," he writes.

3. The ruling party's interests can be harmed by "asking too much of the media." Media that fail to be a credible mirror of reality are more valuable to the opposition, he adds. Moreover, "no Third World country can eliminate foreign information sources, genuine or otherwise." (Nor, he should have added, can any developed country, no matter how large

or totalitarian, any longer effectively—not to say morally—block foreign radio or underground reporting.)

4. The transference of journalistic professionalism to Third World news media has meant primarily an "ideological" transference. Therefore, when domestic journalists have acquired that professionalism but are required to implement political policy, they become confused or leave, and "reinforce unproductive political controls." Perhaps, says Campbell, the answer ultimately is a Third World accommodation similar to that which permits the BBC's broadcasters considerable freedom from the British government.

Campbell's conclusion should be the starting point for editorialists of the *New York Times* to join with Unesco delegates in pondering the problems and opportunities of new communications technologies for all the "worlds." Writes Campbell:

"A new information order [is] not designed to replace the domination of the transnationals by that of national bureaucracies, however well intended; it is not a move toward 'a more restricted press,' but towards a freer one, which would really meet the need to inform and to be informed —one of the fundamental human needs."

What channels has the Third World created to provide alternatives to news from America, Britain and France, the headquarters of the four transnational news services?

Their "monopoly"—a loaded term to describe Western wire-service preeminence—stems from nineteenth-century colonialism. A Nigerian working for Unesco told me in 1977 that only two years earlier when in Lagos, he could not communicate directly with friends in Benin, a few miles away. "We had to phone London, get connections to Paris, and then a line to Dahomey." That linkage followed the precolonial patterns for Anglophone and Francophone Africa. Today, a satellite connection makes every place in the world equidistant. Yet critics maintain that the preeminence of Western journalism still impedes the national emancipation of dependent countries. They argue that "the free flow of information is a one-way torrent, from the capitalist countries, in particular the United States, to the rest of the world."[12] Exposure to Western news reports and analyses are said to conflict with national culture, values and sovereignty.

Alternative news systems

To change this, the Nonaligned Movement sought ways to alter the structure, content and flow of information. They sought to mesh it with the desired political and economic changes. The Nonaligned News Agencies Pool was officially constituted in July 1976. Its stated purpose was to "improve and expand mutual exchange of information and further strengthen mutual cooperation among nonaligned countries." That double objective provides the vital distinction from the Western news media. The pool, by this mandate, must not only exchange information but serve the national interests—however defined—of the constituent sovereignties. This is far different, of course, from the commitment of the Associated Press, the agent solely of its cooperative owners, nongovernmental newspaper and broadcast proprietors.

The pool described itself as "not a supranational organization but a community of free, independent and in every respect equal nonaligned countries' news agencies." The principles of the pool acknowledged there are differences among members, but said cooperation is based on the "broadest democratic principles with the fullest respect of the principle of voluntariness." Each national agency sends reports to the pool at one of more than a dozen regional centers. There are more than seventy-six member agencies and regional centers in Belgrade, Havana, New Delhi, Tunis, Baghdad and Rabat, among others.

Pool centers have little editorial function. They receive copy from a member agency and place it on the circuit for their region, and for transmittal to other regions. Each news agency pays the cost of sending and receiving messages. The pool also created a training school for journalists in Yugoslavia. The pool has developed an intercontinental network of radio teleprinters linking Yugoslavia, Mexico, India, Cuba and Sri Lanka. Satellite links between Yugoslavia and Mexico, and India and Yugoslavia were also established. This enables Mexican news, for example, to be received quickly in Yugoslavia and India where the pool centers can relay the information to the regions.

Is the pool an effective alternative to the major international news services? Clearly not. The pool can supplement but not replace the several thousand trained correspondents stationed in major cities around the world where stories affecting nearly all countries occur with some regularity. The pool, however, does provide an indigenous response, indeed, an official reaction, to whatever event or personality is described. To that extent, pool reports are credible as governmental handouts. They may not be balanced in either the Third World or Western sense of the term. It is unlikely, for example, that Iranian newspapers will publish without alteration or comment a news report from Iraq, or Egyptian newspapers one from Libya. The pool has also helped create more than thirty national news agencies among the nonaligned. The establishment of a satellite tie between Tanjug, the Yugoslav news agency, and the United Nations, however, provides a link with the major global news agencies. That link places reports of the pool into the U.N. system, which, in turn, is monitored by the global services.

Those independent, nongovernmental journalists use pool reports mainly to cite governmental sources, and then attribute the information to those sources. Rarely, do Western wire services treat pool reporting as they would stories coming from their own reporters. Western news services have long avoided retaining, either as permanent correspondents or as stringers, nationals still inside countries with repressive regimes. Such countries may blackmail journalists serving foreign outlets. Western news services are also wary of outright government agencies—at home or abroad—serving up "news." Consequently, Nonaligned Pool reports are likely to be relayed only rarely through the major global channels.

Pool reports fail to gain ready acceptance in mainline news outlets, additionally, because of the early claims for the Nonaligned material. Prime Minister Indira Gandhi, in her inaugural address to the Nonaligned Press Agencies Pool, 8 July 1976, cited her declaration of "emergency" the previous year. "The Western media interpreted it as an onslaught on democracy or an abrogation of our constitution which was not at all correct," she said. "Most, if not all, developing countries understood the position," she added, but did not explain that she had severely censored Indian newspapers and struck harsh blows at their economic base while thoroughly controlling all broadcasting and the national news agencies. Indians themselves understood.

On 27 March 1977, when it became clear that Madame Gandhi had been voted out of office, several thousand demonstrators marched through the streets in the constituency of the Information Minister carrying banners reading, "Victory of the free press" and "Defeat of censorship of the press." It was V.C. Shukla, information minister, who had joined with Gandhi at the creation of the pool saying, "The idea of 'free' flow of information fits insidiously into the package of other kinds of 'freedom' still championed by the adherents of nineteenth century liberalism." Mrs. Gandhi added, "We should be able to get an Indian explanation of events in India...Self-reliance in sources of information is as important as technical self-reliance." Shukla put it more crudely, "The transnational press agencies continue to invade the minds of people living in the nonaligned countries just as multinational corporations invade their economies." These early ideological barks were more fierce than the eventual bites.

For reasons cited above, the pool may become an active competitor to the major news services only if it adopts the journalistic philosophy of the global services; that is, it becomes less the proponent of diverse ideological objectives, depending upon which country is providing the report, and features generally unpoliticized news. Given the history of the pool, that is not likely to happen soon.

Ninety countries now have national news agencies. Of these, two-thirds, or sixty-two, are government-owned. A half-century ago, twenty-eight countries had national news agencies but only one-third were run by governments. We asked twelve journalists from as many Third World countries their opinion of the national news agencies operating in their nations. [See Appendix B for a roster of national news agencies.] Except for one reporter who works for such an agency, all said their national agencies had little credibility at home. They said these agencies did not cover many crucial issues. If they did, it was only from one viewpoint, the government's. The journalists said the national agencies also edited incoming news from the global news services in order to suit government policies.

The one reporter who works for a national news agency in one of the most free countries said the agency tries to cover the country fairly. It eliminates *all* news relating to politics of the country, no matter how much it may be a part of the story. That is done to avoid siding with either of the two political parties. In former years, under another administration, the national agency favored the government party. None of the eleven other journalists said they would work for their national news agency. They said they do not see or hear stories from the Nonaligned News Agencies Pool, though their countries are members of the pool. One said she occasionally sees a feature story in the back of a newspaper.

A similar history, with many of the same actors, is represented in the Third Word nonprofit news cooperative, Inter Press Service. It is headquartered in Rome, and operates through a wholly owned subsidiary, IPS Third World, located in Panama. IPS, created in Rome in 1964, is the sixth largest global news agency (after AP, UPI, Reuters, AFP and Tass). It says it has bureaus or correspondents in sixty countries and distribution or news exchange agreements with more than thirty Third World nations. It has strong technical and ideological links to the Nonaligned Press Agencies Pool. IPS carries much of the pool's daily reports, unaltered but appropriately labelled, over satellite teleprinter links leased to the pool. IPS, moreover, is committed to providing a daily flow of development news, and other Third World stories prepared by its own correspondents. It has also been subsidized by international organizations —FAO, HABITAT, ILO, UNDP, UNCHR, and Unesco—to provide stories about women and other special-interest subjects. Because of these ties to interest groups, including the nations in the Nonaligned Pool, the IPS for years was generally not accepted by Western news media as an independent news carrier. IPS actively fought this negative perception and secured grants from West German party foundations to open offices in Western Europe to service independent newspapers.

IPS makes little attempt to compete with the major global services in covering spot news. Its output may be seen as a daily feature service which covers rural, labor, church and development-related individuals and events. In a 1983 study of IPS, C. Anthony Giffard[13] found that IPS's budget in 1981 was less than $5 million, of which one-third came from the sale of news and other services to national news agencies, and another third from contracts with national agencies for carrier and distribution services. Contracts with several U.N. agencies to provide coverage of their activities, and with nonmedia clients for subscriptions and services, accounted for about 10 percent each.

IPS has had a strong ideological perspective since it was formed in the 1960s to promote the ideas of the Christian Democratic movement in Latin America and Western Europe. IPS seemed to reflect the radicalization of Latin America as right-wing dictatorships seized power. Gifford concluded that the growing internationalization of the agency and the need to adapt its services to northern markets "have had a [politically] moderating effect."

Roberto Savio, IPS' director-general, has devoted more than twenty years to building global interest in development news. He is a perpetual traveler and tireless salesman, both for IPS and because he has the need to tell the south's story for northern consumption. Savio says he met some 110 high-ranking officials throughout Europe in 1987 in order to explain his campaign. I met many times with Savio in the '70s to discuss the opposition to IPS by the major Western news services. They regarded IPS as a spokesman of governments rather than an independent journalistic voice. Savio heatedly, sometimes eloquently denied the charge, and sought to discover who was behind his failure to secure a major U.N. contract.

The U.N. Development Program had agreed to commit about $40 million over six years to a program IPS would direct as a loosely held affiliate. The program would create a network for development-information exchange. The plan aroused American media leaders and the State Department which cabled its mission at the U.N. in New York to discuss the contract with UNDP Administrator Bradford Morse. "We perceive serious problems with the contract," the cable said, "as well as with the possible political ramifications." These were spelled out in talking points for the U.S. ambassador: The program "is largely an initiative of Inter Press Service Third World News Agency (IPS), an organization which has consistently opposed U.S. interests in economic, political and information matters. For example, on Middle East issues the IPS steadily supports PLO themes." The cable adds that IPS "publicizes a standard 'anti-imperialist' line and unabashedly promotes the most radical version of a new international economic order and a new world information order. It is regarded as a stalking horse for Third World press interests and is an object of deep suspicion both of U.S. private media and the U.S. G[overnment]." The State Department cable also states that "the UNDP will be providing U.N. money to build IPS into a world class competitor of the commercial news agencies." The Associated Press responded, "We welcome competition wherever it is. What concerns us is what happens if U.N. money is used to build up an agency that would keep us out of areas where we need to go to do our job." The scale of the project—some $40 million over five years—"makes this one of the most important developments on the world information scene," the cable adds—as indeed it may have become.

Roberto Savio rushed to Washington, and spoke directly to State Department officials. He satisfied them on several accounts, but the UNDP did not support the program.

Savio has always had strong government ties. At IPS' birth, Savio persuaded Chilean President Eduardo Frei to provide a daily 1,000-word bulletin which IPS distributed to Chilean embassies in Latin America and Europe. Frei had just been elected on the platform of "Revolution in Freedom," which included planks to nationalize part of the copper industry, redistribute some Chilean income, and institute housing and development programs. These generated strong opposition, and Frei wanted his ambassadors abroad kept informed. IPS provided radio-teleprinter links. Savio also persuaded the Italian Ministry of Foreign Affairs to retain IPS to distribute news about Italy in Latin America. Slowly the network grew, but it suffered setbacks whenever left-leaning governments were deposed.

When the north-south communications controversy erupted at Unesco in the 1970s, Savio concentrated IPS coverage on Third World issues. Savio told Giffard, "Enemies of Unesco were our enemies, and its friends our friends." Yet Unesco also became the severest competitor of IPS, first, by creating the International Program for the Development of Communications (IPDC). While occasionally receiving funding from Unesco to advance particular projects, IPS was regarded as too independent by the Unesco bureaucracy. Moreover, the IPDC began funding regional news agencies in the Third World which tended to undercut the need for IPS. Now functioning are the Pan-African News Agency (PANA), Asia-Pacific News Network (ANN), Caribbean News Agency (CANA), and the Arab Project for Communication and Planning (ASBU).

All of these agencies, with the exception of CANA, which is controlled by a board of independent as well as governmental members, reflect the objectives of national governments. That is not to say that there is unity. It took some twenty years to create PANA because of strong divisions within the Organization of African Unity (OAU). Along the way, PANA had to withstand such spokesmen as Field Marshal Dr. Idi Amin Daba, VC, DSO, MC, CBE, Life President of the Republic of Uganda—later accused of mass murders, and forced into exile. He served as chairman of the OAU Information Ministers Conference in 1977. The most important item on the agenda was the establishment of PANA. Amin then said that the delay in establishing PANA "has exposed our continent far too much to the unscrupulous machinations of imperialistic and Zionist press media, whose interests have hitherto been to find fault with independent African countries." He said, "The imperialist mass media agents have always tended to twist news items from Africa so as to suit their taste (with) a negative element in it." He announced he had taken steps "to regulate the free entry into (Uganda) of imperialist and Zionist press media agents."

Amin fled Uganda for Libya in April 1979. PANA did not go into operation until 1983. It reports on Africa in English, French and Portuguese and is bound by its charter to distribute—without editing—news items contributed by its member national news services. Though the worst riots in twenty-five years swept Algeria in October 1988, PANA waited for days to receive the official Algerian report before filing its own report. When conflicting reports come from two member agencies, both reports are filed, allowing editors to make up their own minds. This, says PANA, covers Africa "with objectivity and exactitude." Clearly, however, in Africa, nationalism and not class struggle or worker revolt is the overriding ideology. Yet, says Auguste Mpassi-Muba, a PANA director, "It is high time the official, controlled, censored, muzzled or partisan news give way in Africa to news based on diversity of opinions and ideas, with free access to the various sources of official and unofficial information."

As with the Nonaligned Pool, other regional Third World agencies and, to a lesser extent, Inter Press Service, carrying diverse national views on the same subject, do not assure a complete or even a balanced report. Contradictory coverage of the same battle, reported by Libya and Chad, is unlikely to provide an accurate account of what transpired. Indeed, reality may not lie somewhere between the two reports, and both may omit essential facts which officials—for quite different reasons in each country—find inadmissible. IPS, in its own reporting, does not hew as completely to a governmental line. But it is committed to certain general propositions regarding the "constructive" approach to development and governments in the Third World.

The Nonaligned Pool and PANA are, therefore, self-proclaimed extensions of state power. As such, they should not be regarded as competitive with, or supplementary to, nongovernmental news services. But with appropriate identification they are, of course, subjects of coverage by the mainline services.

* * *

IT IS EASY to write off the government-subsidized or wholly owned press, or, indeed, the party press. One tends to forget that the party press was the nature of American journalism from the earliest days of the country. Roberto Savio of IPS defends Third World journalism which does not depend on private enterprise to provide a free flow of information. Newspapers in the West, he says, survive on a mass market for information which industrialization has helped create. News agencies, in turn, are sustained by a well-financed press. But this cannot be the universal model, Savio argues, saying it is "as if the inhabitants of the Sahara, on the basis of their reality, claimed the only efficient quadruped was the camel."

In the Third World, he adds, there is no mass market to sustain commercial media. Since communication is needed for national development, the state must provide the media. He insists, however, that this does not necessarily mean the end of press freedom. He points out that Sweden subsidizes the media, "but it would not occur to the government to use these [subsidies] to control the media."

Give Third World countries time to clarify the borderline between government and media, Savio pleads.

Time may, indeed, be an ameliorating factor. It should be recalled, however, that some Third World countries are nearly forty years removed from colonization, and show no sign of diversifying their news media. Ghana, the first Black African state to become independent, has had some press freedom. But that has been obliterated in recent years. If poverty were the sole determinant of press freedom, Botswana and other poor but free countries would be among the most oppressive. But they are quite the opposite: they are poor and free.

Yet the IPS model deserves further consideration. Giffard found (1983) that IPS "is indeed providing a different kind of news service to that of the traditional agencies. Its claim to be a new source of Third World news is well justified. More than three-quarters of the news carried on the [English and Spanish] networks was sourced in the developing nations, and deals with topics and actors of relevance to them. The large and growing proportion of news it carries from its own correspondents contradicts suggestions that it is merely engaged in distributing news from government-controlled national agencies...There is little to support allegations that the service is biased against the West: far more of its news reports are critical of the shortcomings of the Third World. Although the proportion of national agency material carried on the networks is not high, it is substantial, and is in accord with the IPS philosophy of encouraging a pluralism of voices in international news exchange."

Yet IPS is perceived, properly so, as a spokesman for government ideology, if not representing governments directly. IPS is not solely committed to filling the reportorial gaps in international reporting. It does tell Third World countries about one another, and describes developing nations to the developed world. This is a serious and necessary objective. But IPS also promises Third World clients that it will reflect their governmental attitudes regarding developed-country economic and political interests and relationships. In its 1980 "memorandum of understanding between Inter Press Service, Third World News Agency and the International Coalition for Democratic Action," IPS refers to "a person who will channel information to IPS" to "strengthen G-77 [Third World countries] understanding of differences and divisions between the industrialized countries and build an awareness of common problems of exploitation, dependence and powerlessness between communities in the industrialized and communities in the Third World." The memo then lists nine kinds of critical or negative stories that should be sought within the developed countries and channeled through IPS to the Third World.

This is justifiable as propaganda or polemic; as journalism it is clearly adversarial and far from neutral in intent. IPS thus is suspect among the normative news media of the developed nations. The AP, UPI, AFP, and Reuters also offer their news reports to governments. And, as with nongovernmental clients, the big four will try to provide coverage desired by the purchaser. That is, more news may be moved to or from certain places if there is a particular demand for it by a client. But it is understood that though the news flow may be increased on certain subjects or places, the standard of journalism will not be different from that employed everywhere else in the system. That means that reportorial neutrality will be sought by using diverse sources and reflecting different viewpoints.

Since 1979 IPS has acted as carrier and operator of the eighteen-nation Action of National Information Systems (ASIN), a Latin American network which exchanges official news. From time to time, IPS has also sought to create other networks for government agencies. This is not a journalistic function, and makes IPS a party to events it may report as an observer.

There is a serious and important role for IPS to play. It can provide a more professional report of Third World activities than the Nonaligned News Agencies Pool, which is stultified by strict multi-governmental rules. IPS operates in geographic areas which need more and better coverage. The agency also provides "process news," long-term features of potentially important social and economic development—stories which rarely appear in mainline Western news media. The reality is that by trying to serve two masters—interpretative journalism about the Third World as well as specific governmental interests—IPS loses credibility in the West. One would hope that some Western editors would use IPS, at least as a guide to stories that need to be covered by Western journalists themselves. And perhaps in time Inter Press Service can persuade its funders that their cause is better served by committing the agency solely to reportorial functions without ideological components.

It is no less essential for Third World spokesmen to make the distinction between information and news. Journalism obviously conveys information about all forms of national development. The "news" of such development, however, can be limited to reports of change or progress. The far more detailed coverage, and particularly rhetorical and even polemical appeals for social and economic development should be regarded as "information." The latter is the primary promotional channel of Third World governments. They properly expect their communications media to carry such exhortations. They often expect—but in vain—Western news media to convey such exhortations. Seldom are such promotional issues regarded as "news" in Western journalism. The opening of a Third World phosphate plant may make a small item on the U.S. business pages, though it will be front-page news in the country creating it. Exhorting citizens to learn the uses of the products is useful "information" locally, and probably of no news value elsewhere.

"Small" media

The use of "small" media, particularly radio, can be highly significant for many issues, raising the nutritional level of Third World children, for example. That is a valuable dissemination of information. In Bangladesh, traditional song and dramatic performances were broadcast to popularize the growing and consumption of vitamin A-rich vegetables to prevent blindness among children. Some 60,000 youngsters in Bangladesh go blind every year due to lack of vitamin A in their diet. Through the radio campaigns, blindness in the project areas has been cut in half.[14] Similarly, the assignment of a U.S. satellite to India for a year produced significant educational advances in isolated, rural areas. India, in succeeding years, enlarged its own state radio to reach rural regions in all parts of the vast country.

The next step in Third World development is creating greater access and participation in the communications networks by citizens. The ISDN linkages will enable developing-country citizens to influence the nature of the communications process. They should have access in villages, towns and even tribal areas via electronic terminals tied to normal telephone lines. The International Telecommunications Union (ITU) is determined to see that by the early years of the next century every citizen of the planet will be within easy reach of a telephone. Wherever there is a phone—whether it is a communal installation for a village, or a private instrument—the multi-service information terminal can be added. And there, too, print and broadcast journalists and information specialists can produce first-hand coverage. Some may tie into portable satellite uplinks for rapid transmission a continent away.

All of this requires the Third World elites to redirect their political will. They have slowly accepted radio and other small forms as one-way channels to their countries' masses. Will they accept the hands-on, talk-back capabilities which ISDN would provide? The recent history of Iran, Malaysia and Nigeria suggests that the growth of communications media raises expectations of more consumerism, and greater advantages for specific racial, religious or small groups. Such expectations then raise the elite's fear of political instability. In response, the elites restrict communications resources from the areas where major segments of the population reside.

Such political restriction inhibits the constructive use of hundreds of millions of dollars of development assistance presently available annually to Third World countries. For example, these are current grants and loans to Third World countries for the development of their communications facilities (in thousands of U.S. dollars for the latest year):

Unesco programs	$ 19,423
IPDC (Unesco)	2,500
International Telecommunication Union	27,234
Universal Postal Union	5,540
Food & Agriculture Org. (communications)	11,184
World Bank (loans, hard and soft)	560,100
	$625,981

In addition to $626 million given in grants and loans for communications projects in every sector of the world during an average year, the primary supporter of Third World communications, the United Nations Development Program (UNDP), has contributed $288,650,865 from 1966 to 1987 for a total of 729 communications projects. To this should be added the considerable support provided Third World communications development by the British Commonwealth Fund and the foundations of four political parties in West Germany.

This funding—perhaps $650 million in one year—has mixed blessings. If it were coordinated according to a common development plan it could soon produce highly significant improvements in the connections among the more than one hundred developing countries, and with the rest of the world. But such coordination is presently precluded by divisions over national, ideological, religious and economic traditions and objectives. Yet all profess to want far better communications linkages. Adding to the complexity is the overwhelming predominance of funding directed only to government entities. Only a small percentage assists private entrepreneurs who add the diversity and competitiveness associated with Western societies. Those qualities in the free-market countries have spurred all forms of development, particularly in the field of communications. While it is unlikely that governments, securing grants and loans mainly from intergovernmental agencies, will readily permit the growth of an independent communications infrastructure, there may be a middle-road alternative in the model of those mixed economies which permit various forms of cooperative or private/public ownership or regulation.

It should also be noted that while the Third World clamored for assistance in building a communications infrastructure, the developing countries were spending $180 *billion* dollars a year for the purchase of armaments.[15]

Liberalization of telecommunications—new diversity, broader access, mixed ownership—is taking place mainly in democratic industrialized countries. Public monopolies, meanwhile, remain and are being strengthened in developing countries. In the immediate future this has negative political implications. Third World citizens, who need liberalized telecommunications the most, are suffering more restrictions. The gap between them and users in developed nations therefore widens. At some point, however, citizen demands in the developing countries are likely to coalesce—as we began to see in China in mid-May 1989. Foreign broadcasts and VCRs, whether arriving surreptitiously or by permission, will steadily reveal the thrusts of change in other countries—including such places as South Korea and Taiwan. It may become increasingly difficult for central authorities to deny their peripheral citizens access to empowering telecommunications. That denial delays the day when one-way mass communications will be joined by interactive, two-way personal communications.

6.

Count the Ways of Censorship

IT IS IMPOSSIBLE to forget. In their benighted lands, I have spoken to the best wire-service journalist in Black Africa, fired for his integrity, now driving a truck; the Asian newspaper editor who keeps a small suitcase beside his desk, lest he be suddenly imprisoned; the Latin American editor who must spend every night in prison for more than a year, though released in the morning; the talented black editor in South Africa whose untimely death was hastened by the tension of publishing a newspaper so restricted by censorship that retaining journalistic integrity was an unbearable struggle of mind over conscience; the heart-rending letter from the widow of slain Nicaraguan publisher Pedro Joaquin Chamorro to the president of Nicaragua [see Appendix C].

* * *

"LET ME COUNT the ways," not of love, but of power—word-power, that governs access to ideas. Ideas precede action, including the act of transmitting ideas either by truth and persuasion; or indoctrination and imposition. In a free society, ideas take their chances in open competition. We are concerned here with the rigging of ideas, denying access to facts, imposing "news" and information as truth when they are not, blinding each generation to the larger world of ideas beyond the limited universe set by censorship.

Our generation now knows the horrendous tragedies wrought in this century by enforced noninformation masking unlimited barbarism. More than six million innocent Jews and others were murdered in Nazi death camps under the cloak of information-control. A quarter-century later, two million Cambodians were slaughtered by Maoists in the blackout hiding class-war genocide. And in the 1980s: In Ethiopia, a government wielding information power used the cover of famine to force deadly mass migration, and political reprisals. Censorship further restricted the volume and proper use of foreign aid. Thousands starved to death hideously, and needlessly. In Afghanistan, for eight years atrocities and bombings were committed against civilians, and the Iran-Iraq War took a frightful toll—all endured in virtual silence because professional journalists were barred. "When there is no publicity," said Jeremy Bentham, "there is no justice."

When long-suppressed truth emerges, however, it can produce official embarrassment, and belatedly reveal years of public disbelief of governmental statements. The current re-revision of Soviet history of the Stalin slaughters tells the people what most had whispered for seventy years in guarded privacy. Soviet citizens had masked their disbelief under the euphemism that the official press was "boring." The Marxist leadership only now under Mikhail Gorbachev seems to grasp the irony.

As officials wield nearly absolute control, the words of the government become less credible to the public. Sheer police power prevents any major talk-back. Yet a free press, though small (the samizdat, or unofficial press of Poland, for example) can achieve great credibility among the public—because it is known to be independent. As officials yield some word-power, the government itself becomes more credible, and its statements more believable. The news media then become freer, and gain credibility—provided they act with perceived responsibility. For an independent press that misuses its freedom also loses credibility.

The Chinese are learning the grim lessons of their word control. In August 1983, seven years after one of the deadliest earthquakes in history, the Chinese government finally allowed foreign journalists to visit Tangshan, the devastated city just 100 miles southeast of Beijing. The quake had killed 148,000 people, and severely injured another 81,000. Yet few words were permitted to pass outside China. Because information was held hostage, so were the people. Foreign aid was denied the pitiful survivors. Said Vice Mayor Zhang Qiying in 1983, "We should have accepted aid from other countries. But you know the background of China at that time."

The background: The quake came two months before the death of Chairman Mao Tse Tung, while the powerful "leftist gang" led by Mao's wife controlled the country and its closed-door policy. Tens of thousands of seriously injured and homeless suffered far more because information power was used in devilish ways. In 1988, another quake struck China. This time, word flashed around the world immediately, and help went out to the stricken families.

"Let me count the ways," then, not just in China, Nazi Germany, the Soviet Union or Ethiopia, but universally—ways in which "information sovereignty" is invoked to achieve mind-sealing as well as mind-killing effects.

* * *

THE GLOBAL FLOW of information since 1975 has been repeatedly described in international meetings as "imbalanced," as a blow to "information sovereignty." This was invoked as if it were as sacred as internationally recognized borders and citizenship. When a message is forcefully projected by a government, however, police power backs up word-power, and can impede or silence an independent communicator

That struggle to speak or write freely is the ultimate imbalance in the global flow of information. The old childhood chant, "sticks and stones can break my bones, but words will never hurt me," is not true; it never was. Not only sticks and stones, but the full armory of the state—ideological as well as police power—can be deployed to influence words and ideas. And most countries calling for a new information order lose their standing and credibility when they use police power to control news and information within their own jurisdictions.

In the struggle between state and press, state has the overwhelming power. It is a war, usually bloodless, but not always. In 1988, thirty-eight journalists were murdered because of their occupation, in 1987, thirty-two. At least 156 journalists were killed from 1982 to 1988 (sixteen died in 1985 in the Philippines, twenty-five in Mexico over six years). [For the full tally of physical and other assaults on journalists worldwide in 1988 see Appendix A.]

What are the weapons of the state in controlling or influencing what the citizen will read, hear or see at home (and what others abroad may learn about the country)? [See box on pages 161-165.]

Outright murder is only one of several forms of physical violence that journalists face in many countries. Others include assassination, kidnapping, torture, "disappearance" (seizing and killing a victim, and secretly disposing of the body), beating. Shooting is conducted not only by government or political-party death squads. Drug traffickers have also taken a major toll of journalists (as of judges and attornies-general) in Colombia. Some thirty-five Colombian journalists have been killed the past twelve years for covering the drug story. In August 1987, a death list fingered nine prominent Colombian reporters and newspaper columnists, among other prominent citizens. The list was taken seriously when two important politicians on the hit list were murdered. Several journalists fled the country.

Left-wing guerrillas and right-wing goons have included journalists among their victims in El Salvador. And reporters have been killed in

Weapons of the state versus the press

Physical, against journalists

Assassinate journalists by government- or party-associated death squads.

Force the disappearance of newspersons. In Argentina for several years before 1983, 82 journalists were "desaparecidos."

Kidnap reporters. Maltreat them in captivity. Hold them hostage to political or financial demands. Return them, sometimes, with threats to their colleagues, and all other in-country journalists.

Torture in private, beating in public.

Crossfire victims in wartime ambushes and bombing, and finger informers in civil wars and insurgencies; treat as traitors if covering one side, and attempting to report from the other.

Punch in the nose, the unique tactic of a Nicaraguan official used against a journalist.

Physical, against the media

Attack, raid, set mobs against a media plant.

Destroy press and broadcast equipment.

Impound media equipment.

Jam broadcasts.

Occupy newsrooms by the military.

Cuts by government of water, electricity lines to newspapers.

License typewriters, control photocopiers.

Psychological, against journalists

Threaten physical harm.

Threaten loss of job.

Detain without charge.

Psychological, against the media

Threaten to shut down the print or broadcast facility.

Threaten to imprison management.

Expel management from the country's leadership circle.

Hamper the media by supporting or withholding language in a multilingual country.

Editorial, against journalists

Government controls domestic news agency; establishes the facts and tone of media coverage.

Government sets guidelines, mandates the areas of coverage, the slant, and the "responsibilities" of the news media to advance political, economic, devel-

opmental, socialist and other objectives.
   Favoritism in controlling access to official news.
   Leak to control the "spin," and thereby influence coverage.
   Government training schools for journalist indoctrination in ideology as well as occupational skills.
   Out-of-country training for indoctrination in international affairs.
   Deny access to the place where government news is released.
   "Press clubs" of government and journalist.
   House foreign journalists in a separate enclave for surveillance.
   Monitor foreign journalists as suspicious adversaries.

Editorial, against the media

   Information Ministry or other government official releases news and information.
   Favoritism in releasing news to certain media.
   Incoming foreign news admitted only through government agency.
   "Secrecy" invoked to withhold information or avoid embarrassment.
   Confiscate certain editions.
   Full-text publication of certain speeches or statements required by the government
   "Calls," guidelines to editors and publishers giving direct orders to cover or not cover certain events.
   Government gazette providing the lead and tone for other media, government or independent.
   "...no one may come" to permitted press conference by a nongovernment medium.
   Government press releases, by volume, overwhelm the independent media.
   Disinformation, the use of known falsity, or planned distortion.

Legal, against journalists

   Official censorship.
   Legislation re: abuse of publishing, contempt, security, confidentiality, official secrets, arms control, anti-terrorism, anti-protest, military codes, anti-communism, defense of socialism, defense of the revolution, demeaning the president or his family.
   Contempt citations.
   Forced corrections and retractions.
   Libel laws, particularly for criminal libel.
   License journalists, and withdrawal of licenses.
   Monitor content against a journalist code.
   Imprison for long terms, including year-long, nightly incarceration.
   Detain
   Ban (no visitors, house arrest, no meetings, unquotable).
   Expel from profession.
   Expel from the country.

Deny access to geographic or sensitive areas.
Demand sources under threat of imprisonment.
Bar entry to country; deny multiple visas for easy reentry.
Refuse or delay satellite feeds.
Surveillance of foreign journalists.

Legal, against the media
Suspend a publication or broadcast station.
Confiscate a print or broadcast facility.
Ban for an indefinite period.
License the medium, and withdrawal of license.
Monitor output against a code of performance.
Government ownership of the media.
Ruling party ownership of the media.
Independent party ownership of the media.
Ban opposition-party papers.
Source disclosure made publisher's, editor's responsibility
Search editorial office for documents.

Financial, against journalists
Bribes by government or others, of newspersons.
Firing, loss of career or demotion for unwanted coverage.

Financial, against the media
Newsprint prices or allocation controlled by government.
Circulation and distribution of independent publications, including foreign media, controlled by government.
Price of independent publication controlled by government.
Foreign exchange, needed to purchase newsprint or equipment for independent media, controlled by government.
Subsidies to favored independent media provided by government.
Loans to favored independent media provided by government.
Government advertising placed in favored independent media.
Tax rate adjusted to favor or harm press.
Subscriptions to favored independent newspapers, magazines, news agencies purchased by government for distribution to its offices at home and abroad.
Ownership of major independent media by government-related industrialists.

Technical, against the media
Deny satellite use for domestic or foreign feed.

Consequences of governmental pressures
Self-censorship by journalists and managements.
Muzzling of journalists by editors and managements.

Media councils created by government.
Labor union pressures, including strikes, to influence content of the media, as well as secure job improvement.
Financial dependence on government of weak media.
Linkage of state and media power through government-owned media, producing tension for the weaker, independent media where they are permitted.

WEAPONS OF THE PRESS VERSUS THE STATE

Legal
Reveal *corruption* in government in the media.
Secure *injunctions* from courts to get information from government.
Access to many government files through Freedom of Information Acts.

Economic
Press chains acquire cumulative financial and political power in market-economy countries.
Broadcast networks, rarely privately owned as in Brazil, *acquire similar power.*

Political
Supporting or *opposing* government policy.
Editorial and broadcast commentary on candidates or officials.
Exit polling discouraging some voters who have not yet voted, and undercutting politicians' interpretation of election results.
Publicizing financial and political support or opposition for leaders/candidates.

Editorial
Investigative reporting of hitherto closed aspects of a candidate's or official's career, or the system.
Publish or broadcast sensitive secrets—may create diplomatic, military or political problems.
Denial of coverage to political figures.
Immediate critical replies to major speeches of officials.
Sensational reports which may distort or, even if true, embarrass leaders.

Policy
Agenda-setting for the public and government by the manner in which news and commentary are selected by the news media.
Campaigning by the media to set policy or change policy.

Psychological
Character assassination of political leaders and bureaucracy.
Blackmail—threaten to run a bad story, unless other information or payment is provided
Questioning public confidence in a policy or the leadership.

the crossfire between government and guerrilla combatants. Sometimes, it appears, journalists are lured to their death in Salvador, for to cover one side almost certainly arouses the deadly suspicion of the other.

In November 1987, journalists covering the election in Haiti were not caught in the crossfire; they were targeted by gun squads, including the Haitian military. Alfredo Mejia, a soundman for ABC, described his attackers:

> They stalked us...I watched (one) aim and fire, and saw the flash from his pistol. He fired from point-blank range. Carillo (an ABC cameraman) was hugging his camera to his body, and I hugged my sound recorder to my middle. I closed my eyes hoping he would stop shooting. My right arm jumped. I was hit twice by 45-caliber bullets in the arm. We were still rolling [filming], and afterwards Carillo said he had filmed me being shot. We hardly had time to speak when we saw them returning. We played dead...[They] tore off my gold neck-chain, our watches, and tore out our wallets, and then took away the camera and sound equipment.

How to stay alive

To help protect journalists on dangerous missions, the Inter American Press Association in 1985 issued a series of survival tips. They begin with the reminder that "no story is worth your life." If the authorities cannot "guarantee your safety, leave the country," the IAPA declares. But that is easier said than done. A government facing an insurgency is likely to warn a journalist he/she cannot be protected outside lines clearly held by the central authority. The truth is, even that cannot be guaranteed. Journalists have been killed or wounded well inside territories supposedly dominated by the central government in El Salvador, the Philippines, and Colombia. That warning, reasonable though it sounds, carries an occupational hazard. Remaining only within government-stipulated bounds makes it more difficult to cover "the other side," whether that be an insurgency or a drug-ridden sector.

The IAPA gives this useful advice to journalists in danger: Never carry a gun or weapon, or point your finger so it can be mistaken for a gun. Carry a white flag. Use extreme care in selecting local drivers or guides. Mark your car "press" in the local language (not always a protection if the insurgents are angry with their press coverage; I saw newsmen and -women targeted by the special forces in Chile in October 1988). Never wash your car. Unwanted tampering can more easily be detected on a dirty car. Talk to local residents, and listen to their advice. Dress appropriately: Sometimes you may want to blend into a crowd, and other times you may not want to be seen as part of a group. Never wear olive green or anything that makes you look like a soldier. If guerrillas at roadblocks ask for your "war tax," give something. It shouldn't be much, but it can avoid unpleasantness. When confronted by hostile persons, identify yourself in their language and attempt to convey what you are doing. Always carry complete identification papers. Do not cross the line between journalist and active participant. Make sure your employer carries adequate insurance in case of injury or death. Pieter Van Bennekom, with the UPI in Mexico City, puts it tersely: "Press passes don't stop bullets."

Journalists may become victims without even leaving the newsroom. They may be at their desk when government troops occupy the premises. In this form of physical violence against news media, troops, police or the paramilitary may remain on the premises to prevent further reporting. Or they may attack newspaper and radio facilities in order to destroy or fire the press and broadcast instruments. In April 1986 the Stroessner government sent a mob to attack Radio Nanduti, the last free voice in Paraguay. After breaking windows, firing pistols and sounding shrill blasts, government goons later destroyed the transmitter, and finally used jamming devices to force Nanduti from the air. This jamming was rather imaginative, however; unlike the continuous screech of Soviet jammers over America's Radio Liberty and Voice of America, the Paraguayans used selective jamming. It erased Radio Nanduti's signal whenever the broadcaster discussed certain issues. These were always matters deemed prejudicial to the government.

In other places, notably Eastern Europe and the Soviet Union, where all the information facilities are state-owned, control of the word begins at a lower level. Telephones are sharply restricted (in the USSR there are no phone books for public use), photocopiers are strictly controlled, and typewriters must be licensed. All of this makes journalism more difficult, particularly for foreign correspondents.

In Lebanon, official censorship and subtle controls are the least fearsome aspect of journalism. After thirteen years of war—and sometimes "bloody peace"—journalism is unique. One newspaper, *An-Nahar*, continues to publish amid constant turmoil, bombings and slaughter from many directions. Ghassan Tueni, the dean of Arab editors, describes[1] his newspaper: "It is," he says, "the only thing that crosses the dividing lines of the warriors when nobody else dares. There were times when it rolled from one part of the (Beirut) demarcation line to the other wrapped into tires. Birds, of course, were frightened away from the crossings a long time ago. So are the journalists of *An-Nahar* of whom many were kidnapped, killed or intimidated." They now live two blocks away from the *An-Nahar* building. To reach the newspaper they cross, one at a time, nine different crossing points, most of them non-Lebanese. Says Tueni, "The game is to print whatever is possible, not only what fits. The game is also to survive the destruction of everything civilized in what was once the cradle of early civilizations." His correspondents, says Tueni, "are not only witnesses to what we have been through, but also victims." There is a "ruthlessness about wars," he adds, "and journalism is usually its best target."

A government war against the word can be psychological as well as physical. Threats, too, are powerful weapons. Warning a journalist he faces physical harm is excruciatingly credible when there is a history of such attacks. Responses to the questionnaire conducted for this book in seventy-four countries produced the surprising result that killing of journalists did not influence the content of news and commentaries in most places. The exceptions were those few countries in which assassinations of newspersons were endemic. Colombia was a major example of this. More journalists have been killed in the Philippines in recent years, yet their surviving colleagues, heroically, return to their microphones and typewriters. Detention without charge, provides another form of psychological attack on journalists. Once in custody without formal charge, usually in a climate of "emergency" or martial law, the victim knows he cannot expect the rule of law to prevail—even to the limited extent an authoritarian government may have earlier permitted. Isolation, perhaps torture, at least deep uncertainty, for an indefinite period, produces a severe psychological trauma which may never be fully erased. Zwelakhe Sisulu, editor of the *New Nation*, a South African weekly newspaper, was detained without charge in December 1986. More than two years later, the former Nieman Fellow at Harvard University was released from a South African prison but held in house arrest, and still not charged with any offense.

Not only domestic reporters and editors are targeted for psychological assault. Kenya, which has the largest concentration of foreign journalists in Africa, threatened in March 1987 to begin censoring all news reports from foreign correspondents. The ruling Kenya African National Union, the country's only political party, made the threat. Governments also threaten to imprison domestic press management, or expel managers from the country's leadership circle. Officials also use state power to influence the content of the news media by threatening to shut down the print or broadcast facility. Or government may undermine an independent press by ruling that the vernacular in which it is printed is a latent threat to public peace.

More subtle are governmental intrusions in the editorial processes of journalism. Some governments place themselves in the news business. There are ninety government-run news agencies [see Appendix B for a roster of government-run news agencies]. A half-century ago there were thirty-nine national news services in twenty-eight countries. Some 70 percent of these were at least nominally independent of governments. Today, 56 percent of all countries have a government news agency. Of the countries designated in the Freedom House survey as "least free," 72 percent have government news agencies. Of the "most free" nations, 9 percent own their news agencies. There is a high correlation between the freedom a government grants its own citizens—their right to choose leaders, approve policies, associate at will, and have access to diverse press and broadcast reports—and the operation of an independent national news agency.

A government news agency may be the country's sole receiver of news from abroad. In such cases, the Associated Press, Agence France-Presse or other foreign services feed reports only to the government agency. It edits the incoming file, and passes on to domestic press and radio editors only that information deemed fit for local consumption. Many newspapers in Asia and Africa which in years past bought daily reports directly from the international wire services, can now get the day's foreign news only from their government agency's editors. Frustrated domestic editors may hear on the BBC news censored by their national agency.

Most government news agencies also distribute domestic news to the outside world. Most government agencies of developing countries, since 1975, have joined the Nonaligned Press Agencies Pool. The pool was created to end the complete dependence of Third World countries on the four major Western news services headquartered in the United States (AP, UPI), United Kingdom (Reuters) and France (Agence France-Presse). The pool sought to provide—in the words of the late Indira Gandhi—"news of India by Indians." The pool features articles about the economic and social development of Third World nations, with a heavy overlay of "protocol" stories. These describe visits and speeches by officials of pool countries. Only rarely does the pool carry investigative stories which the Western media are accused of featuring. The daily pool file is composed of stories submitted by the affiliated government news agencies. The pool cannot edit or eliminate these stories. It transmits them, as received, through regional centers in Yugoslavia, India and Cuba. The pool thus magnifies the disadvantages of each government agency. But, for supportive officials, the pool is a safe conduit, if boring and repetitious. Each government news agency in its domestic operation sets the tone for government-approved news and information; the pool automatically follows those guidelines.

More explicit guidelines are provided by governments to "independent" domestic media. These guides, wielding state power, ensure official control of words and ideas reaching the public. Yet a semblance of press freedom is preserved. The press, technically not run by government, thus enjoys more credibility than if it were an official mouthpiece. Classic examples are provided by Poland and Hungary on the political left, and until 1987, South Korea on the right.

The Polish censor

The secret "recommendations" of the Polish censors reveal the totalitarian mind as rarely seen. In January 1978 Freedom House released the translated text culled from 700 pages prepared by Poland's Central Office for Control of the Press, Publications and Spectacles. These classified documents had been carried out of Poland by Tomasz Strzyzewski, a defecting censor. After working in the censors' office for nearly two years, Strzyzewski said he "then realized the enormous scope of destructive opportunities, and the reach of censorship's ruinous influence on our national culture and on the Poles' social awareness."

The documents were in three categories. The *first* includes directives for the censors. They are told in great detail what information to kill, and what to transmit through the mass media. Nothing was to appear about limitations on Communist parties in Arab states, conflicts between Arab nations, Polish contacts with South Africa, or use of terms such as "military dictatorship" for states with which Poland has diplomatic relations—except for "Chile, Paraguay, Guatemala and the Dominican Republic." While prohibiting all references to environmental problems in Poland, the censors thereby revealed the prevalence of contamination, accidents and alcoholism. References were forbidden to the purchases in the West of machinery and foodstuffs. Suggestions were prohibited that Soviet goods are inferior. Opportunities to emigrate to the United States were not to be revealed. "Any material critical of the situation of various religious denominations in countries of the socialist camp, must be eliminated." Obituaries or announcements of meetings in connection with the Warsaw Uprising (against Nazi extermination of Jews) cannot be passed for publication. "No information can be published about the nondeliveries by the farmers to the (compulsory) purchase centers of pigs, cattle and fodder." Kill references to the "re-exports of coffee." "No potential criticism can be allowed of decisions relating to wages or of current social policies." No information about the "bribery affair in Sandomierz."

The *second* category consists of examples of texts already censored. This reveals cuts in every form of communication: the press, books, films, public meetings, invitations, personal advertisements, diploma certificates and memorial plaques. Extensive samples are culled from politics, economics, religion, education, history, culture, arts and literature. Censors cut statements by Pope Paul VI, material from the Bible, and reports from the thirty-five-nation Helsinki Conference (on security and cooperation in Europe), among countless other references.

The *third* category deals with censors' "oversights"—material published which should not have been approved—the work of the "double-checking department."

The Poles of KOR, the Social Self-Defense Committee, which first received these documents from the defector, said that the directives "unmask...the anti-national activities of the censorship enforced by the State and party authorities in the Polish People's Republic: the murder, according to plan, of the nation's culture and the crippling of individual and national personality through the methodical suppression of freedom of speech, of the printed word, of information." That was 1978 (one year, incidentally, after Poland ratified the international covenants on civil and political rights, and on economic, social and cultural rights!).

At the beginning of the 1980s, during the legal period of the Solidarity union, some alternative publications were briefly permitted, and censorship was eased. With the imposition and later abrogation of martial law, alternative publications managed to survive, but penal proceedings and extra-legal acts of police violence were directed in 90 percent of the cases against people suspected of being active in the "alternative circulation" (editing, printing and distribution of trade union papers, leaflets, etc). To put the "alternative circulation" outside the law, authorities use detention, refusal to issue passports, searches of homes and confiscation of typewriters and manuscripts, and general harassment of writers and editors. In the words of the Polish Helsinki Committee (1986), intellectual life in Poland is "surrounded by a framework of preventive political censorship."

The Hungarian censor

An equally significant censors' document emerged from Hungary late in 1986. This analysis and directive from the Hungarian Politburo was labeled "strictly confidential." Using cold-war terms of the 1950s, the directive termed domestic critics "enemies," not opposition. "Enemy groups" must be prevented from "seizing the initiative in any genuine problem, including environmental questions," said the directive. These "enemies" were said to be linked to the "imperialist" West. Party organs hereafter must devote greater attention to "work among the intelligentsia, the press and mass media." In particular, "leading cadres and communists employed in the press and the mass media, cultural and scientific institutions must be obliged to heed political and cultural-political principles and norms." The directive urged, "where justified...disciplinary action" should be imitated. For, "widening of the influence of enemy groups must be prevented; illegal publishing activity and the circulation of illegal publications must be suppressed." To accomplish this, the directive laid down this command:

> The principles and norms of our publishing policy must be applied consistently. Editors offending against them must be held responsible. It cannot be tolerated, that enemy viewpoints spread by illegal publications should seep into the press. Individuals, who play a role

in illegal publications and hostile press organs and radio stations abroad, must be prevented without regard to person from publishing in the press or having their works put on by the mass media or cultural institutions. The appropriate State organs must take the necessary measures to this end. The Central Committee's Agitation and Propaganda Department, as well as the Scientific, Educational and Cultural Department must give them the necessary assistance.

The directive stated the objective of bringing about "the disintegration of the environment of the enemy groups, isolation of those sympathetic to them, as well as the detaching of those with honest intentions." The directive acknowledged that, "if justified, we must accept direct argument with the views formulated in these publications." This secret document orders the full power of the State to "disintegrate" dissenting views in the news and information media. Most significant, this order came at a time when Hungary was perceived in the West as adapting Marxist economic principles to a market-like system.

By mid-May 1989, both Hungary and Poland permitted their underground media to act as "illegally legal."

Korean "guidance"
In June 1987 South Korea ended its elaborate system of guiding the country's independent newspapers. Just before the guidance system ended, Korean journalists increasingly ignored it. Helping end the guidelines was the leaking of hundreds of the government's "requests for cooperation" in 1985-1986. The leakers were arrested and convicted, but their sentences were later suspended. The "requests" were published in Mal (Words), the organ of independent journalists dismissed under pressure of an earlier government. Some examples:

"Do not report on...the alleged torture of [two politicians named] and others...the self-immolation of [two named], and the mysterious death of [a student]...the fact that a total of 15,000 farmers staged demonstrations on thirty-two occasions this year, the largest resistance movement of farmers since the revolt of the Tonghak Party in 1984, and reports that 95 percent of the people support a civilian government and not a military government."

"Do not report the *Financial Times* story which says that South Korea and Communist China are setting up a joint venture...do not carry any photos of Kim Dae Jung, Kim Young Sam and Lee Min Woo (opposition party leaders) at today's meeting...do not write that the import of cigarettes was forced on South Korea by the USA...In connection with the sit-in of the Council for the Promotion of Democracy, do not report the sit-in, the moves of the three (opposition) Kims, or the street blockade in the neighborhood either in a straight news story or in a sketch-type story." "Do not report Jack Anderson's column about the civil war in Nicaragua in which he says that the Taiwan government and the Unification Church are supporting the contras." "Election in the Philippines—do not give prominence to foreign dispatches, especially those written from the U.S. standpoint."

Detailed, explicit instructions were given regarding the contents of year-end or new-year special editions. Even the placement and prominence of certain stories was described: "The news report on the Philippine situation (24 February 1986) should be at the side of the front page, and the rest of the report should be dealt with in the middle pages." "Put on the front pages (6 March 1986) a four or more column story about the statement to the effect that South Korea differs from the Philippines."

When I interviewed, on the same day in April 1988, the minister of information of South Korea and the editor of *Dong-A Ilbo,* the leading daily newspaper, both confirmed that the guidance system had ended—and both expressed relief.

Guidance and government news agencies are not the only editorial weapons used by officials against the news media. Controlling access to information is crucial. Officials control the "spin" on a story by leaking documents or statements to favored journalists or publications. Leaking may also give both journalist and official a favorable symbiotic relationship for future coverage. Journalists in disfavor may find themselves barred from Government House, as in democratic Costa Rica where official releases are handed out. In other countries, particularly Japan, the "old boy" press clubs nurture a year-round relationship between officials and selected journalists. Even while Japanese reporters compete among themselves for most stories, they generally accept guidance from the top official on particularly sensitive issues. The Japanese "familial culture" thus serves political power through sophisticated editorial persuasion.

There are less subtle means by which governments exert editorial control. Foreign journalists requesting accreditation in Israel must sign a statement that they will submit in advance to the censor all stories "which contain reference to the defense establishment of the State, the security of the State in the territory administered by the Israeli Defense Forces." Some countries demand that certain speeches or statements be published in full in the nongovernment press. Others publish a government gazette which provides the lead and tone for all other media to follow. Some simply create great volumes of government press releases to make it easy for independent journalists to cover their beats. Still others tackle press control at the source. They send aspiring journalists to government-run schools of journalism. There, political ideology and occupational skills are taught. Where domestic journalism is inadequate, men and women are sent for training to the elaborate journalism cum indoctrination centers supported by the Soviet Union in Moscow, Prague, East Berlin and Budapest (the last three operated by the East bloc's International Organization of Journalists). Other developing country journalists may attend U.S. schools where they are likely to learn about adversarial journalism. Editorial pressure is placed on foreign journalists by the Soviet Union, which houses them in a separate enclave and treats them as suspicious adversaries requiring physical surveillance and occupational monitoring.

Then there are highly secret forms of governmental intrusion in the flow of information. Government agents, at home or abroad, may act as journalists, using the profession as a front for police action. This compromises journalism by casting suspicion on all legitimate professionals and it inevitably wipes out sources of information needed in accurate reporting. A far more grievous distortion of the news flow is the use of disinformation —the careful placement of "facts" known to be erroneous or, at best, half-true. Disinformation generally takes the form of faked or doctored documents purporting to reveal heinous acts planned by an adversarial government. In recent years, the Soviet Union has studiously planted anti-American propaganda by infiltrating non-Communist publications in the West and the Third World. Some newspapers, such as the *Patriot* in India and *Ethnos* in Greece, have repeatedly served as first publishers of Soviet disinformation. The stories were then carried by the Soviet press as originating in Third World newspapers—and republication of the stories began anew.

Examples: an East German scientist working in Cuba blamed the origin of the AIDS epidemic on an American germ-warfare experiment in Maryland; Indian "sources" claimed Prime Minister Indira Gandhi was assassinated by the CIA; a forged letter from President Reagan to the King of Spain, in November 1981, urged the King (in terms calculated to offend Spanish sensibilities) to crack down on Catholic pacifists and the "left-wing opposition." Spanish papers exposed the hoax, and attributed it to the Soviet disinformation campaign.

At a libel trial in London in March 1987, a KGB defector and a former Greek employee of Radio Moscow stated they had participated in high-level meetings in the Soviet Union on the means of spreading Soviet disinformation. The defector, Ilya Dzhirkvelov, said he was personally involved in the use of several journalists and newspapers in the West as channels for "planned propaganda, often without the publications being consciously aware of it." He said major newspapers in the United States were "actively used" by the Soviets. He testified that he personally participated in financing and setting up the *Patriot* newspaper in India. Mr. Dzhirkvelov was speaking as a witness for Britain's influential magazine, the *Economist*, during a libel case with *Ethnos*, Greece's largest circulation newspaper. The Greek daily sued the *Economist* when it alleged that *Eth-

nos was launched with a $1.8 million Soviet subsidy. The British magazine counter-sued *Ethnos* for saying that the *Economist* served CIA interests. Both cases were tried simultaneously, without a conclusive decision.

Even with glasnost—perhaps especially to demonstrate glasnost—a new form of disinformation has surfaced. Twice in 1988 Western reports credited to Moscow sources said Alexander Solzhenitsyn, the leading Soviet writer in exile, was negotiating with the Soviet Union for the publication in the USSR of his banned novel *Cancer Ward*. The report said Solzhenitsyn would return to the Soviet Union in 1988. Mrs. Solzhenitsyn told Freedom House she and her husband had never been approached by the USSR since their citizenship was withdrawn in 1976. By such reports, however,—in the form of disinformation—glasnost seems to reach out even to the USSR's Nobel Prize-winning critic.

Law, the courts and penalties figure high in most governments' power plays against journalists and the news media. Official censorship, formerly the mainstay of governmental press-control, has been largely superseded by scores of other control techniques recounted in this chapter. In some places, censors still sit at the elbow of journalists. *La Prensa* in Managua felt the heavy hand of Sandinista censorship daily until the paper was closed down in 1987. When it reopened in 1988, the censors were gone, but other governmental threats hang over the lone dissenting paper in Nicaragua.

South Africa instituted measures in 1988 which could assign censors to newspapers formally warned they have broken emergency regulations. "The really worrying thing is the pre-censor," said Anton Haber, co-editor of the liberal *Weekly Mail*. "This is a vicious new form of censorship," unprecedented in this country, he added. The result was dramatically shown by Anne Taylor Fleming on the MacNeil/Lehrer News Hour, 29 December 1987. "The South African government imposed censorship and took away our pictures," she said. "What do you see?," Ms. Fleming asked, as the television screens of everyone watching the program went black. "What do you imagine? Can you conjure again the old troubling images [of riots and police action]? Or have they faded in time, too? It's as if an entire country has simply been lost to us. Withdrawn from our view, from our TVs, from our minds."

Precensorship worked in South Africa, and to a lesser extent in Israel. But mostly other legalities and legalisms now bolster State power challenging the nongovernmental news systems.

In civilian societies, targeting the news and information media are acts protecting official secrets and security (Malaysia and the United Kingdom repeatedly invoked these acts in 1987-1988).

Such acts are designed to protect confidentiality and arms control, penalize "contempt" or "abuse" of publishing, or demeaning of the president or his family. Some acts make it a crime to publish "false news" (as in Hong Kong), with the government determining what is truth. There are acts opposing communism in Western countries, and defending socialism in the Eastern bloc; e.g. acts opposing aid to terrorists, and acts opposing public protest. In time of limited war, such as the British campaign in the Falklands/Mariannas, and the U. S. in Grenada, journalists are either kept out of the early fighting, or isolated and denied communications to home offices. There are acts in defense of the revolution and, in governments run by the military, special codes of law and punishment.

That is just the beginning.

Laws provide contempt citations, detention, imprisonment, or expulsion from the profession. Foreign correspondents may be barred from entry or reentry to the country. Journalists (in South Africa) may be banned, the equivalent of being placed under house arrest with no visitors and, as an unperson, left unable to be quoted by other journalists. Laws there and elsewhere deny newspersons access to sensitive geographic regions; others demand they reveal their sources, under threat of imprisonment. Still others deny them satellite feeds, place them under police surveillance, or subject their writing to regular monitoring against a government-approved code.

The newspaper, radio or television station is also subject to harsh reprisal from government: suspension for a definite or indefinite period; confiscation of equipment or the entire plant; monitoring the output against a government code; banning of opposition party papers by a ruling party; police or military search of an editorial office.

Libel laws intended to protect the public from false and damaging statements in the news media have become a significant threat to press freedom, in some instances. Governments sometimes invoke libel laws to protect officials or their policies from sharp criticism in the press. In February 1988 the Honduran military demanded that the government file lawsuits against the *New York Times* and *Miami Herald* for linking high-ranking officers to Colombia cocaine traffickers. The papers, said the Superior Council of the Armed Forces (fifty colonels and generals), should present proof or accept legal responsibilities for attacking the "dignity of Honduras." No further action was taken.

In South Africa, Zulu leader Mangosuthu Buthelezi, also the head of Inkatha, the largest black organization, filed a lawsuit against a small, articulate magazine, *Frontline*. Buthelezi charged he was the target of a concerted international campaign of vilification. *Frontline* had quoted a negative assessment of Buthelezi from the *Spectator* of London. The Zulu leader chose the nearby target of opportunity, and sued for some $10,000. He won the suit, and seriously crippled the small anti-apartheid magazine. It is a rare publication in South Africa. Its articles regularly seek a sane, nonviolent solution to South Africa's seething racial issues. *Frontline*'s editor, Denis Beckett, and a remarkably able black columnist, Nomavenda Mathiane, provide passionately clear thinking—a unique contribution in southern Africa.[2] They frequently support Buthelezi, and particularly his effort to create a democratic legislative model. How shortsighted, then, was Buthelezi's recourse to libel law, and its punitive action. Just months later, almost as a tribute to *Frontline's* objectivity, the government banned the magazine.

Similar abuses of libel law may be found in the United States. American libel law, since *New York Times v. Sullivan* in 1964, insists that the journalist must be shown to have conveyed untruth with malice before a public person can secure redress. A private citizen has greater protection. Yet libel suits against American newspapers and broadcasters have dramatically increased in number. Juries provide million-dollar verdicts which threaten to bankrupt some newspapers. Almost routinely on appeal, however, penalties have been greatly reduced in 75 percent of the cases. The cost of litigation when a small paper wins the suit, however, can be crippling. Lawyers' fees average $150,000. The Alton (Illinois) *Daily Telegraph* was made bankrupt by a libel verdict of $9.2 million which it could not pay. The "punitive" part of the verdict was $2.5 million. The protection added by the *Sullivan* ruling may make it harder for the plaintiff to win, but there is no ceiling on the cost of losing to the defendant.

Demonetizing libel

Giant news organizations are not immune either. Two—*Time* magazine and CBS—suffered costly libel-law engagements in 1985. Cash was not the greatest price paid in either case. Indeed, General Westmoreland withdrew his $120 million suit against CBS before it went to the jury. CBS had pictured Westmoreland as conspiring to falsify estimates of enemy troop strengths during the Vietnam war. But CBS's reputation as a fair, balanced reporter was damaged by revelations during the five-month court battle. The telecast, for example, had ignored evidence that did not fit the serious charges made against Westmoreland. *Time* magazine suffered similarly. Its accuser, Israeli General Ariel Sharon, maintained he had been libeled when the magazine wrote he ordered the attack on southern Lebanon to "take revenge" on those responsible for the assassination of Bashir Gamayel, an Israeli ally. In the subsequent fighting, several hundred civilians in Palestine refugee camps in Lebanon were massacred, and this was linked to Sharon by the magazine. *Time* was found negligent and careless in reporting Sharon, but not legally liable. Quietly, a year later in an Israeli courtroom, *Time* admitted there had been no proof that Sharon had spoken of revenge. *Time* agreed to pay some of Sharon's legal expenses. It had been a $50 million damage suit.

In both cases, two factors dominated: the legal question of what constitutes libel, and the financial issue of how expensive it has become to answer the first question. Under *Times v. Sullivan*, CBS came very close to revealing malice against General Westmoreland, but he seemed to run out of funds and lacked adequate legal representation. Both were needed to press the matter through to a determination by the jury. *Time* magazine also made a serious editorial mistake and refused to admit and correct it.

Clearly, libel suits should be demonetized. It is easy for plaintiffs to sue. Lawyers usually agree to a contingency fee: no payment unless they win. The integrity of the journalist—as with CBS and *Time*—should be at stake in deciding whether a subject has been harmed. Massive court costs and punitive threats are not necessary to establish for the public the rightness or wrongness of a journalistic report. If a journalist's or a news organization's integrity is on the line, punitive financial penalties become far less significant. Only actual financial loss to a plaintiff should be considered, beyond determining the truth of the contested statement.

Although the mass media can libel a citizen massively, it is folly to argue that crippling a channel of public information is the intention of the law. Yet present libel litigation supports that fear. Most media do not have the resources of *Time* or CBS. The never-ending tension between press freedom and the check on license which libel law provides is a necessary function of a free society. No matter how high a price tag one places on an alleged defamation, the ultimate reward should be the clearing of one's name, not inordinate profit. Actual financial loss should certainly be recompensed, but that is all. This calls for demonetizing the libel-law process.

Litigation should be severely streamlined. Perhaps the small-claims or family courts in the United States suggest the model. They provide low-cost, relatively lawyer-free procedures in fields of specialized law. A "libel court" would be run by a particularly qualified judge with a staff able to deal fairly with the content and procedures of published or broadcast material. The libel court could remove the financial burdens that pose a threat to First Amendment rights. The person who believes he has been dealt with maliciously would have far less costly recourse. Since most libel suits are brought by persons seeking to restore their reputations, the adequate correction of misstatements should be the principal objective. The public would learn the truth. And where libel has been proved, editors and employers may be expected to deal appropriately with the erring journalist. For credibility, not dollars, should be the news medium's and the plaintiff's most precious possession. Proper reporting of the truth, not punitive settlements, should be the objective of the libel court.

Without demonetization of libel suits it is likely that journalists, and particularly editors and publishers, will increasingly avoid risky investigative coverage. Critical city and state investigations are likely to be the first evaded by the smaller, less-endowed newspapers. Well before the need for litigation, however, should be the preventatives: self-correction by a news organization, and mutual criticism by the media generally. In their absence, increasing government pressures in even the freest societies are inevitable.

The Singapore case

Such pressures became most visible in 1987-1988 in Singapore, where Prime Minister Lee Kuan Yew insisted that his government had the right to publish in the American and European press the full text of his government's responses to earlier press reports about Singapore. When the prime minister's "right to rectification" was not acknowledged, he invoked new legislation to greatly restrict the circulation in his country of *Time, Asiaweek*, the *Asian Wall Street Journal* and the *Far Eastern Economic Review*.

The right to reply is not a new issue. Under an 1881 French law, any citizen mentioned in a newspaper or magazine has the right to an unedited reply not longer than the original article (and no more than 200 lines). This right may be invoked whether the original piece was true or false. In 1986, according to Professor Franklyn S. Haiman of Northwestern University, *Le Monde*, France's most prestigious newspaper, received only fourteen requests for replies, of which only three were printed (the others did not conform to the law). Another professor, Jerome Barron, of Harvard, proposed a law mandating the right to reply. This was passed by the Florida legislature, and later declared unconstitutional. The right of reply, however, is increasingly demanded at Unesco and other international forums where countries such as Singapore stridently demand access to the Western news media which, they assert, malign their countries. Singapore withdrew from Unesco in 1986, but persists on its own to advance the right of a government to enter, on demand, the pages and broadcasts of independent news media.

Singapore's two-year struggle with the news media over the government's "right" has been impressively boring, but highly instructive. Of the four publications gazetted, and made to suffer severe financial losses, the case of the *Far Eastern Economic Review* sets the issues most clearly. On 26 December 1987, Singapore declared *FEER* "to be a newspaper engaging in the domestic politics of Singapore, and restricted its circulation to 500 copies per issue" (from a government advertisement appearing 28 January 1988, in *FEER*). Singapore invited *FEER* to nominate a local distributor who would select the 500 readers "in such a manner as the Minister [of Communications] may direct."

FEER responded by saying that it had suffered censorship and banning elsewhere in the past but that Singapore's "gazetting is invidious." The magazine declared, "As the *Review* is prevented from serving all its Singapore readers, it prefers to serve none. Gazetting, in effect, places the distribution of the publication into the hands of the Singapore authorities, allowing them to pick and choose the institutions or readers which the *Review* reaches. This is unacceptable. The act of gazetting also enables the Singapore government to claim misleadingly that the publications affected have not been banned."

The specific reportorial issue here is of little consequence outside Singapore. The story that inspired the government's wrath involved coverage of the arrest and detention without trial of twenty-two young men and women said by the Singapore government to be part of a "Marxist conspiracy." *FEER* covered several sides of the continuing story, including the claim by eight detainees that their supposed confessions had been false, and made under duress. The government struck at the magazine, declaring: "Foreign publications are allowed to circulate freely provided they do not engage in Singapore's domestic politics. They should report Singapore as outsiders for outsiders. Refusing the government the right of reply constitutes engagement in Singapore politics."

The *Far Eastern Economic Review* had, in fact, published many letters on this subject from Singaporean officials. By early 1988, when the prime minister took the magazine to court, readers of *FEER* were subjected to repeated, boring letters from Singapore, and responses from the magazine. *FEER* sought to end the detailed minutiae of the government's argument. Back in Singapore, however, the London *Economist* quietly moved its regional office to Hong Kong, *U.S. News and World Report* closed down its Singapore bureau, and *Asahi Shimbun* of Tokyo took its Asian bureau to Bangkok.

After more than two decades of wooing foreign correspondents, and profiting from generally favorable reportage, Singapore seems unable to accept the downside of Western-style reporting—that is, the straightforward coverage of events and personalities that may be perceived as flawed. Writing "as outsiders for outsiders," to meet the prime minister's dictum, would mean writing with far less insight, and without the details and nuances that reveal good reporting. Indeed, the strange charge of writing as an insider is an unintended compliment. In debates at Unesco it has often been claimed that Western journalists cover the Third World superficially, without sufficient understanding of the insider's views. Attributing fault to an "insider's" writing implies that Singaporeans cannot be given their own story as it truly exists. And that, in fact, is the prime minister's point: he doesn't want his countrymen to know some of the realities in their own country. Demanding the right of reply is mainly a dodge. He has had that right, repeatedly, but he will not be satisfied until he can alter the content of future coverage. And that is the ultimate fear of those who have resisted the right to rectification at Unesco and in other forums for so long.

Licensing of journalists

Often in these debates, rectification has been linked to the licensing of journalists by governments. The extreme argument: Journalists should operate under a governmental umbrella in order to perform according to a code, also officially inspired. The moderate position: Newspersons have a quasi-public mission; like doctors or engineers, they should be licensed to protect the public from quacks. The self-serving factor put forth by some journalists themselves: Anyone who can write his name may compete as a reporter; therefore, to reduce competition, particularly from the unskilled, journalists should be formally certified.

The campaign for degree-granting *colegios* in Latin America began in the 1960s. Costa Rica passed the Colegio de Periodistas (Guild of Journalists) Law in 1969. It was intended to increase professionalism among journalists. The law requires that anyone working as a reporter on domestic media must be a graduate of the University of Costa Rica's School of Journalism. Though other schools appeared, including one with more professional journalists teaching than at UCR, only graduates from UCR could enter the colegio and, under the law, become accredited journalists. This tends to limit the number of practicing reporters and grant members of the colegio monopolistic control over the profession. Since Costa Rica is one of the most democratic countries in the world, the government has not taken the step most feared by Western journalists. Costa Rican officials do not dictate those to be accredited or to be removed from colegio membership. In less democratic places, the licensing of journalists, newspapers, magazines and the broadcast media serves as a power mechanism that influences the words and pictures put before the public.

Yet the darker side of licensing appears even in Costa Rica. For, after all, the government yields some of its power to the colegio, and permits that body to determine what information the people of Costa Rica may or may not receive. One man challenged that law. He is Stephen B. Schmidt, an American who had worked for several years as a journalist in Costa Rica. He holds a masters degree in journalism from the Autonomous University of Central America in San Jose. He was not licensed, however, because the colegio was open only to graduates of the University of Costa Rica. Schmidt, nevertheless, practiced journalism and officials ignored his lack of license (as they ignored others) until in 1980 he challenged the colegio to enforce the law. A lower court ruled in January 1983 that the law violated the Costa Rican constitution's guarantee of "freedom of opinion and expression." Five months later, however, the Supreme Court reversed that decision and gave Schmidt a suspended three-month sentence. He was, nevertheless, a convicted felon. The case then came before the Inter-American Commission on Human Rights, an arm of the Organization of American States. The seven members from as many countries heard arguments in October 1984 and voted five to one in support of the Costa Rican Supreme Court's decision. (The Costa Rican member excused himself from participation.) The lone dissenting vote was cast by R. Bruce McColm, my colleague at Freedom House who was the only North American on the Commission.[3] The Schmidt case moved in August 1985 to the Inter-American Court of Human Rights, also situated in San Jose. The government of Costa Rica, prodded by the Inter American Press Association, sought an advisory ruling on whether the press-licensing law violates existing inter-American and universal human rights conventions.

The court's unanimous opinion, 13 November 1985, was a startling rejection of the licensing of journalists. The finding was particularly surprising because thirteen Latin American states by then were licensing journalists—several for as long as two decades. By then, too, the licensing of journalists had attracted global interest. Unesco's International Commission on the Study of Communication Problems (1980) provided the pros and cons of licensing. The commission, headed by Sean MacBride, concluded:

> Experience shows that the granting of professional licenses and all complicated accreditation procedures tend to foster government intervention in the national and international flow of news. [More explicitly:] We share the anxiety aroused by the prospect of licensing and consider it contains dangers to freedom of information.[4]

Not only was licensing frequently discussed at Unesco, but countries in Africa and Western Europe (Spain, for example) were seriously considering licensing of journalists. So serious were the possibilities, the International Federation of Journalists (IFJ) in June 1986 completed a study on the Licensing of Journalists and the Colegios. The IFJ, an association of journalist trade unions on five continents, headquartered in Brussels, found it difficult to support or reject fully the idea of journalist licensing and the colegio. The IFJ said they "should not have a compulsory nature." Yet that is, indeed, the nature of the foremost colegio, particularly that in Costa Rica. The Inter-American Court's ruling, therefore, is of great importance. It said:

> The compulsory licensing of journalists is incompatible with Ar-

ticle 13 of the American Convention on Human Rights insofar as it denies some persons access to the full use of the news media as a means of expressing themselves or imparting information.

The court further held that "When an individual's freedom of expression is unlawfully restricted, it is not only the right of that individual that is being violated, but also the right of all others to 'receive' information and ideas." That second aspect, said the court, "implies a collective right to receive any information whatsoever and to have access to the thoughts expressed by others."

Freedom of expression goes still further, the court continued. It includes "the right to use whatever medium is deemed appropriate to impart ideas and to have them reach as wide an audience as possible." This means that "restrictions that are imposed on dissemination represent, in equal measure, a direct limitation on the right to express oneself freely. The importance of the legal rules applicable to the press and to the status of those who dedicate themselves professionally to it derives from this concept," the court declared.

Newspaper publishers through the Americas immediately petitioned Latin American countries with journalist-licensing laws to consider repealing them in light of the court's ruling, which was advisory and without sanctions other than moral suasion. Four years after the ruling, no country, including Costa Rica, had altered its licensing law or the colegios. In 1988, however, Costa Rica began studying ways to bring the country into compliance with the inter-American covenants.

Diverse legal powers of the state, then, are deployed to control the words and ideas of the news media; and, as we have seen earlier, editorial, psychological and physical powers as well. Financial and technical instruments are also used by governments to control or influence the flow of news and information.

The crudest form is outright bribery, the threat to fire, or the giving of favors such as the promise of a job, the offer of assignment to a desirable location, junketing, or public association with an important official. A government's financial assaults on a newspaper can be serious, indeed. These may include raising the price of newsprint, generally set by the state's import body (as in India), controlling the allocation of paper to favored publications (as in Mexico), restricting the distribution of the newspaper (as in Singapore), setting the sales price of the publication, denying foreign exchange to the publication so that it cannot purchase newsprint or parts for the presses (as in Paraguay and Guyana), providing subsidies or loans to weak papers (boons that may be just as easily removed if reporting does not please officials), adjusting the tax rate to favor the friendly media, placing or removing government advertising depending on editorial content of the publication (Nicaragua), and taking a variable number of paid subscriptions to independent newspapers, magazines and even news agencies for distribution of the product to government offices at home and abroad.

The state has a limited number of technical weapons. Perhaps the latest bit of press-control ingenuity is Mexico's legal requirement that "from the moment [new] telecommunications equipment is put into operation... in general, everything that conforms this equipment, will henceforth be the property of the nation." Thus, if an independent newspaper or broadcaster buys the equipment and installs it, it becomes nationalized. Governments may also deny a domestic publication, a broadcaster, or a foreign news medium access to a satellite. And, as Colombia has done, it may seek to charge another country (generally the U.S.) for "parking" its satellite in geostationary orbit 22,300 miles above Colombia's land mass.

The government's newspaper

It is routine for Marxist states to own and operate, directly or indirectly through cooperatives, all the means of communication, including newspapers and magazines. Many non-Marxist governments also own one or all the newspapers in a country. Nelso Etukudo, a Nigerian communications scholar, discusses "the forces militating against the editor of a government newspaper" and the "conflict between the professional demands of the editor and the private commands" of the governmental proprietors. This requires, he says, a "balancing act" to determine how far the editor should go in serving as the "inspector-general" of the government's policies.[5]

His colleague Etim Anim[6] describes the usual objectives of governments in creating a newspaper: to use it as an "instrument of nation-building or communal harmony [and] for disseminating information about government programs, actions, and activities." But Anim also quotes an advocate of government newspapering. He says, "above all, the law of self-preservation and survival dictates that every government should arm itself against the possible ravages of political newspaper lords and against the rampages of capitalist controllers of the mass media in their bid to subdue the people to their service, and to bend the nation to their will." Anim responds that government ownership is not necessary to fulfill these objectives. A private newspaper, he says, can be "an agent of integration and an advocate of change," and also provide the "dynamics...for development and social harmony." He concludes there are other reasons—"overt and covert"—for government ownership of newspapers. Anim says "the burden of editing a government newspaper, then, is tied up with resolving the conflicts which arise between the overt reasons and the covert ones; the conflict between what is and what ought to be. The one impinges on the performance of the journalist as a professional, and the other on the performance of the journalist as a government employee." Editing a government newspaper, he adds, is like "walking a tightrope," balancing the "demands of one's profession and the demands of political realities, between the right and the just, and expedient; between fair play and a closed conscience." Anim then drew on a survey of editors of government newspapers to list, in descending order of pressure, the following sources of conflict and pressure to which they are subjected:

The Executive
  1. *Commissioner/Adviser on Information*
  2. *Governor's office officials (not the governor directly)*
  3. *Other commissioners*

Board Members

Chief Executive of the newspaper

The Legislature
  1. *Assemblymen*
  2. *Committeemen/Chairmen*
  3. *Deputy Speaker/Speaker*

Interest Groups
  1. *Ethnic organizations of people in the government*
  2. *Friends of government officials in the news*

Political Parties
  1. *Majority party*
  2. *Minority party*

The Bureaucracy
  1. *Permanent Secretary (Information)*
  2. *Head of Service*

Anim concludes that "the editor's balancing act is successful to the extent to which he can persuade his superiors to encourage the easier flow of communication between the government and the people. Success in that task is a function of the editor's professionalism."

Nigeria has perhaps the freest journalism in Africa, though there are government-owned and independent newspapers [a fuller description is provided in Chapter VIII.] This analysis, however, may be compared to the far less hopeful plight of journalists elsewhere in Africa. Elie Abel, former dean of the Graduate School of Journalism of Columbia University, tells of a young African student he advised some years ago. Said Abel:

> His work was extraordinarily good and he went home in a rather triumphant glow to take charge of a rather important news organ-

ization in his own country. Less than two years later I received from this young man a letter that must have been as painful for him to write as it was for me to read.

"You taught us all too well," he wrote, "I've discovered unhappily that my country is not ready for American standards of press freedom. This, as you know, is a one-party system and the things that I've learned in America have made it difficult—probably impossible—for me to function in such a system. It's not just that criticism, exposures of inefficiency and official misconduct have no place in our press. The system lays upon us an affirmative obligation to praise the regime and above all to glorify the head man. The dilemma of adhering to professional ethics and at the same time trying to satisfy the whims of politicians is a headache. But as you know there are no answers to these difficult questions in developing countries."

This former student did not dare to write such a letter from his home. He waited until he was out of the country attending an international conference in Europe. And the last time I heard from him he had given up journalism, his chosen profession, as well as his country, to prepare himself for a new career in another field, one somewhat removed from Third World politics in which he could hope to function as the honest man I know him to be.

I think that's a very sad story.

Press power

What, then, are the countervailing forces the independent news media may use to survive, and perform their functions in the service of the public? In every category—legal, economic, political, editorial, policy-influencing, and psychological—the media of news and information are comparatively weaker than government. The power of the media varies from country to country. Even in this hypothetical listing of all possibilities, however, the strongest media are no match for the weakest state. Often, the weaker the state in comparison to other countries, the more firmly will it crack down on journalists and news media within its borders.

The media have some legal instruments at their command, however. They may get court injunctions to secure some government information. Or in a few countries they may acquire extensive government files through freedom of information acts. By revealing corruption in government, the media use their primary weapon—truthful information—to restrain government power. By regular coverage of government operations, the press provides the context within which not only the government but the society functions.

News media, print and broadcast obviously acquire economic power in those places where independent media are permitted to function with relative freedom. In such places the tendency to form newspaper chains, broadcast networks, and other media conglomerates creates significant economic power. While that cannot fully balance state power, press power can influence the naming of officials and the policies they pursue. For economic power translates into political power—the opportunity to support or oppose governmental policies, provide editorials and news coverage for candidates or officials, publicize a public person's financial and political support or opposition, and influence the electoral process at every stage—including too-early exit polling on election day, or campaign polling generally to provide an authoritative demographic profile of the society. That can be as indispensable to government as to its opposition.

Editorial power can touch the innermost secrets—strengths or weaknesses—of government and officials. Even if the scale and impact of the Watergate revelations could not be duplicated outside the U.S.—and probably, with some luck, not soon again in the U.S.—investigative reporting is now a regular function in the American press, and is slowly being taken up in other countries. It is recognized that investigative reporters wield power when they reveal sensitive secrets. They may also create diplomatic, military or financial problems. But even by routinely covering public figures, or denying them and their statements print space or broadcast time, journalists wield considerable influence. When they provide instant replies to officials' speeches, journalists, particularly television anchors, become part of the country's political-power system. If a journalist turns sensationalist in his reporting or commentary, the distortion may create official embarrassment, inspire reprisal, and probably distort the message for the public. But sensationalism is power, nevertheless.

So, too, is the function of the news media to help set the agenda for the public and the government. By selecting news and issues for dissemination, the print and broadcast media informally join with government in debates over policy-setting. The news media may initiate campaigns to preserve wildlife and defeat commercial developers or deride or clamor for military action abroad. Independent news media may get ahead of, or stand behind, government policies. In either case, officials and journalists are separate players, with different sets of ground rules, though serving an overlapping constituency.

Of all the powers of the press over government, perhaps none is as potent as the psycho-political: the daily ability, where permitted by constitutional writ, and not abrogated by officials, to question and even undermine public confidence in a policy or the leadership of the government. This question influences not only short-term issues, but the general functioning of the ruling party, and, if such questioning is deep and sustained, the legitimacy of the system of governance itself. With that pos-

Power, the Press and the Technology of Freedom

sibility always present, weak governments crack down on the news media (or withhold permission to criticize) well before the critical challenge to governance is reached, no matter how justified may be the questioning by the press, and no matter how much such criticism may be in the longer-term interest of the people and the state.

Involvement in such interplay may not always be a noble purpose in the performance of the news media, or even an ethical procedure. Character assassination of political leaders and the bureaucracy may be engaged in by the news media for their own political, commercial or other objectives. Outright blackmail—the threat to run a bad story about a subject, unless he provides other information, or even favors or payment—is infrequently practiced, but deeply disturbing when it appears. Most often, an "irresponsible" journalist is one who is so accused after he accurately quotes a jailed official. John Tusa of the BBC recalls an irate citizen on the streets of Belfast thrusting her fist over the camera lens, and delivering the immortal verdict "You are showing things that aren't even happening." That, says Tusa, could be the motto of the official censor down the ages.

[For a listing of governments which own all or most of the press, license the press, license journalists, practice censorship or provide "guidance" see Appendix H.]

7.
How the News is Reported and Distorted

A Survey of State versus Information Power in 74 Countries

This is a study of thirty-six most-free countries with 1,340 million citizens, and thirty-eight least-free countries with 2,223 million population. The designation of most- or least-free is based on the 1988 Freedom House survey of political and civil rights in all countries.

The findings in this separate press-freedom study show the reality in each of seventy-four countries. But the patterns formed by the responses reveal the larger relationships of state versus information power that are likely to prevail through the '90s. These broad patterns are not easily altered, though individual states may display notable movement along the least-free to most-free spectrum. It is worth noting that press freedom is established now as a universal norm even by those regimes which most harshly deviate from the standard. They generally do so by self-serving rationalizations which pay lip service to the right of their people to know what they need to know.

Our press-freedom study provides these analyses for the present and the short-term future:

Political pressures influence newspapers everywhere. Political influences are felt strongly in nearly one-third of the most-free nations, twice as strong in the least-free. Strong commercial pressures on the press, surprisingly, occur only half as frequently as political influences in most-free countries. And no less surprising, commercial and ideological influences on the print press are twice as strong in least-free as in most-free nations. All pressures considered, in 68 percent of the most-free places print journalists exert "strong" editorial control, but broadcast journalists maintain full control in only 44 percent of most-free nations.

More than one-third of the least-free countries strongly control news content by limiting access to newsprint and broadcast materials. India is the exceptional most-free country, with political influences strongly restricting newsprint and controlling prices of independent newspapers.

Licensing of publications and journalists strongly influences the content of news reporting in nearly half of the least-free countries.

Even in one-quarter of the most-free states, however, restrictive laws of many kinds influence newspaper and magazine content. In three-quarters of the least-free nations, such laws affect news reporting.

Newspaper journalists regarded by governments as unfriendly are penalized to some extent in 18 percent of most-free countries, and in two-thirds of least-free nations. Unfriendly broadcasters are penalized in 95 percent of least-free nations.

More than half the most-free countries somewhat limit access to news in order to control the information flow. This is done in nearly all least-free nations. About two-thirds of the latter countries also strongly engage in the practice of sending to newspapers, information known to be false. Denunciation of journalists to destroy their credibility or cost them their jobs is practiced by up to 80 percent of least-free governments.

Harshest measures—murder, imprisonment, harassment, banning, confiscation—are endemic in least-free countries. Murdering journalists, however, influences the content of the press less than arresting or harassing journalists.

Surprisingly, private ownership of the news media and market competition inhibit the free flow of news more in the least-free countries than in the most-free.

EARLY IN 1987 I asked correspondents in seventy-four countries to tell at year-end how the news media function in their lands. The monitoring was updated to December 1988. The correspondents are journalists and academics who specialize in communications studies. I had known many of these men and women for more than a decade, during which I travelled to forty-five countries. On nearly all of these visits I met privately with local journalists, and at national and international conferences discussed press freedom and the uncertain fate of journalism.

For this study, we divided the countries into the most free (MoF) and least free (LeF). No MoF country totally meets an ideal standard for full public access to diverse information, or for a free flow of news and data, unhampered by bias or unjournalistic pressures. And no LeF country is totally restrictive. All countries fall somewhere along the continuum from absolute freedom to absolute restriction.

Yet, we must be clear: the highest standard provides the greatest freedom for the individual citizen, within the limited rules needed for all to live together in peace. The quality of life in a most-free nation is patently more satisfying than life in a repressive regime. Similarly, the quality of the flow of news and information is an integral aspect of *every* country's level of freedom. Where a country appears on the freedom-of-information spectrum is, therefore, very important. That placement tells us much about the nature of the society, and the quality of life of its citizens.

Quality of life is primarily determined by how a society is organized or governed. By employing two-dozen criteria based on the system of governance in each country, it is possible to judge which are more free or less free countries. The systems of governance are assessed by the seventeen-year-old Comparative Survey of Freedom, conducted by Freedom House. The degree of freedom of the news and information media was examined independently for this book, based on actual journalistic performance in seventy-four countries in 1987 and 1988. Our study of journalistic practice fully sustains the separate assessments of systems of governance. There is, then, a distinct correlation between a country's style of governance and the manner in which it permits journalists to report and comment. Governmental influence, control, or censorship—the struggle to use the *word* in the exercise of power—is the primary determinant of news media freedoms in every country. Our study, therefore, focuses primarily on government/media relations, but examines as well the market forces which also influence press freedom.

The most-free group includes thirty-six countries with a population of 1,340 million, or 40 percent of the people in this study. The least-free include thirty-eight countries with a combined population of 2,223 million, or 60 percent of the population total. (This parallels the 1989 Freedom House survey of comparative freedom which finds 39 percent of the world's people free, and 61 percent either partly free or not free.[1])

Fully one-half of our correspondents requested anonymity. We list the names of the countries represented (see Appendix G) but, regrettably, none of the respondents. The plea for anonymity, I believe, is a dramatic indication of the fear factor in domestic and international journalism.

The questions

In ten basic categories, we asked a total of 194 questions of each participant (see Appendix F for a sample questionnaire). We received tallies of the number of newspapers, magazines, radio stations, television channels and national news agencies which are state- or privately owned, or share private-public ownership, or some other form of control. We were told that 1987, for the news media, was a year of great improvement in three *most-free* countries and one *least-free*. "Some improvement" appeared in eight *most-free* and seventeen *least-free* places, no change in twenty-six *most-free* and nine *least-free* nations, "some deterioration" in one *most-free* and two *least-free* nations, and "great deterioration" in seven *least-free* countries. (For listing of countries by the degree of change see Appendix D.)

Percentages are used no matter how small the number in a category. Each number represents a country. No projection is made beyond the actual countries responding in any category. This, then, is *not* a poll of a representative sample of a larger universe. It *is* the universe itself.

The information we sought

1. the *kind* and *degree* of influence (political, economic, social/racial, commercial, ideological) over the editorial content of newspapers, magazines, radios, television

2. the degree of *formal* governmental influence on each medium

3. the degree of *informal* governmental influence on each medium

4. the kinds of news "management" *practiced* by the government over each medium

5. the degree to which *harsh governmental actions* (murder, imprisonment, harassment of journalists, banning of media, confiscation of product) actually influence news reports or commentaries in each medium

6. the degree to which the content in each medium is influenced by *private ownership,* such as special interests, advertisers' interests, mergers, pressures from competing media

7. the degree to which *self-censorship* and other pressures affect news reporting

8. the forms of governmental control practiced over *international news.*

Kind and degree of influence

What kinds of influences do newspapers, magazines, radio and television experience routinely in most-free (MoF) and least-free (LeF) countries?

How strong are these influences?

Our tables show the percentage of the *countries* responding under each of the influence categories, and for each of the degrees of influence.

For example, *political* influence is *strongly* directed to *newspapers* in eleven countries (or 30 percent of the responding countries) regarded as *most-free.* Twice as many *least-free* governments, twenty-two (or 55 percent of those replying), exert *strong political* pressures on *newspapers.* It may not be surprising that newspapers in 89 percent of *least free* nations face significant political pressure. But the degree of political influence in free nations—apparent in 86 percent of the countries, though showing only *some* influence in 56 percent of the places—is nevertheless greater than may be generally supposed.

Far stronger political pressures are exerted on radio and television in *least-free* (LeF) countries. These are the primary means of communication with the public in most countries. The more government seeks to control or influence a society, the more it employs radio and television as its principal political channel to "the people."

(The January 1988 Freedom House study of press freedom[2] shows that, worldwide, the print press is free in 34 percent of the countries, the broadcast media in only 24 percent. "Free" here does not refer to absence of government ownership of the media, but rather to the diversity of content and degree of balance in presenting dissent or competing ideas.)

Television channels in 88 percent of LeF countries receive strong political pressures. Radios in 74 percent are also strongly influenced. Even one-third of *most-free* (MoF) countries, however, report strong political influences on radio and television, and 68 percent of MoF-country radio has "some" political pressure—far less than the "strong" influences on broadcasting in LeF countries, but significant nevertheless.

The political pressures on the media in MoF and LeF countries are generally stronger than economic, social, racial, ideological and even commercial influences. This is particularly significant in the MoF countries where market systems predominate. Commercial influences on newspapers are one half as strong as political pressures. Weaker commercial influences are about the same as weaker political pressures on newspapers in MoF countries. Strong commercial influences in MoF-country radio and television approximate the level of strong political pressures in MoF countries.

In LeF countries, the strong influences on newspapers from economic, commercial and ideological sources are about double those forces in MoF countries. Among social or racial influences, however, the four kinds of LeF media suffer strong influences from three-and-a-half to five times as great as in MoF countries.

We asked, finally, the degree to which the judgments of editors and journalists prevail over other considerations. Nearly one-third to one-half of the LeF countries report that editors and journalists in newspapers, magazines, radio and television do not provide the determining judgments over the content of the media in which they operate. The media managers and journalists in MoF countries exert their own judgments strongly in half to two-thirds of the countries, depending on the form of media. Newspaper journalists exert the strongest freedom (68 percent of the places) and radio journalists the least (44 percent). But when "strong" and "some" journalistic freedom are taken together, in most-free countries, the prevailing nonjournalistic judgments are minimal. (See Table 1.)

Formal governmental influences

Governments use formal institutions and practices to influence or control the news media. The ministry of information (or its equivalent under different names) provides the most obvious tool to wield governmental

Power, the Press and the Technology of Freedom

I. Kind & Degree of Influence Over Editorial Content

			A. Newspapers			B. Magazines			C. Radio			D. Television		
			Strong	Some	None	Strong	Some	None	Strong	Some	None	Strong	Some	None
1. Political	MoF	%	30	56	14	14	47	39	32	68	0	37	60	3
		#	11	20	5	4	13	11	11	23	0	11	18	1
	LeF	%	59	30	11	48	40	12	74	21	5	88	6	6
		#	22	11	4	15	13	4	25	7	2	25	2	2
2. Economic	MoF	%	15	50	35	4	48	38	9	47	44	14	54	32
		#	5	17	12	4	14	11	3	16	15	4	15	9
	LeF	%	38	48	14	40	42	18	49	34	17	50	34	16
		#	14	18	5	13	14	6	17	12	6	16	11	5
3. Social/Racial	MoF	%	3	32	65	4	21	75	6	18	76	7	21	72
		#	1	11	22	1	6	21	2	6	25	2	6	21
	LeF	%	26	48	26	15	56	29	30	35	35	31	43	26
		#	9	17	9	5	18	9	10	12	12	10	14	8
4. Commercial	MoF	%	17	57	26	21	43	36	19	47	34	25	46	29
		#	6	20	9	6	12	10	6	15	11	7	13	8
	LeF	%	30	42	28	30	45	25	29	46	25	38	42	20
		#	10	14	9	9	14	8	9	15	8	11	12	6
5. Ideological	MoF	%	23	49	28	25	29	46	12	51	37	27	35	38
		#	8	17	10	7	8	13	4	17	12	7	9	10
	LeF	%	52	31	17	38	43	19	48	26	26	50	25	25
		#	18	11	6	12	14	6	16	9	9	16	8	8
6. Judgments by Journalists	MoF	%	64	36	0	48	33	19	44	47	9	50	47	3
		#	23	13	0	13	9	5	15	16	3	15	14	1
	LeF	%	32	38	30	32	36	32	17	40	43	21	31	48
		#	11	13	10	9	10	9	5	12	13	6	9	14

II. Formal Government Influences on Media

			A. Newspapers			B. Magazines			C. Radio			D. Television		
			Strong	Some	None	Strong	Some	None	Strong	Some	None	Strong	Some	None
1. Information Ministry	MoF	%	9	49	42	9	18	73	28	34	38	27	31	42
		#	3	17	15	2	4	16	9	11	12	7	8	11
	LeF	%	50	44	6	34	56	10	84	10	6	90	6	4
		#	19	17	2	11	18	3	32	4	2	29	2	1
2. Military	MoF	%	3	9	88	4	7	89	3	6	91	3.5	3.5	93
		#	1	3	30	1	2	26	1	2	30	1	1	27
	LeF	%	32	34	34	30	27	43	40	23	37	35	30	35
		#	11	12	12	10	9	14	14	8	13	11	9	11
3. Government Press Club	MoF	%	0	9	91	0	11	89	0	10	90	0	11	89
		#	0	3	29	0	3	25	0	3	28	0	3	24
	LeF	%	13	19	68	7	14	79	10	16	74	14	20	66
		#	4	6	21	2	4	23	3	5	22	4	6	19
4. Newsprint Control	MoF	%	3	6	91	4	8	88	0	3	97	0	4	96
		#	1	2	30	1	2	24	0	1	29	0	1	25
	LeF	%	37	26	37	31	28	41	20	16	64	20	20	60
		#	13	9	13	10	9	13	6	5	19	6	6	17
5. Price-Setting	MoF	%	3	13	84	4	8	88	14	7	79	17	9	74
		#	1	4	27	1	2	22	4	2	23	4	2	17
	LeF	%	21	27	52	16	25	59	11	18	71	19	14	67
		#	7	9	17	5	8	18	3	5	19	5	4	18
6. Licensing Media	MoF	%	0	9	91	0	8	92	35	17	48	42	21	37
		#	0	3	29	0	2	24	10	5	14	10	5	9
	LeF	%	47	17	36	47	18	35	61	3	36	59	3	38
		#	16	6	12	15	6	11	19	1	11	17	1	11
7. Licensing Journalists	MoF	%	3	9	88	0	15	85	6	6	88	7.5	7.5	85
		#	1	3	28	0	4	22	2	2	26	2	2	23
	LeF	%	29	18	53	29	22	49	29	18	53	32	13	55
		#	10	6	18	8	6	13	9	6	17	10	4	17
8. Restrictive Laws	MoF	%	3	21	76	4	19	77	9	18	73	14	24	62
		#	1	7	26	1	5	21	3	6	24	4	7	18
	LeF	%	38	38	24	39	42	19	51	18	31	48	17	35
		#	13	13	8	13	14	6	17	6	10	14	5	10

power over the media. The information ministry may be an innocuous distribution point for government handouts on better farming methods, and the comings and goings of local officials and foreign visitors. Or the ministry may be the choke point for controlling the domestic and foreign press. The ministry may determine which journalist is permitted access to information, and which bureaucrat is authorized to provide it. Through such controls the ministry may mount campaigns for or against certain policies or individuals, domestic or foreign. The military often serve clear-cut censoring functions. Less frequently, the government "press club" provides an old-boy network that includes journalists and department bureaucrats. They discuss pending developments, sometimes those which officials do not wish to be publicized. As good "club" members journalists will generally comply. This tool has been finely honed in Japan. Though journalists there are wildly competitive, and the news media are generally free, the press clubs associated with ministries practice self-censorship.

Governments also rely for power on the control of newsprint, setting the price of newspapers, licensing the media or individual journalists, or passing a wide variety of restrictive laws.

Only three (9 percent) MoF countries report strong influences on newspapers from information ministries, but fully half the LeF countries are strongly influenced. Half the MoF, however, show "some" ministerial influence. Such influences are three times as strong in the broadcast as in the print media of MoF countries. In LeF countries ministerial pressures are almost universal.

The military provides little influence on newspapers in MoF nations. Israel was an exception even before the 1988 uprisings in Gaza and the West Bank. In fully two-thirds of LeF countries, however, the military influences the content of newspapers, magazines, radio and television.

Government-associated press clubs are a factor in only a small percentage of the countries. Their strongest influence occurs in some LeF places, but the most sophisticated use of this system appears to be in Japan; a variation, "the lobby," serving as an unofficial leaking mechanism, is traditional in Britain.

Strong controls over newsprint or broadcast materials are practiced infrequently in MoF countries. An exception is India, with substantial political influences over both the price of, and access to, newsprint. More than one third of LeF countries, though, report strong or some influences on all the media from governmental controls over newsprint or broadcast materials.

Similarly, government price-fixing directly influences the size of the audiences for news media. Only one MoF country reported strong governmental price-setting practices affecting newspapers, but four MoF nations influence the price of broadcast materials. Half the LeF countries influence the cost of newspapers and magazines. Broadcast media are affected somewhat less because they are generally government-operated.

Licensing of newspapers and magazines, though unevenly practiced in MoF countries, does not appear to influence publication content strongly anywhere among the MoF. Broadcast media, even in MoF countries, however, are strongly or somewhat affected by licensing practices in nearly two-thirds of the places. Nearly half the LeF suffer strong influences over their newspapers and magazines from governmental licensing of these media.

The licensing of individual journalists—the government determining who is or is not a journalist, and often monitoring his/her practices—is a growing procedure in Latin America, despite the 1985 unanimous ruling of the Inter-American Court of Human Rights. It concluded that licensing of journalists is a violation of hemispheric human rights covenants and the Universal Declaration of Human Rights. Licensing of journalists has some influence on the content of newspapers in 12 percent of the MoF countries sampled and nearly half the LeF countries. Magazines were similarly influenced in half the LeF, but only marginally affected in the MoF nations. Journalist licensing influences the broadcast media in nearly half the LeF countries, but only 12 percent to 15 percent of the MoF.

Restrictive laws of many kinds—among them, outright censorship, official secrets acts, closure of certain geographic areas to press coverage, limitations on press travel—influence newspapers and magazines in fully 25 percent of the MoF countries (MoF broadcasters are slightly more pressured by laws). LeF governments reverse the percentages. About 75 percent of the LeF apply legislation to affect the content of news media. (See Table 2.)

Informal governmental pressure

In addition to the formal governmental influences on the media there are the subtle or informal pressures: favoring friendly journalists, penalizing critics, or threatening them with physical harm or loss of employment. Such negative pressures also cast a pall over journalists that stimulates wider self-censorship. While 13 percent of MoF countries report the favoring of friendly newspaper people by governments, almost no strong use of the negative informal pressures is indicated in MoF nations. They are not entirely above using these subtle pressures, however: "some" MoF governments threaten newspaper journalists (3 percent) or penalize unfriendly press men or women (18 percent). The LeF countries, however, show strong use of these pressures one-third to two-thirds of the time directed

How the News is Reported and Distorted

III. Informal Governmental Influences on Media

			A. Newspapers			B. Magazines			C. Radio			D. Television		
			Strong	Some	None	Strong	Some	None	Strong	Some	None	Strong	Some	None
1. Favor Friendly Journalists	MoF	%	13	62	25	13	47	40	9	64	27	16	58	26
		#	4	20	8	4	15	13	3	21	9	5	18	8
	LeF	%	65	30	5	57	33	10	74	21	5	77	13	10
		#	24	11	2	19	11	3	25	7	2	23	4	3
2. Penalize Critical Journalists	MoF	%	3	18	79	3	17	80	3	19	78	7	20	73
		#	1	6	26	1	5	23	1	6	25	2	6	21
	LeF	%	40	43	17	33	45	22	44	32	24	54	23	23
		#	15	16	6	11	15	7	15	11	8	16	7	7
3. Threats to Harm Journalists	MoF	%	0	3	97	0	3	97	0	3	97	0	3	97
		#	0	1	32	0	1	29	0	1	32	0	1	30
	LeF	%	32	32	36	28	34	38	26	37	37	26	37	37
		#	11	11	12	9	11	12	8	12	12	8	11	11
4. Threats to Deprive Journalist Jobs	MoF	%	0	13	87	0	4	96	3	6	91	3	7	90
		#	0	4	28	0	1	27	1	2	29	1	2	27
	LeF	%	42	29	29	40	28	32	52	21	27	48	26	26
		#	15	10	10	13	9	10	17	7	9	15	8	8

IV. News "Management" Practiced by Government

			A. Newspapers			B. Magazines			C. Radio			D. Television		
			Strong	Some	None	Strong	Some	None	Strong	Some	None	Strong	Some	None
1. Limiting Access	MoF	%	9	59	32	7	38	55	12	52	36	14	46	40
		#	3	20	11	2	11	16	4	17	12	4	13	11
	LeF	%	58	37	5	53	44	3	43	37	20	44	34	22
		#	22	14	2	17	14	1	15	13	7	14	11	7
2. Selective Leaking	MoF	%	12	62	26	14	45	41	12	50	38	23	50	27
		#	4	21	9	4	13	12	4	17	13	7	15	8
	LeF	%	39	50	11	25	56	19	35	47	18	34	50	16
		#	15	19	4	8	18	6	12	16	6	11	16	5
3. False Information	MoF	%	3	21	76	3	18	79	6	14	80	6	17	77
		#	1	7	26	1	5	23	2	5	28	2	5	23
	LeF	%	19	46	35	22	38	40	29	38	33	28	44	28
		#	7	17	13	7	12	13	10	13	11	9	14	9
4. Favorable Information	MoF	%	17	57	26	10	41	49	18	59	23	27	47	26
		#	6	20	9	3	12	14	6	20	8	8	14	8
	LeF	%	61	29	10	55	36	9	68	18	14	75	16	9
		#	23	11	4	18	12	3	23	6	5	24	5	3
5. Denunciation of Journalists	MoF	%	0	38	62	0	31	69	0	24	76	0	23	77
		#	0	13	21	0	9	20	0	8	26	0	7	23
	LeF	%	37	43	21	32	45	23	22	39	39	23	45	32
		#	13	16	8	10	14	7	7	13	13	7	14	10
6. Restricting Size	MoF	%	0	6	94	0	4	96	0	0	100	0	4	96
		#	0	2	31	0	1	26	0	0	32	0	1	26
	LeF	%	20	22	58	13	23	64	10	20	70	10	13	77
		#	7	8	21	4	7	20	3	6	22	3	4	23

to newspaper persons. Broadcasters in MoF countries fare no worse than their print colleagues, but in LeF countries radio and television news persons work under harsher pressure than print journalists. Ninety-five percent of the countries report governmental favoring of friendly journalists (three quarters "strongly"), about half penalizing unfriendly journalists, and a quarter to two-thirds threatening radio or TV people with physical harm or loss of employment. (See Table 3.)

Governmental influence by news management

Governments which set out to influence the news and information channels use many levers other than institutional controls or more subtle, informal pressures. They employ outright news management. This may take the form of limiting journalists' access to government information, leaking of information to selected news carriers, dissemination of false information, denunciation of journalists by officials (intended to reduce their credibility, or cause employers to dispense with their services), or restrictions on the size of the print or broadcast audience.

Nine percent of MoF governments strongly limit access to news, but another 59 percent use this method occasionally to affect newspaper coverage. Radio and TV are strongly affected in 12-13 percent of MoF countries, and somewhat influenced in 43 to 52 percent of MoF places by limiting the access to news. In 95 percent of the LeF countries, print media suffer strong or some limitation of access, with broadcast media slightly less overtly restricted (mainly because they are already government controlled).

Leaking of news to favored journalists in all the media is strongly practiced in an average 15 percent of MoF nations. About half the MoF countries, however, exhibit "some" degree of selective leaking to journalists. Some 80 percent of LeF countries employ this practice (most strongly directed to newspapers, while MoF-country leakers tend to favor TV outlets).

Only one MoF government was reported to employ strongly the practice of disseminating to newspapers or magazines information known to be false. Seven other MoF countries (21 percent) were said to employ this practice some of the time. Two MoF governments were reported to direct information known to be false "strongly" to the broadcast media. About two-thirds of the LeF countries, however, were reported sending to newspapers known-false information (20 percent of these as a "strong" practice). The broadcast media in LeF (as in MoF) countries were more strongly targeted than the print press to disseminate false information.

The other side of the coin, the governmental use of favorable information to influence the media, is practiced in three-quarters of the MoF countries. Print and broadcast media are about equally targeted for this purpose. About 90 percent of the LeF governments disseminate favorable information to influence all the media.

Denunciation of journalists to destroy their credibility or their jobs is not strongly employed in any MoF country, though between seven and thirteen MoF nations report "some" use of such practices, directed to television and newspaper journalists respectively. From 60 to 80 percent of LeF governments denounce journalists. One-quarter to one-third of these countries "strongly" use this practice.

Restricting the size of a publication or limiting broadcast equipment is an indirect form of censorship. No MoF country "strongly" uses this technique, though two MoF nations impose "some" size restrictions on newspapers. One limits broadcast media in this fashion. About 44 percent of the LeF countries restrict the size of newspapers to influence content. Magazines are slightly more targeted in this manner. Broadcast media, already under governmental influence, fare better. (See Table 4.)

Harsh measures to control news

It would be hard to persuade journalists who are victims of institutional, informal, subtle or clever news-management techniques practiced by governments that these are "soft" measures. Yet there are harsher news-control techniques—far harsher—directed with the same intent. These "censorship" techniques are:

Murder of journalists.
Imprisonment of journalists.
Harassment of journalists.
Banning or closing of print or broadcast media facilities.
Confiscation of newspaper or magazine plants, or broadcast facilities.

Murder was found to influence the content of newspapers somewhat, but never strongly, in two MoF countries (6 percent), and "somewhat" but not strongly in five LeF nations (14 percent). Killing of journalists was not a factor in pressuring broadcasters in MoF countries. Yet, thirty-eight journalists were killed in 1988, eighteen in free countries. Drug traffickers and insurgents, not governments, mainly accounted for these deaths. In three LeF nations (8 to 10 percent) murder influenced news content. For all the victims, however, murder was the ultimate censorship.

Imprisoning journalists was not a strong factor in influencing any of the media in MoF countries. In only one MoF place was jail said to be "some" influence, and only on newspapers. In 58 percent of LeF countries, however, newspapers were influenced by the imprisonment of journalists (in 22 percent "strongly" influenced). The broadcast media in LeF countries were only half as targeted as the print press in those countries.

V. Harsh Measures to Control News

			A. Newspapers Strong Some None	B. Magazines Strong Some None	C. Radio Strong Some None	D. Television Strong Some None
1. Murder of Journalists	MoF	%	0 6 94	0 0 100	0 0 100	0 0 100
		#	0 2 33	0 0 28	0 0 32	0 0 29
	LeF	%	0 14 86	0 16 84	0 8 92	0 10 90
		#	0 5 30	0 5 27	0 3 31	0 3 27
2. Imprisonment of Journalists	MoF	%	0 3 97	0 0 100	0 0 100	0 0 100
		#	0 1 35	0 0 30	0 0 34	0 0 31
	LeF	%	22 36 42	16 38 46	15 11 74	11 9 80
		#	8 13 15	5 12 15	5 4 25	4 3 28
3. Harassing of Journalists	MoF	%	0 20 80	0 13 87	0 9 91	0 16 84
		#	0 7 29	0 4 26	0 3 31	0 5 26
	LeF	%	27 46 27	19 59 22	18 30 52	10 34 56
		#	10 17 10	6 19 7	6 10 17	3 10 16
4. Banning, Closing of Media	MoF	%	0 3 97	0 0 100	0 0 100	0 3 97
		#	0 1 35	0 0 30	0 0 34	0 1 30
	LeF	%	26 37 37	13 48 39	18 15 67	10 14 76
		#	9 13 13	4 15 12	6 5 22	3 4 22
5. Confiscation	MoF	%	0 0 100	0 0 100	0 0 100	0 0 100
		#	0 0 36	0 0 30	0 0 33	0 0 31
	LeF	%	14 20 66	9 25 66	15 3 82	14 3 83
		#	5 7 23	3 8 21	5 1 27	4 1 24

VI. Private Ownership Influences on News Content

			A. Newspapers Strong Some None	B. Magazines Strong Some None	C. Radio Strong Some None	D. Television Strong Some None
1. Owner's Order	MoF	%	31 50 19	27 46 27	12 68 20	20 57 23
		#	11 18 7	8 14 8	5 27 8	6 17 7
	LeF	%	50 20 30	39 39 22	45 10 45	45 5 50
		#	13 6 9	11 11 6	10 2 10	9 1 10
2. Special Interests	MoF	%	6 58 36	3 55 42	3 66 31	7 58 35
		#	2 21 13	1 17 13	1 23 11	2 18 11
	LeF	%	38 44 18	34 48 18	39 32 29	48 31 21
		#	13 15 6	11 16 6	12 10 9	14 9 6
3. Advertisers	MoF	%	3 50 47	14 52 34	8 47 45	13 55 32
		#	1 18 17	4 15 10	3 17 16	4 17 10
	LeF	%	14 36 50	21 29 50	15 15 70	13 16 71
		#	5 13 18	7 10 17	5 5 23	4 5 22
4. Mergers	MoF	%	3 9 88	4 4 82	3 6 91	3 11 86
		#	1 3 30	1 1 26	1 2 30	1 3 25
	LeF	%	9 17 74	6 16 78	7 11 82	7 11 82
		#	3 6 26	2 5 25	2 3 22	2 3 22
5. Competition	MoF	%	14 54 32	14 52 34	9 47 44	10 37 53
		#	5 19 11	4 15 10	3 16 15	3 11 16
	LeF	%	6 53 41	10 42 48	0 28 72	3 19 78
		#	2 18 14	3 13 15	0 9 23	1 6 24

VII. Self-Censorship and Other Pressures

			A. Newspapers Strong Some None	B. Magazines Strong Some None	C. Radio Strong Some None	D. Television Strong Some None
1. Security Restrictions	MoF	%	0 15 85	0 11 89	0 9 91	0 14 86
		#	0 5 29	0 3 25	0 3 30	0 4 25
	LeF	%	44 36 20	47 41 12	53 21 26	63 13 24
		#	16 13 7	15 13 4	18 7 9	19 4 7
2. National Dev. Needs	MoF	%	3 36 61	0 17 83	6 32 62	7 30 63
		#	1 13 22	0 5 24	2 11 21	2 9 19
	LeF	%	28 64 8	30 61 9	29 59 12	37 50 13
		#	10 23 3	10 20 3	10 20 4	11 15 4
3. Self-restraint by Journalists	MoF	%	3 72 25	0 47 53	3 65 32	3 68 29
		#	1 26 9	0 13 15	1 22 11	1 21 9
	LeF	%	49 49 2	43 45 12	49 24 27	57 23 20
		#	17 17 1	14 15 4	16 8 9	17 7 6

Harassing newspersons is more prevalent than jailing them in MoF countries. Though governments are usually the transgressors, drug rings or political terrorists often harass journalists. (Harassment may take many forms: sending threatening mail, beating, kidnapping, charging a crime without imprisoning, etc.) Though no MoF place strongly does, 20 percent of MoF nations "somewhat" harass newspaper persons; radio (9 percent) and television (16 percent). Strong harassment of print press journalists occurs in 27 percent of LeF countries; radio (18 percent), TV journalists (10 percent).

Temporary banning or shutting down print or broadcast facilities has an immediate impact on news coverage. No MoF country uses this as a major weapon, but one employs it "somewhat" against newspapers and television. In 63 percent of LeF countries, however, banning is strongly used against newspapers and magazines; in one-third to one-fourth of the places, respectively, against radio and television.

Confiscation is the permanent censoring weapon. No MoF country uses this tactic to influence news and information media. But 14 percent of LeF nations resort to newspaper confiscation "strongly." Another 20 percent confiscate press facilities somewhat. (See Table 5.)

[For a complete enumeration of physical and other harsh attacks on journalists, world-wide, in 1988 see the Journalism Morbidity Table, Appendix A.]

Private-ownership influences on news content

Not only governments or terrorists seek to influence the content of news media and the journalists who operate them. Owners and managers of news media, advertisers, and other free-market forces influence cut access to newspapers and broadcast channels. Codes of ethics of journalists and publishers call for a distinct separation between the advertising and editorial departments of a newspaper or broadcast facility, and between the owner/manager and the journalists. As a matter of high ethic, the writer or editor should be free of pressure from a publisher or advertiser to satisfy commercial, political or ideological interests. How real is this separation?

One-third of the MoF countries report strong publisher's influence over the content of newspapers. Another 50 percent of MoF nations show "some" such influence on the press. Half of LeF countries show strong publisher influences, another 20 percent, some such pressures. In about three quarters of the MoF nations broadcast media feel some ownership pressures on editorial content.

Special interests—corporate, labor, commercial, etc—tend to influence editorial content of all the media. Perhaps not surprisingly, such pressures are greater in the LeF countries. There, the print media are financially weaker, and more vulnerable. In 38 percent of LeF countries newspapers are strongly influenced by special interests; in another 44 percent they are somewhat pressured. In only two MoF countries (6 percent) is the press strongly pressured by special interests. Weaker special-interest influences on newspapers are felt in 58 percent of MoF places. Broadcasters in MoF countries feel about the same degree of special-interest pressures as the print press.

Advertisers, too, have stronger impact on the press in LeF nations than in MoF countries. This may be due in part to the importance of advertising by governmental agencies. A majority of LeF countries are developing nations with financially weak publications. Governmental advertisers often expect editorial content to mesh with paid public announcements. In 14 percent of LeF countries, then, strong advertising influences are felt in newspapers, 21 percent, in magazines. Another 36 percent report some advertising pressures. Only 3 percent of MoF societies feel strong advertising pressures, but 36 percent record some influence from advertisers.

Corporate mergers of formerly competitive media systems, and growing networking of communicators—particularly through the linking of new technologies—generate fear of monopolization. This phenomenon is visible among government-owned as well as independent communications media. Surprisingly, however, little influence on editorial content is noted in either the MoF or LeF countries. Only 3 percent of MoF nations report strong influences on newspaper, magazine or broadcast content from mergers or networking. Another 9 percent say there are some pressures on the press, less on other media. While merger/networking pressures on the press are stronger in the LeF countries, the percentage (9 percent) is small.

Other forms of commercial competition sometimes influence editorial content of the news and information systems. Market competition on newspapers and magazines is a strong influence in 14 percent of MoF countries, and somewhat weaker in another 53 percent. In 20 percent of LeF countries, the press also faces pressures from market competition.

When all forms of market pressures on the news media are considered, it appears that the journalists and the media in the more-free countries suffer fewer commercial pressures than their less-free counterparts. This may be explained by the general ethic of separation of power—the government versus the independent media—in the more-free nations, the linkage of commercial and political interests, particularly among the elites in the less-free places, and the greater financial strength of more-free media. But the degree of freedom, while higher in the more-free countries, is neither absolute nor without challenge.

In countries where economic, social and political development are primary national concerns, pressure is placed on the mass communication systems to promote development. Such influences are strongly felt by newspapers, magazines and broadcasters in about 30 percent of LeF countries. In another 60 percent of LeF places there is some developmental pressure on the mass media. Virtually no strong pressure to promote development is reported on the print press in MoF countries, though in about one third of the MoF nations some such pressure is directed to the print press. (See Table 6.)

Self-censorship and other pressures

In the face of all these influences—economic, market, political, governmental—the journalist develops a sixth sense. He selects a subject or edits his own material with possible objectors in mind. Call it self-restraint, or, if you will, self-censorship. The LeF countries report that self-restraint is practiced in 98 percent of the places. The phenomenon is strongly practiced in about one half of LeF nations, less strongly in the other half. The practice is by no means limited to the LeF. In 72 percent of the MoF countries "some" self-restraint exercised by newspaper people is reported. The same level of self-control is practiced by broadcasters in MoF countries.

Finally, particularly in LeF countries, there is the pervasive pressure of security restrictions on mass media. In 44 percent of these nations, strong security restraints are felt in newspaper offices, 47 percent in magazines, 53 percent, radios, and 63 percent, television. Weaker security restrictions influence the mass media in most other LeF countries. No strong security pressures are felt on the print or broadcast media in MoF nations, but about 10 percent of these countries report some national security pressures. (See Table 7.)

Control of international news and foreign journalists

We examined the forms of governmental control over international news, and foreign journalists. The thirty-eight less-free countries reported:

Channeling incoming foreign news through a government news agency (thereby controlling what their own people will learn): 28 percent of the countries.

Channeling outgoing domestic news through a government news agency (thereby influencing what the rest of the world will know): 23 percent.

Barring or limiting foreign publications to the general public: 20 percent.

Restricting the movement of, and access to information by, foreign journalists: 18 percent.

Expelling foreign journalists: 13 percent.

Jamming of incoming foreign broadcasts: 10 percent.

Of the thirty-six more-free countries surveyed, only two jam incoming broadcasts and one (not always the same) practices each of the other five controls over international news and foreign journalists.

Countries Examined in Chapter 8

Afghanistan	Kenya
Albania	Kiribati
*Algeria	Korea (South)
Antigua	Liberia
*Argentina	Luxembourg
Austria	Malaysia
Bangladesh	Malta
Barbados	Nauru
Belgium	Nepal
Bolivia	Nicaragua
*Brazil	Nigeria
*Bulgaria	Norway
Cameroon	Panama
Canada	Papua New Guinea
Chile	Paraguay
China	Peru
Colombia	Philippines
Costa Rica	Poland
Cuba	Portugal
Czechoslovakia	St. Kitts-Nevis
Denmark	St. Lucia
Dominica	St. Vincent
Dominican Republic	Senegal
Egypt	*Singapore
*Ethiopia	Solomon Islands
Fiji	South Africa
Finland	Soviet Union
France	Spain
Germany (Fed. Rep.)	Sri Lanka
Ghana	Sudan
*Greece	*Sweden
Grenada	*Switzerland
Guatemala	Taiwan
Guayana	Tanzania
*Haiti	Togo
*Hungary	Tongo
Iceland	Trinidad & Tobago
India	*Turkey
Indonesia	Uganda
*Israel	*United Kingdom
Italy	Vanuatu
Jamaica	Western Samoa
Japan	Zaire
Jordan	Zimbabwe

(*Not included in questionnaire responses)

8.

The Information-Power Struggle
88 Country Reports

THESE ANALYSES OF government/press relationships in 1987-1988 are based on private reports to us from correspondents in eighty-eight countries. The countries are rich and poor, industrialized and developing. They have market, socialist or mixed economies, and political systems and ideologies that may roughly be regarded as democratic (in Western terms) or centralized, following Marxist, military-controlled or authoritarian-nationalist patterns.

Many correspondents specifically requested anonymity. I respect those requests, and appreciate the authors' contributions all the more. The correspondents' reports were prepared toward the end of 1987, covering that year's events. In most cases, however, the reports have been expanded to cover 1988 through December. The country reports in this form are the responsibility of the author of this book, and not of the individual correspondents.

The line graph below the country name indicates, in the judgment of the author of this book, the relative approach to a high standard of freedom by the country's news and information media. The criteria for this standard are the political, economic, social and other pressure points addressed in responses to the questionnaire. They form the basis of discussion of this chapter. It should be repeated here that government ownership of news and information media does not automatically reduce the degree of freedom in a country, unless government control blocks dissent and diverse reporting and commentary (as all too frequently it does). The more free the news media in a country below, the closer the graph moves toward MoF (Most Free).

Afghanistan
[IIIIII_____MoF

This is a country at war. Some 1,000,000 Soviet troops were rotated in and out of the fighting in Afghanistan, according to competent estimates. Until April 1988, when the Pakistan-Afghanistan agreement was signed at Geneva, the country's borders were open only to foreign journalists from countries friendly to the Kabul government. Other reporters entered at their own risk, usually accompanying a group of the mujahedeen resistance. After Soviet troops began withdrawing, 15 May, Western reporters were escorted through the limited Kabul-controlled areas. Most of the country remained under guerrilla domination. Each mujahedeen commander continued to control admission of Western correspondents to the area he dominates. For most of the 1987-1988 period, therefore, the coverage of the war for foreign consumption was strictly controlled either by Kabul or the mujahedeen, with crossing of the lines virtually impossible.

Censorship began with the first newspaper published in Afghanistan in 1912. This newspaper became too outspoken and was replaced by the ruler. From 1965 to 1973 Afghan journalists enjoyed some freedom. During that eight-year period, eleven widely circulated and politically important papers were published. Another nineteen minor publications appeared. These included diverse dailies and weeklies. One was described as "non-party, no ideological information, conservative"; another, "anti-communist, anti-Soviet and strongly Islamic"; others, "non-party, supporter of constitutional monarchy and its processes"; "organ of Progressive Democratic Party, pro-West"; "organ of pro-Chinese communists"; "non-party, Islamic religious oriented, belonged to mujadeheen family"; "no particular trend, commercial."[1]

Since the 1978 coup, however, domestic and foreign coverage of Afghanistan has been strictly controlled by the government in Kabul. The domestic news media became a tool of state for propagating Communist ideology and Soviet objectives. No criticism of the Soviet Union is permitted. The mass media are credible, however, as a reflection of the thinking and policies of the party in power, and of the Soviet Union. Most domestic news is supplied by the Ministry of Information. The chief source of news for the public is the National Radio-TV operated by the Ministry of Communications. Loudspeakers are installed in all the major towns. The Afghan news media have access only to foreign information provided to the state-owned Bakhtar news agency from Tass, Novosti and other Soviet sources. From these, Bakhtar receives only selected items from non-Communist media. Many speeches delivered by Afghan government or Party leaders are presented fully. Western radio broadcasts in domestic languages are generally jammed. As the Soviet withdrawal approached, the press became more subservient to Communist objectives.

The Bakhtar news agency describes itself as committed to "strengthen the friendship of the Democratic Republic of Afghanistan with the Soviet Union." Soviet advisers through 1988 held key posts in the press, broadcasting and telecommunications. The Geneva agreement does not provide for the withdrawal of these or other civilian advisers. On the contrary, a little-noted clause of the agreement (Article II, paragraph twelve) states that the parties shall "prevent the use of mass media" to create "subversion, disorder, or unrest." Any print or broadcast message can thereby be interpreted as objectionable, leading to demands for the banning of the communicators and their facility.

The training of Afghan journalists in most of 1988 was conducted in the USSR, East Germany and particularly through the International Organization of Journalists in Prague. A Soviet-installed satellite system stepped up Soviet broadcasts into Afghanistan. Kabul radio, through this installation, can now reach Europe, the Middle East, Pakistan, Iran and elsewhere in Asia. The same satellite system expanded television coverage throughout Afghanistan, and across the border into the refugee camps in Pakistan. The transmissions all strongly support Soviet policies.

In 1985 the U.S. Congress appropriated $500,000 to train Afghans as television cameramen and reporters. Since Western TV crews rarely moved inside Afghanistan, Afghans would be shown how to use the equipment so they could cover the war in their own way. American opponents charged that public money was used to generate one-sided propaganda. Supporters of the project countered that Americans who were contributing $600 million that year supporting the mujahedeen should be able to see what is happening in an area closed to normal press coverage, and open only to reports filtered through the Kabul-Moscow systems. By mid-1988, seventy missions by resistance camera crews on the battlefields produced more than 300 hours of film, and 20,000 photographs and slides, used in 122 countries.

The fates of Western journalists in Afghanistan underscored the dangers. They faced ambushes, bombing, mines, informers, and a Soviet threat to "liquidate" journalists who enter the country "illegally" with the resistance. Since 1982, at least seven journalists were killed and five captured. Two foreign journalists were arrested and sentenced in 1987. The United Nations human rights inquiry in Afghanistan in February 1988 urged that captured newspersons be treated in accordance with U.N. resolution 2673 (XXV) on protecting journalists on dangerous missions. Two journalists, French and Italian, were released as the Soviet troops began withdrawing in 1988.

A personal tragedy and a major casualty in the coverage of Afghanistan was the assassination early in 1988 of Syed Bahauddin Majrooh. He had been dean of the faculty of Kabul University. Shortly after Soviet troops entered Afghanistan, he joined the refugees in Pakistan. There, Professor Majrooh created the Afghanistan Information Center. Amid all the rumor, disinformation and propaganda circulated by all sides of the Afghan war, Professor Majrooh's reports and analyses were (as far as humanly possible) objective and authoritative. This was no small feat in the bitterly-contested arena of Afghan resistance, as well as the Kabul-Moscow entente. The professor was so committed to journalistic integrity that he angered many committed to single-minded domination. Those who gunned him down may have come from Kabul, or an extremist resistance group.

Professor Majrooh contributed to this book, and was a friend of this writer. He was a warmhearted, quiet, modest man, with considerable academic qualifications. He seemed ill-suited to any role outside the academy, hardly on the frontline of a bitter and complex war. We first met in 1980 in Madrid, when we appeared together to urge compliance, particularly by the Soviet Union, with the human rights provisions of the Helsinki Accords (Conference on Security and Cooperation in Europe). That was the first time an Afghan resistance leader had spoken publicly of Soviet soldiers being held by the resistance. That conversation began an eight-year struggle—particularly by Ludmilla Thorne of Freedom House—to persuade the mujahedeen to release Soviet troops who sought asylum in the West. Until then, captured or defected Soviet soldiers were generally killed. Rone Tempest, a *Los Angeles Times* correspondent, provided Syed Majrooh's fitting eulogy: "His death is a serious blow to the sparse information network used by Western diplomats and journalists who monitor the war."

Albania
[IIIIII_____MoF

Journalism has never been free in Albania since the first printing press appeared in 1563. The Romans, Goths, Vandals, Byzantines, Turks, and the Soviet and Chinese Communists saw to that. Since the end of 1944 the Albanian press, radio and television have operated under the strict control of the Communist party. The media serve as the Party's principal instruments of political indoctrination and mass education. The Albanian constitution forbids private ownership of news media. The constitution also supports freedom of the press, but says such freedom may not be used to oppose the socialist order. The media, consequently, are by far the dullest and least informative in all of Eastern Europe. Their tone is preachy, exhortative, patronizing and self-righteous. No form of censorship is necessary. Tight control is held by the party through its agitation and propaganda units. Straying from the party line brings instant dismissal.

Coverage of domestic and foreign news is highly selective and tendentious. Important domestic events such as party and government reshuffles, political trials and purges, and even ordinary crime are hardly ever reported, or if they are published, they are made to serve some specific propaganda purpose. International news is consistently relegated to the back pages of newspapers and consists of carefully selected items, heavily slanted to suit the current official line. Such news is accompanied by crude propagandistic commentary on international affairs. It is not surprising, therefore, that for more than forty years the people of Albania have been ignorant not only of major events in the outside world, but also about developments in their own backyard.

Since the death in 1985 of Enver Hoxha, the long-time Albanian Communist leader, there has been a modest improvement in the press and other news media. This was prompted not by editors and journalists themselves, but by the decision of the country's leaders. They felt the need —a kind of strongly-managed glasnost—to speak more frankly than before about serious economic, industrial, farming, cultural and other national failures and shortcomings. Since 1986, consequently, the press has dealt with some important issues it had completely ignored before.

It has been admitted in the press that many of the country's young people are work-shy, parasitic and violent, and that the young generation is thoroughly bored by the poor quality of newspapers, periodicals, books and leisure activities. There have been many exhortations to improve matters, but only a few signs of better press coverage of many domestic problems.

On the other hand, there has been no change in the press coverage of foreign news. The Albanian Telegraph Agency, an arm of the government, has a monopoly on the distribution of foreign and domestic news. The agency receives news from ANSA (Italy), as well as Tanjug (Yugoslavia), and Hsinhua (China). One significant development in 1987 was Albanian press attacks on Mikhail Gorbachev's reforms in the Soviet Union. They were called revisionist and capitalist in character. It would appear the Albanian leadership wants to avoid at all costs the serious ideological and political risks implicit in Gorbachev's reforms.

Foreign correspondents were not allowed to visit Albania for four decades. This was relaxed somewhat the past two years. Some Western European journalists have been allowed in as tourists. Such visits are generally strictly supervised, but some journalists report a greater readiness on the part of officials to discuss the country's difficulties.

Algeria

The mass riots in Algeria in October 1988 were an explosive indicator of the true nature of Algerian society and its problems. The press, its mode of operation and its relations with the government and the public, did not escape this cruel x-ray.

The Algerian press, which is politically and economically linked to the state that closely controls it, has come through this test in disorder. While press structures remain intact, its convictions have been shaken. Most important, the credibility of the press in the eyes of the public is now a direct function of the press's attitude to the October events. Newspapers and journalists who had the courage and independence to condemn the repression and then take advantage of the new spirit of their professional freedom in Algeria have now gained a new degree of credibility.

There clearly exists in Algeria a great thirst for information. It is the result of the deep silence which has prevailed for more than twenty-five years in Algeria. Listening to the foreign radio, common before 1988, has now become almost an obsession. Articles on the riots from the foreign press, banned in Algeria during this period, were circulated in Algiers in the form of photocopies. Some Algerian papers, especially those which dealt somewhat freely with official statements, have had spectacular increases in sales.

It would be false, however, to exaggerate the impact on the public of the foreign press and the "liberated" or "now being liberated" Algerian press. There is a silent majority which condemns the "October rioters" and which accepts the language of the government and the system, although perhaps without enthusiasm. Government officials and party functionaries were not the only ones to accuse the foreign press—in particular the French press—of "distorting" Algerian reality. Algerian nationalism is strong. Moreover, it often takes the form of Poujadism or chauvinism which do not allow criticism. And the exaggerations and even complete fabrications by some special correspondents supplied excellent examples for spokesmen denouncing the "neocolonialist" press. The Algerian press, nevertheless, like Algerian society as a whole, can no longer be as it was before. Algerian journalists and newspapers must now be seen as two currents.

The *first*, the minority, but becoming stronger every day, consists of journalists who decided to break with a past of which they are ashamed. They have begun to explore the areas of freedom which are open to them, or which they have glimpsed. Their work contributed material to the various commissions of inquiry on torture, arbitrary arrests, kidnappings or massacres in October. Their reflections served as a basis for the "Summary

Report on the Distortions and Defectiveness of Information Related to the Events of October 1988" which was published in November by the Algerian League of the Rights of Man. Several examples of investigations of corrupt officials, policemen guilty of brutality, or abuse of authority were made public. An Oran newspaper (in Arabic) has had considerable popular success by publicizing what some observers called "the first Algerian Watergate": a scandal involving lying and corruption in which local personalities were implicated. An Algiers newspaper (in French) was the first to undertake—and publish—an investigation of a series of kidnappings carried out by policemen. Articles on striking factories and attempts at worker control of factories were published in several papers. In general, in most newspapers the staff discussions are much more lively and journalists hesitate much less when proposing subjects which were completely unacceptable before October 1988.

Even for journalists and newspapers most active in this direction, however, there are obstacles, more cultural than political, that are still very hard to overcome. Everything, for example, that concerns the position of women is practically untouchable. Energetic partisans of democracy and liberty become fierce defenders of the established order when the principle of equality between men and women is mentioned. For the future, there are two positive factors:

1. The weakening of self-censorship by journalists themselves seems to be an irreversible trend which is even contagious. Whenever one publishes an article of investigative journalism which has been freely researched, another piece of the wall of silence has fallen. Particular attention should be paid to the professional pride of younger Algerian journalists who are eager to extend the boundaries of their freedom a little more each day.

2. The opening of new areas of freedom in a press which is still far from free is favorably received by the public. If an article attacks the failings of the system, with examples, the paper is widely bought. As another consequence of the events of October, Algeria during the next few years will move further along a path, if not to capitalism, at least to a relative liberalization of the economy and rehabilitation of commercial competition. Today it is satisfying to be widely read. Tomorrow it will also be financially profitable. This type of motivation for independence should not be overlooked.

The dominant *second* current of the Algerian press is still a tendency to be conformist. Many Algerian journalists who have been the docile favorites of the regime do not have any tradition of press freedom. They have been educated and pampered in a system of the press which has accumulated all the failings of socialist regimes—authoritarianism and Manicheanism, and the failings of the Third World—poverty and nepotism.

Today they are ready to adopt the new attitude just as they docilely transmitted the previous one. They simply drop one dead language and take up another, which is a little more supple and better constructed than the preceding one.

The journalists and papers (the best example of which is *El Moudjahid*) are ready to criticize a functionary, an institution or a decision if the party or the government thinks that the criticism is appropriate or useful. They do not question the system which has generated errors or crimes, and they prefer editorials or readers' letters to investigations or stories that are harder to control. The fact is that the entire Algerian press has openly had to face its failings and responsibilities. Should it speak or be silent? Should it let others speak? Should it defend the state or the citizen? The press has come out of the crisis deeply troubled.

There are proposals that a new information code be rapidly formulated. Yet the Algerian press does not need codes but only freedom itself. "Anything could come out of the present effervescence of ideas—either the best or the worst," our correspondent writes.

Antigua
[IIIIIIIIIIIIIIIIIIIIIIIIIIIIIIIIIIIIIIIIIIIIIIIIIIIIIIIIIIIIIIIIIIIIIIIIIIII_____MoF

This country has a free and diverse press—no daily newspaper but four weeklies attached to parties and labor unions—but less independent broadcast services. The government, therefore, dominates the daily news flow through its preeminence in the electronic media.

Among the newspapers, the *Outlet* represents the left-wing Caribbean Liberation Movement whose leader, Tim Hector, is widely believed to be pro-Communist and pro-Libya. The *Trumpet*, the organ of the Antigua Worker's Union, was the voice of ex-Premier George Walter who was removed ten years ago after he required a costly bond for newspapers, and was discredited by widespread corruption. The *Worker's Voice* represents the Trades and Labour Union and is the source of power of the prime minister's family. The *Nation's Voice* is the paper of the governing Antigua Labour party. The main opposition, the United Democratic party, has no newspaper.

In radio, there is the government's Antigua Broadcasting Service, and two private stations, one owned by the ruling Bird family. Similarly, in television there is the government's ABS and the private cable TV company owned by the Birds.

There has been no overt danger to press freedom, but the *Nation's Voice* came under heavy criticism in 1987 for allegedly racist articles against Indians, and for encouraging demonstrations against the government. In the latter case, an inquiry revealed that Vere Bird, Jr. had been manager of construction at the airport while he was also minister of aviation. His father, the prime minister, refused demands for his resignation. Vere Bird, Jr. is now chairman of one radio station and owns cable TV.

The region-wide Caribbean News Agency (CANA) and Radio Antilles provide some news coverage not available from the limited Antiguan services. The greatest impediment to the free flow of news and comment is the absence of a daily press, and the inadequacy of the local newspapers.

Argentina
[||||||||||||||||||||||||||||||||||||||||||||||||||||||||||||||||||||||||||||||||||||||||||||_____MoF

The once globally-distinguished press of Argentina does not yet enjoy the eminence it had before the Peron crackdown. Harsh military rule, during which eighty-two journalists disappeared, ended in 1983. Yet, in 1985, bombings of press and broadcast facilities were frequent. Violence has ended, press freedom prevails, and six major national newspapers reflect a wide political spectrum. Criticism of the government is freely and broadly expressed. Yet the government dominates the country's newsprint industry, a veiled economic threat. Many provinces also have a "right to reply" law. This places the government in the role of mediator between the press and anyone in the public who feels he has been badly reported. While not an immediate threat to press freedom, the concept can be used by government to insist on certain coverage in newspapers. A court in mid-1988, however, cast doubt on the right to reply. It sustained that right in defense of one's character, dignity and honor (as a matter of libel), but not in order to impose ideas on the press.

Clouding the new political democracy, however, is the government's near monopoly on television news, and its ownership of most provincial radio stations. This domination began during the forty years of authoritarian regimes. During the brief military rebellion in 1987, the government was accused by the print press of using its broadcast power for propaganda purposes. During the crisis, government TV urged citizens to go to the Plaza de Mayo in Buenos Aires or the plazas in the provinces to demonstrate support for the government. Officials also appealed to the print press not to report the positions of the rebellious officers. The private radio did, however, carry their views. For that, the radios were publicly criticized by the mayor of Buenos Aires. But, wrote the country's largest newspaper, *Clarin*, this disagreement over coverage reflects the conflicting views over the role of the press in a democracy.

Yet Argentine TV also provided in August 1987 the country's first televized political debate. The basic decision—whether private ownership of broadcast services should be increased—lies ahead. The state owns

40 percent of the radio stations, has franchises on others, and on television channels it does not yet own. Three of four major TV channels in the capital are government-owned, and syndicated throughout Argentina. Clearly, full freedom of news and information cannot be achieved until the issue of government versus private ownership of broadcasting is resolved. Private ownership may not be the sole model for the full freedom of the electronic media. But then clear limitations are needed for the government's use of the media it owns, and formal commitment to provide access to the channels for opposing and competing viewpoints.

Austria
[|||||||||||||||||||||||||||||||||||||||||||||||||||||||||||||||||||||||||||||||||||||_____MoF

Concentration of news power is striking. In a national population of 7.5 million, 40 percent of Austrians read the tabloid *Neue Kronenzeitung;* 16 percent, *Kurier*; and 10 percent, *Kleine Zeitung*. The three papers, published in Vienna, account for three-quarters of the total newspaper circulation. They tend to be conservative in such issues as state-run industries and abortion. *Die Presse*, a "quality" daily, has 2.8 percent of the readership. This situation has been a disappointment to Austrian journalists who hoped for an independent outlet with wide readership. During the 1987 controversies over Austrian President Kurt Waldheim's service under the Nazis in World War II, *Kronenzeitung* staunchly supported Waldheim, and other papers backed the president. Consistently objective and in-depth reporting is found in the weekly newsmagazine *Profil* and, on Waldheim, in the popular weekly *Die Ganze Woche*.

The Austrian Broadcasting Company (ORF) controls both radio and television, but is protected by law from political pressures. Since 1974, "editor's statutes" were enacted to shield broadcast editors and journalists from pressures by ORF management. This was the first European broadcasting system to install this protection. The trustees of ORF include representatives of different political parties, federal and provincial governments, churches, education, the arts, sports, the sciences, and ORF staff. The purpose of this representation is to provide balances of content at several levels. In addition, a council of viewers and listeners, without executive power, provides further oversight from unions, business groups and the general public. This complex structure is designed to provide a balance between social responsiveness and political independence.

The courts generally protect the newspapers as well from political and other interference. Politicians have gone to court, unsuccessfully, when attacked by the media. When a lower court allowed the searching of an editor's office and private house, journalists and publishers associations protested, and a higher court reversed the decision. While a publication may be barred from distribution if it violates the law concerning morality or public security, such actions are extremely rare.

Though the political-party orientation of the press is recognized, the credibility of both print and broadcast media is high. The press is regarded as the political and moral watchdog of the nation's affairs.

Bangladesh
[IIIIIIIIIIIIIIIIIIIII_____MoF

Throughout 1987, disruptions in Parliament—a walkout by the opposition and government efforts to avoid debate on major issues—were coupled with mounting demonstrations in the streets of Bangladesh. This led, in turn, to the declaration of a state of emergency, sweeping arrests, and the closure of several privately owned newspapers. The government of Lt. Gen. H. M. Ershad owns four major dailies in several languages, as well as the Bangladesh News Agency. Three of four English-language dailies in the capital are privately owned, as are most of the thirty other newspapers. Under the Special Powers Act, however, a ban was placed on foreign and domestic journalists. They could only report news that came from the government. Domestic violators faced a three-year prison term; foreign journalists, expulsion.

Four newspapers were banned and seven journalists arrested in mid-May. Another daily newspaper, owned by an opposition-party member, was closed in August. Some twenty-five newspapers carried a blank spot on their front pages one day in November to protest curbs on anti-government news, and the ban on one opposition paper. In December, the local BBC offices were shut down and its correspondent expelled. Government crackdowns continued into 1988, as did political unrest. Four weekly publications were suspended by the government in January and February, and a daily opposition paper shut down indefinitely.

Even when permitted to publish, the print press is under strong pressures from the government. This generated a twenty-four-hour strike by newspapers in March. They protested the government's manipulation of the press during the country's general election. The papers were forced to ignore violence at the polls, during which some thirteen people were killed.

Governmental pressure begins before the first issue is published. The government licenses newspapers, controls newsprint and places government advertisements in the press (about three-quarters of newspaper revenues). Applicants for press licenses are subject to police investigation of their political affiliations and activities. Although the government denies that censorship is practiced, varied forms of press control have been exerted since Bangladesh became independent in 1971.

Perhaps the most persistent form of news management are the nighttime telephone calls from the ministry of information to the newspapers. Examples of "calls" to the press: "Pictures of the Chief Martial Law Administrator (CMLA) visiting the national monument should be placed on the first page...Picture of an adviser should appear on the bottom of page one...Importance should be given to the new CMLA talking to primary teachers, and the headline should be: 'Primary teachers are government servants, and will continue to be so,: Ershad.'..." He was the prime minister. Other "advices" concern stories to be omitted, as well as those to be played down.

While journalists complain about government pressures, officials counter that journalists frequently accept bribes for their coverage. Some journalists use blackmail, charging subjects to withhold publication of negative information. The public as well as officials come to mistrust journalists. In 1985, on the very day when it became publicly known that the editor of one of the largest vernacular dailies had joined the Ershad government, the circulation of that paper dropped precipitately.

Eight cabinet ministers directly or indirectly own a newspaper. Hardly any uncolored news may be found in these papers. They carry mainly press releases, other handouts, and ministry "calls." The government simply fails to distinguish between news and information. It seeks to manipulate both. A case can be made for any government providing information on consumer subjects, agriculture, climate and industrial development, but hardly news analysis and comment. Apart from governmental interference, self-censorship is a perennial factor in Bangladesh journalism.

The loss of credibility of the print media may account for the wide audience for the BBC and the Voice of America. The government owns and operates all radio and television in Bangladesh.

Our correspondent writes: "Despite all these shortcomings and drawbacks, the majority of our mediamen are engaged in service to the nation, as well as the profession of journalism. Journalists in Bangladesh have often been brilliant and inspiring. In the face of diverse trials and tribulations, they have upheld the cause of democracy, social justice, equality and human rights. Even in the darkest hours of pessimism, journalism has been the torch-bearer of hope."

Barbados
[||||||||||||||||||||||||||||||||||||||||||||||||||||||||||||||||||||||||||||__MoF

The news media enjoy a high level of acceptance here, and play a traditional role as defender of the rights of individuals. Freedom of the press is guaranteed in the constitution. This right is generally respected and practiced. Two privately owned daily newspapers, the *Nation* and the *Advocate-News*, play a constructive role in national development but do not hesitate to take issue with the government. The high public regard for freedom of the press has insured the protection of the media from any meaningful interference.

The government does not attempt to control the flow of news and information, though the government-owned radio (three of the five radios are independent), and government television (CBC) are used to ventilate government positions on all issues. Even on the CBC, however, views taking issue with government policies are frequently aired.

Journalists generally practice their profession in an atmosphere of freedom.

Belgium

[IIIIIIIIIIIIIIIIIIIIIIIIIIIIIIIIIIIIIIIIIIIIIIIIIIIIIIIIIIIIIIIIIIIIIIIIIIIIIIIIIIIIII__MoF

Newspapers have been published here for more than 300 years. There is no governmental censorship. Nearly every paper, however, emphasizes opinion and has clear political-party affiliations. These stem mainly from family ownership of the papers. In recent years, financial exigencies and competition from television have caused the newspapers to emphasize less serious matters and provide more reading for relaxation and amusement. There are government prohibitions on publications for undermining "public order," but these are rarely imposed. The Belgian constitution (1830) stipulates that "censorship may never be established [and] no deposit in earnest of good faith may be demanded from writers, editors or printers. When the author is known and is resident in Belgium, the publisher, printer or distributor may not be prosecuted."

The constitution makes clear that only the usual civil and penal regulations binding on all citizens apply as well to journalists. To comply with the constitution, the responsible editor must publish his name and address in every issue. There is competition in the press. Twelve dailies are published in Brussels, six in Antwerp and three in Ghent. Neither newspapers nor journalists are licensed, nor must bonds be posted in advance of publication. Journalists themselves pushed through legislation in 1963 intended to protect the title of those in the news profession. The law specifies that the title of professional journalist applies only to people older than twenty-one with full civil and political rights whose main income comes from journalism. The individual citizen is protected by a right-to-reply ruling (1961). An individual may respond to errors published about him, or statements which impugn his honor. The reply must not be longer than twice the original, and the paper may refuse to publish the response if it is not relevant to the original text, or is contrary to law or in other ways defamatory. If a paper does not publish a legitimate reply it may be fined.

The obligations as well as rights of journalists are covered under law. Obligations are set forth by the professional journalists' association. This includes respecting truth, supporting the free flow of information and commentary, refraining from plagiarism and libel, carrying information from known sources, and avoiding service on behalf of commercial or political interests.

The rights of the professional journalist include confidentiality, access to all sources of information, and the right to refuse to express views opposed to his own convictions or conscience.

The role of broadcasters is more complex. Private radio stations operate with government licenses. By 1988 there were still no privately owned television stations. In radio broadcasting, the French-, Dutch-, and German-language stations are government owned under grants by the Parliament. The board of each station determines programming, but there is no direct governmental control of program content. The programs are supervised by representatives of the main political, language and opinion groups. A government representative sits on the board without veto power.

Belgium has become a major global news center. The country serves as a platform for the European Economic Community, NATO and other major international organizations. Many international press groups, consequently, are headquartered in Brussels. Journalists, particularly foreign reporters, do not need special accreditation. Belgium, therefore, is a European model for press freedom.

Bolivia
[llllllllllllllllllllllllllllllllllllllllllllllllllllllllllllllllllllllll_____MoF

A citizen older than forty would probably consider this the golden era of Bolivian journalism. Since 1982, the country has been a multiparty democracy, and the press has been free. In more than 160 years since Bolivia achieved independence, however, only during a brief period in the mid-1960s did the government permit the newspapers to criticize public policy, and serve as a relatively free voice. The 1980 military coup (after eleven years of earlier military rule) brought an immediate crackdown on the press. Troops were placed in editorial offices, and twenty-five journalists were arrested. Today, journalists and citizens generally exercise freedom without government interference.

Journalists have survived left- and right-wing terrorist attacks (including destruction of a radio station in 1986, and a regional newspaper and a labor radio station in 1987). The present government reversed its predecessor's limitations on press ownership and restrictions on foreign correspondents in Bolivia. All newspapers are privately owned and reflect a broad spectrum of views in a country with some sixty political parties. The government permits full criticism by the media, though officials still have authority under the constitution to impose censorship.

The Organic Statute of the Journalist, enacted in May 1984, requires that practicing journalists hold a professional academic degree issued by a Bolivian university, or granted because of seniority in the profession. Late in 1987, the Bolivian journalists' association expressed concern over the possibility that Parliament might consider certain amendments to the law on press freedom. This, writes our correspondent, was a warning to the legislature that journalists would vigorously oppose even the consideration of restrictive legislation.

Brazil
[IIIIIIIIIIIIIIIIIIIIIIIIIIIIIIIIIIIIIIIIIIIIIIIIIIIIIIIIIIIIIIIIIIIIIIIIIIIIIIIIIIIIIIIIIIIIIII____MoF

In March 1985 the government announced the end of political constraints on Brazil's extensive print and broadcast systems. By late 1988 a new constitution abolished censorship of films and television. For more than twenty years military rulers did not tolerate opposition views in the mass media. Censorship—including the placing of the censor in the office of a major newspaper—also generated self-censorship. At the peak, ninety censors maintained stringent surveillance over Brazil's 3,000 newspapers, magazines and radio stations. Editors were informed by "calls" from the government what they could not print. Though few newspapers were subjected to prior censorship, the possibility altered coverage. That ended dramatically in 1975 when the country's leading newspaper, *O Estado de Sao Paulo*, received a phone call from the censors saying, "We won't be coming any more." The paper had frequently defied the government and become a regular defender of press freedom. *O Estado*, as other newspapers, receives lucrative advertising or printing contracts from the government. Many journalists draw paychecks from "add-on" jobs. Yet investigative, and politically independent, journalism is emerging.

Just as *O Estado* is one of the leading newspapers in the hemisphere, Globo TV is Brazil's (and one of the world's) largest television network. Both are vigorous, independent voices with considerable power in their country. Newspapers are privately owned and vigorously question government policies and performance. Radio and television stations, with the exception of two, are also privately owned.

Globo TV draws ratings of 70 to 80 percent of the viewing audience. The private owner wields information power for clear political purposes. He generally supports the administration, but also presses for improvements, as he sees them. Reporters and editors of Globo are attuned to the owner's views. Globo owns seven television channels in major cities, partly owns six others and has thirty-six other affiliates.

Bulgaria

This author visited Sofia in 1985 and lunched with the president of the Bulgarian Union of Journalists, a close associate of Bulgaria's president. The union leader is also the editor of the leading daily newspaper *Rabotnichesko Delo*. He insisted that journalists here are "free" to write and edit, and even criticize the ministers. They do, as part of "guided journalism," which is used further to support the monitoring of government works and focusing on low-level inefficiency. He said there are only three legal prohibitions for a journalist: he may not espouse war, chauvinism or racism. "In thirty years since that law was promulgated," he said, "not a single journalist has been punished for these reasons." That does not account for self-censorship, or removal for "other" reasons.

Since then, there has been published criticism of the media by the media, though all are government owned and operated. The constitution provides severe punishment for anyone convicted of criticizing the state or spreading "untruthful remarks which might increase distrust of state power." Nevertheless, the Communist party daily *Rabotnichesko Delo* published an editorial on 16 September 1987, accusing the trade union daily *Trud* of "smugness and self-promotion," and a lack of self-criticism. This was only three days after *Rabotnichesko Delo* strongly criticized Bulgarian television. *Trud* had generated the angry criticism by publishing a poll showing that it received more letters from readers than the larger daily; particularly, that *Trud*'s readers complained they found it difficult to buy the paper. The implication: *Rabotnichesko Delo*'s large readership may be due to government-controlled distribution. *Trud*'s editor told the journalists' association earlier that circulation should not be restricted and "readers' demands" should govern distribution of newsprint. Set circulations, he said, caused journalists to lose initiative and interest in improving their work, prevented "noble competition," and resulted in stagnation.

This is Bulgaria's version of Soviet glasnost. The Party daily's criticism of government television and its labor paper suggests an internal conflict over glasnost not unlike that in the USSR itself. Similarly, restructuring (perestroika in Soviet terminology) links media power with larger state power. The Party paper in September underscored the struggle by stating that "the bureaucracy understands that the principles of restructuring can be held up, at least temporarily, exactly at the cost of taking away glasnost." "Bureaucracy," here, refers to negative attitudes that hamper progress.

As a gesture to its northern neighbor's glasnost, Bulgaria at the end of 1988 stopped jamming Radio Free Europe, the American government's broadcaster. RFE could then be heard in every country of the Eastern bloc.

Cameroon
[IIIIIIIIIIIIIIIIII_____MoF

The 1972 constitution guarantees freedom of the press, yet under Cameroonian law and practice freedom is sharply restricted.

Press censorship prevails for private publishers. When an issue is ready for publication, a copy of each page must be read first by the government censor. The ministry must stamp each page for approval. Printers cannot print the material without the stamp. After printing, copies must be returned to the ministry for final approval and another stamp. The former practice of leaving white space where censored material was removed is no longer allowed.

Not subject to overt censorship, but to clear governmental strictures, is the principal newspaper, the government-owned *Cameroon Tribune*. As elsewhere in developing Africa, the media here have set roles. Independent newspapers provide a forum and play a watchdog role though they cannot undertake investigative or adversarial journalism such as the *Nigerian Punch* can. Through satirical articles, papers here sometimes highlight corrupt and unethical practices, and act as gadflies which prick the Cameroonian conscience. They also provide a forum for discussion of some issues even if they are just flights into academic fantasy. The *Cameroon Tribune* is mainly a government gazette. It publicizes government decisions and polices, and, whenever the need arises, rationalizes them. Most lead stories focus on top government officials. Arguably, therefore, the *Tribune* is the authoritative paper, and a major organ of government publicity and propaganda.

Television broke new frontiers when it was introduced in 1986. As elsewhere in the developing world, television is an urban phenomenon. Owing particularly to its prohibitive cost, access to it is restricted to the urban elites in the major towns such as Douala, Yaounde, and Ebolowa. The focus, too, is on the activities of government elites in the major cities.

The transistor radio is the close companion of most Cameroonians in both rural and urban areas. Cameroon's mass medium, twenty-seven years after independence, is radio. The price is low. It is portable, and regional stations carry programs in local languages. Radio broadcasts announcements of births and deaths, and entertainment programs with wide appeal.

News is another matter. Whether in radio or the press, the major role of the media is telling what government officials say or do. News is "la raison d'état." Typically, news coverage will begin with "the head of state..." and at the provincial level with "the governor of..." Inextricably linked to the news, and the focus on political and bureaucratic elites, is the phenomenon of praise-singing. In fact, Cameroonian media men, like most others in Black Africa, are more or less modern-day "griots." They used to recite the oral history of the tribe. Today, they specialize in lavishing praise on those in the corridors of power. Special programs are broadcast extolling government leaders and authorities. The media in Cameroon specialize in praise-singing.

Since Cameroon is less than thirty years old, its emphasis on nation-building is understandable. A major role of the mass media is forging national unity and concord among more than 250 ethnic entities which owe loyalty to their tribal grouping rather than to the nation-state. The government is, therefore, highly sensitive to everything broadcast.

Sports and music dominate radio and television. Unlike his Nigerian neighbor, who will spend his last koboo on a newspaper, the Cameroonian may use it for music or beer. Sports and music are helping to unify this diverse country. Makossa—the Cameroonian pop beat, widely broadcast—has created a "we-feeling," a sense of common identity and belonging. Wherever in the world you meet a Cameroonian, the sound of Makossa produces a feeling of nostalgia, and the national identity grips him. Some cynics, however, attribute this excessive emphasis on sports and music to a political strategy. They argue that the political elite has stifled free speech and substituted the fanatic promotion of sports, music and drinking. This tactic, they say, diverts public attention from crucial political and economic discussions. The mass media, in this view, are a narcotic.

Apart from entertainment and public announcement programs, the media's excessive focus on government issues and personalities raises credibility questions among the public. People are generally skeptical of what they hear over state radio and television. Consequently, papers that employ satirical styles in Francophone and Anglophone Cameroon have a wide audience. In such a situation, rumors run wild.

In Cameroon, the journalist works in a closed society, in contrast to nearby Nigeria where citizens speak and comment freely. Journalists fear harassment and incarceration. They can readily be arrested and detained. Self-censorship is the first professional impediment. Consequently, journalists carefully limit what they broadcast and report. Punishment of colleagues has taught them lessons in cautious journalism. The government monitors their activities closely. Before the news is broadcast, a top Ministry of Information official must vet it. If a broadcaster takes a unilateral decision or adds a piece perceived as inimical to the State's interest, he could be suspended immediately. Within the media, especially those government-owned, editors instruct subordinates on certain angles to be stressed or issues to be avoided. Outside the media, government officials and security agents watch carefully. At provincial levels, government censors must approve articles before publication and can order stories killed. Journalists who doggedly pursue an independent path can be arrested, detained and ruthlessly manhandled by law enforcement officers. Legally, ad hoc laws can be invoked by officials to arrest or detain journalists.

Three radio journalists were detained and charged with subversion in 1986. The government can ban a paper, as it did the *Cameroon Times* in 1985. The *Times* persisted in publishing investigative and interpretive reports on national issues, and the popular Anglophone paper was closed. Most Cameroonian journalists, even those working for the state, would like to be as daring as their Nigerian counterparts. However, even economic sanctions can be imposed on them. The private media can be refused import licenses for the purchase of ink, paper and other printing materials. Those working for the government can have their salaries slashed or their jobs ended. With few alternatives in a poverty-ridden environment, most Cameroonian journalists tend to toe the government line. Yet, 1987 was a comparatively better year for journalists than the previous twelve months. In 1988, however, one journalist was killed in line of duty.

Canada

There are very few impediments to the effective work of the news media in Canada, except rather stringent libel laws and a tradition of journalists not digging too deeply, and government not being required to reveal much.

Television and radio are subject to regulations of the CRTC—Canadian Radio and Television Commission—which is in charge of licensing and content. Stations making application before it are expected to try to live up to certain commitments such as 60 percent Canadian content (primarily offsetting the influences from the United States), news balanced with music, and other factors. Television is stricter, with two private networks, Global and CTV, competing against the state-financed network, CBC-TV. Private radio stations are more numerous and not in direct competition with CBC Radio, which is state-financed. Because Canadians generally prefer to watch U.S. programming, laws are passed to favor the presentation of Canadian shows: cultural nationalism. There is clearly a tension between the ideal of an unrestricted press, and fear of dominance by U.S. media; between a libertarian press, and government's aim to protect and stimulate national media.

Our correspondent writes:

CBC (radio and television), without dependence on market preferences, can program without regard to audience choices. Critics say the radio tends to attract leftwing ideologues on its staff, particularly among its producers and editors. The CBC has token programs periodically exposing Soviet violations of human rights, but it presents a constant, almost daily, flow of negative programs about Chile, Central America, South Africa, etc. Cuba is generally praised for literacy and health. Guests and commentators seem chosen for their similar orientation. Yet extreme conservatives find it easier to get exposure on the CBC than moderate conservatives, partly, one suspects, because extremism of any sort offends Canadians and turns them against the cause espoused by extremists. Communist party officials rarely get hearings on the CBC, though Marxists often do. The CBC bias affects most radio and TV outlets in Canada, which tend to copy rather than initiate.

Newspapers are independent, but conservative columnists, commentators or editors are tolerated as aberrations or tokens. Owners may think of themselves as conservatives, but their papers, generally, uncritically accept a conventional liberal line.

Advertisers do not use economic power to influence content; big business may try, but rarely succeeds. Politicians influence papers by demanding equal time. The Canadian media are generally respectful of authority. Racial or minority groups have considerable influence on all the media. There is a tradition of journalistic independence, so the media insist, when they succumb to pressure, that they do so of their own free will.

The media push for enlarged freedom-of-information legislation, which they think would force governments to reveal all. The government agrees, but does little.

Canada has an official secrets act which is almost identical to Britain's, but unlike Britain, it has been used only once to prosecute journalists and a newspaper (Peter Worthington, his publisher and the *Toronto Sun*, 1978). It is still on the books, however, and serves as a deterrent. The obscenity and seditious-libel laws have a similar effect. Prior restraint is rare, but denying access to publishable material is more common. Even a freedom of information act will not be effective unless the Official Secrets Act is modified. Several major cities have voluntary press councils. They use the power of reprimand to spotlight irresponsible journalism.

Canada is a country of one-newspaper towns—most of them profitable monopolies, most owned by the Thomson or Southam chains. The only cities with two separately owned English-language (as distinct from French) newspapers are Toronto, Calgary, Edmonton, Winnipeg and Ottawa—where the *Sun* provides the opposition. Toronto is the only four-newspaper city in Canada (There are only two in North America!).

The newspapers are usually docile, and oriented towards domestic politics and liberal-left causes. The French media in Quebec are perhaps even more political and ideologically to the left (separatist) than the English. French journalists are more likely than the English to accept a guiding rather than an informing function. Pierre Trudeau, when he was prime minister, accused the CBC of being a haven for ideologues and separatists trying to undermine Canada. Several times in the past, the RCMP Security Service found Soviet KGB links leading into the CBC. But when raised in Parliament this is usually refuted by accusations of McCarthyism.

Chile
[IIIIIIIIIIIIIIIIIIIIIIIIIIIIIIIII_____MoF

At 21:30, 7 October 1988, a half-million Chileans crowded Avenida Alamada to celebrate the victory two days earlier over the rule of General Augusto Pinochet. The celebration, like the earlier plebiscite, had been undisturbed by police or the military. Suddenly, on the evening of the seventh, the special forces under Gen. Pinochet's Interior Ministry swooped down on journalists, mainly photographers. We witnessed the result: wounded, beaten, bloodied journalists met to tally the injuries. White-faced, some still in shock, they checked on colleagues taken to hospitals. Twenty-three, most of them foreign, were injured, four hospitalized. Cameras were shattered. Our formal protest at 7:00 A.M. the next day noted that we had heard rumors earlier that week that once the plebiscite ended just such an assault would target journalists. We discounted the rumor when the police and military acted with great professionalism, indeed obvious courtesy, during the voting 5 October.

The plebiscite was remarkable because of the role television had played. For fifteen years the Pinochet forces controlled all television and radio broadcasts, and influenced the press by repressive measures. But for twenty-seven nights before the plebiscite the opposition was given fifteen minutes each evening to state its case. Almost 70 percent of the TV sets were turned to the "No" program. A skillful mix of politics, music and humor damaged many of the taboos created by Pinochet. Wives of "disappeared" prisoners danced the national dance—without their missing partners. As one spokesman of the Command for the No put it, "In fifteen minutes' television time we destroyed fifteen years of government publicity for the dictatorship."

Though all states of exception were lifted in Chile on 27 August in preparation for the plebiscite, constitutional limitations on dissent and the key law used to imprison journalists (making illegal some criticism of the armed forces) remained in force. And, indeed, death threats and harassment of journalists increased as the plebiscite neared. After the plebiscite, the government said it would privatize its television stations. This was apparently a step to sell TV channels to political friends in the event the opposition wins the 1989 election.

At a meeting in Santiago which this author directed in November 1987, one of Chile's leading editors, Emilio Filippi, stated that news media must "maintain their commitment to democracy, without wavering in the face of difficulties," as a daily task; they must "refuse to be tempted, ever, by those who either consciously or unconsciously denature democracy." Democracy, he said, must be "rooted in respect, dialogue, tolerance and truth"—and there cannot be any of this without freedom of the press. He went on to say several days later, "freedom of the press does not exist in Chile, despite the fact that there are communications outlets that exist independently of the government, criticize the way it operates, or even act as its tenacious adversaries."

Existence of these media "does not suffice," he added. "The restrictions now in force frequently expose them to the risk of having to stand trial (usually in the military courts), make them targets of constant discrimination, and continually deny them legal standing." Despite, this, he said, "these communications outlets are regularly paraded forth as proof that Chile enjoys freedom of the press."

It is not possible to establish, edit and operate communications outlets without express permission from the authorities. The Ministry of the Interior, acting on behalf of the president of the republic, grants or denies the requisite permission as it sees fit. The establishment of new publications may be barred by declaring a "state of disturbance of the domestic peace," under Transitory Article 24 of the constitution. It was necessary to appeal to the courts to permit the publication of the daily newspapers *La Epoca* and *Fortin Mapocho*. Though this gave the political opposition two daily newspapers, journalists writing for both publications in 1987 were indicted.

Coverage of activities by the opposition increased since March 1987, but access to other sources of information is limited to the official version. Authorities either do not permit journalistic investigation of their action, or disqualify it without comment. Neither is there openness in the legislative process. The government has established a system of accrediting journalists for access even to official sources of information. The executive branch reserves the right to grant or refuse such accreditation. Weekly papers cannot gain accreditation for Government House. Among the opposition dailies, only *La Epoca* has an accredited journalist and he is bound by discriminatory limitations that prevent him from covering the president of the republic. One journalist assigned to the governing junta was expelled from the area when a report in his paper displeased the chief of the legislative branch. Generally, preference is given to those media considered to be favorable to the administration.

Economic pressures further limit media freedoms. Several major media enterprises are deeply in debt to government financial institutions. This gives the administration some control over the content of reports and commentary. The largest television network is operated by the government, and the high cost of advertising makes it virtually unattainable for opposition political parties. They were given very limited access to the three university channels in January 1988.

Opposition publications and radio stations are often strongly critical of the government, but they face legal harassment. Chile now has thirty-four sets of legal strictures regulating the print and broadcast media. The constitution guarantees "freedom to publish opinions and to report, without prior censorship, in any manner and using any communications outlet, without prejudice to any accountability for crimes or abuses that may be committed in the exercise of these freedoms in accordance with the law. In no case shall the law establish a state monopoly over the mass communications media." In exceptional conditions the press may be restricted, censored or even closed down, as has happened on several occasions during the past fifteen years. But not only the constitution imposes limits on press freedom. There are the Abuse of Publicity Law, the State Security Law, the Arms Control Law, the Anti-Terrorist Law, the Anti-Protest Law and the Military Code of Justice. Under the last, "offenses against members of the armed forces as individuals" significantly expanded the "offenses" and the penalties.

Perhaps the most threatening of recent acts was the legislation gazetted 29 October 1987 which expands censorship to specific information and subject areas. This law No. 18,662 supplements the restrictions in Article 8 of the constitution. Journalists have dubbed this the "new muzzle." (I called the Chilean embassy in Washington, in mid-June 1988, to verify that this law was still in effect. "Yes," I was told, "but it is rarely used." It is sufficient to have it, however, to cast a pall over daily journalism in Chile.)

Article 8 prohibits anyone from propagating doctrines that endanger the family, incite to violence, or promote a totalitarian system based on the notion of class struggle. It not only declares unconstitutional all groups having such objectives but imposes political and civil sanctions retroactively on those who have been so engaged. These sanctions include a prohibition for a ten-year period against serving as a journalist or acting as a manager of any communications outlet. Those who passed the new law believe that Article 8 was mainly rhetorical and that a statute was needed to regulate its application. In effect, the new law revokes the right to hold political opinions if a person has been sanctioned for "unconstitutional" behavior. Moreover, it penalizes those who use communications media to express opinions or impart information concerning groups declared unconstitutional, or concerning representatives of such groups.

During a three-month period in 1987, more than twenty journalists were tried for a variety of political-professional reasons. An unprecedented sentence was imposed on Juan Pablo Cardenas, editor-in-chief of *Analisis* magazine. He was sentenced to a year and a half of nightly imprisonment for supposed offenses against the president of the republic. This may be the only case anywhere of a journalist having to sleep in prison to atone for an alleged political crime. In another case, two editors of *APSI* magazine were detained for more than two months for publishing a humorous supplement that was never distributed because it was seized by the police. The journalists were tried by a military court and detained for sixty days awaiting an unprecedented "psychopolitical" report. The journalists were found innocent, but this procedure serves to punish those considered guilty even before proof has been established.

Under Article 8 of the constitution, editors of four publications were brought to trial for publishing an announcement paid for by the Communist party. Another reporter spent two weeks in jail for publishing an interview with a political leader who alluded to General Pinochet in a way the government deemed injurious. Though the politician accepted full responsibility for his statement, the government insisted the journalist published it with intent to harm. In mid-1988, more than forty journalists faced charges that could lead to prison. At year-end, the government dropped all charges against journalists brought in civilian courts. But some eighteen journalists still faced charges in the more severe military courts.

The assassins of one journalist in September 1987 have never been found. Some journalists have received death threats. A journalism student and a magazine editor abducted in March 1988 were released to convey death threats to other editors, and two journalists from *Analisis* received death threats through the mail.

Emilio Filippi concluded his remarks in November 1987: "We need an informed, conscious democracy which proves its effectiveness in the maturity of its institutions; and we believe that it is absolutely necessary to have communication media for that purpose, media for whom democracy is a creed and which they support to ensure its stability."

One year later, an appellate court upheld the indictment against Filippi and three other journalists.

People's Republic of China
[IIIIIIIIIIIIIII_____MoF

During a visit to China in 1983, I heard the first rumblings of the revision of Marxism to conform to post-Mao ideology. This was glasnost before Gorbachev promoted the term. The impetus was the need to rationalize the ideology with such pragmatic success as had been achieved, the preceding eighteen months, in vastly increasing agricultural production. This had resulted from permitting personal enterprise tied to a limited market system. Yet Marxism was being revised because it was regarded in its current form as not sufficiently in the Chinese tradition; as too Western.

In 1988, "reform" was distinguished from the "revolution," which led to the founding of the PRC. Reform, said the *World Economic Herald* of Shanghai, means dropping the Marxist emphasis on the class struggle. It was "distorted" during the cultural revolution of the 1960s, said the *Herald*, while "developing the economy was largely neglected." In the 1980s, reforms in many areas of Chinese society have been attempted, and occasionally reversed.

During these policy changes, Liu Binyan, the muckraking journalist, was twice expelled from the Communist party, and suffered twenty-one years in labor camps and internal exile. Winds of reform sustained him briefly, and then came attacks on "bourgeois liberalization." Liu believed that changes in the press of the Soviet Union in 1986-1988 were envied in China. "No mainland Chinese newspaper," he said, "has been allowed to match the way the Soviet media boldly expose political mistakes, stimulate discussion of political questions, and explicitly challenge conservative views." The paper for which he wrote, the *People's Daily* of Beijing, was unusually frank and courageous from 1978 to 1980, he said, but later reverted to a more cautious style. Then came attacks on "bourgeois liberalization," starting in January 1987. Journalists, academics and party officials were dismissed from their jobs. Some were detained, and maltreated. One journalist was sentenced to seven years in prison for speaking out. Party General Secretary Zhao Ziyang said in mid-May 1987 that "the corner has been turned on anti-bourgeois liberalization." In 1988 Liu believed that the experience of the cultural revolution had "cured many Chinese citizens of the superstitious belief in personal authority, and freed them from political dogmatism."

The struggle—revisionists versus Marxist traditionalists—continues in the field of information as in all other fields. Liu Binyan was expelled but permitted to travel to the United States for a year. Meanwhile, in November 1987, the *China Legal Journal* proposed a law covering journalists' activities. It recommended legislation to define and protect journalists' rights, and punish violations of journalists' rights and illegal activities by journalists (particularly in an attempt to make profits—such as asking money for "better stories," advertisements in disguised form). The journal stated that "just as corrupt journalists cannot be expected to protect the party and people's interests with their reports, reporters without rights can hardly do well." So, the proposal concluded, "reporters' activities should be legalized in order to consolidate the role of the press in the development of socialist democracy and supervision."

Promulgation of a press law is expected sometime in 1989 or 1990. Meanwhile, defining of news media performance continued. At journalism seminars in Beijing and Shanghai toward the end of 1987, it was agreed that the press should not act as the "mouthpiece" of the Party, but as the "overseer" of both Party and government. The press should help the "supervisory role of public opinion, and support the masses in their criticism of shortcomings in work." This is not unlike the controlled managerial function of glasnost as a critical channel in the Soviet Union.

The reform of journalism should be made the top priority in 1988 of the Xinhua News Agency, China's main international as well as domestic news link, its managers said early in the year. The president of Xinhua expressed a "pressing responsibility to meet the urgent need of China's political reforms" by making the news agency more useful at home and more "respected" abroad.

A poll carried in the party daily *Renmin Ribao*, March 1988, showed that 200 top officials and other leading figures would like to see extensive reforms of the media. Those responding repeated the demand for a press law, and urged more dialogues with readers, discussions involving different viewpoints on current topics, replies from those criticized, and analysis and commentaries on policies and important leadership changes. The poll reported 62 percent of those interviewed saying they were "dissatisfied" or "not very happy" with the Chinese media. They called for more openness, more consideration of public opinion, more responsibility, and more criticism of the Party and government. The press, in a word, should serve as watchdog. This is a concept which press-control advocates elsewhere have attacked as untranslatable, too Western.

Such discussions in China do not guarantee the liberalization of the press, temporarily or for a longer period. The ebb and flow of press reform may continue. Journalists have become accustomed to such zigzagging. Yet the new press law is expected to include rights as well as duties. A test will be whether such reforms will permit investigative journalism without fear of reprisal, the dropping of taboos on the activity of top government leadership, and questions of foreign policy. All such subjects also remain under tight control in the glasnost of the USSR

Yet one can see signs of unusual libertarianism in China. Wrote the daily *Gongren Ribao*, 29 January 1988:

> Freedom of thought is originally a right that every person is endowed with by nature. It is also the most basic and most elementary freedom for all people. If this freedom is not ensured,

we can hardly imagine what other personal freedoms people can have.

Such aspirations must still cope with Chinese reality. Despite the growing ideological support for limited expression of differing opinions, there are pitfalls for extending criticism of government or party beyond certain still-undisclosed boundaries.

Meanwhile, about 600 million Chinese now watch about 120 million television sets. About 200 million Chinese are illiterate. As 1988 began, there were 4,609 satellite-receiving earth stations in China, one-third of which retransmit by cable. China has program exchanges with Japan, Visnews and Asiavision.

In the print field, *China Daily*, published in English, now has 200,000 readers, many in Europe and the U.S. It is simultaneously published in New York, Hong Kong and London. A nongovernmental magazine, *Nexus*, began publishing in 1987. It has English and Russian editions, all wholly supported by advertising. *Nexus*, which publishes 50,000 copies, is the only medium in China not associated with a party or government organization. Officials say "intellectuals" are the publishers.

China had a landmark libel ruling in September 1988. A court in Beijing rejected a factory's plea that the Beijing *Evening News* had caused financial losses by damaging the plant's reputation in a news report. In Shanghai alone, however, some thirty-five libel suits involving twelve newspapers and eight magazines had been handled by the courts in 1987-8. The increase in lawsuits, said *Legal News* of Beijing, stems from a growing public awareness of individual rights. Interestingly, libel suits are regarded as a good omen by foreign observers in China, a bad one by journalists facing libel actions in America.

The greatest test for China's journalists came in mid-May 1989 when 1 million supporters of the student strike jammed Beijing's Tiananmen Square. The government radio and press, momentarily, described the objectives of the prodemocracy movement. The foreign television cameras simultaneously pictured the extraordinary counterrevolution. But then the plug was pulled on both foreign and domestic journalists, and government control restored. For how long, remained the question pondered world-wide.

Colombia
[IIIIIIIIIIIIIIIIIIIIIIIIIIIIIIIIIIIIIIIIIIIIIIIIIIIIIIIIIIIIIIIIIIIIIIIIIIIIIIIIII_____MoF

More journalists are gunned down here than in Lebanon or Afghanistan. They are targeted by the global drug cartel headquartered here, and by political death squads. In January 1988 the government decreed tough anti-terrorist laws under state-of-siege powers. Yet four journalists were murdered during 1988, and death threats were received by another eleven editors and reporters. Nine reporters and editors were kidnapped, and later released. Three journalists were killed in 1987, and others threatened. In four years, the reign of terror has cost the lives of at least fifteen print and broadcast journalists. Terror has had its intended effects. Six whose names appeared on a late-1987 hit list left for exile abroad. Journalists who remain in Colombia face the daily choice of risking death or self-censoring.

In a country with a strong democratic tradition and full press freedom, this is a cruel dilemma. The best-known Colombian reporter faced this test in October 1987: "Tell Daniel Samper that he has twenty-four hours to leave the country," said the message delivered to *El Tiempo*. Other phone calls and a telegram convinced Samper he was being closely watched. He decided he must flee the country, as had many others before him. Colombian journalists are the target of a variety of criminal gangs, Samper says. The Interior Ministry lists 137 rightist death squads in addition to the drug mafia and the leftist guerrilla groups. "It is sad," adds Samper, "that a common question among journalists is, 'Do you know who is threatening you?'" It may be "several of the above," at once, he believes. Kidnapping reporters as hostages further damages the credibility of the profession, says Samper. "If journalists are to become hostages, and the ransom to be paid for their freedom is access to the media, then we have taken a new backward step for Colombian journalism."

Press freedom in Colombia dates from 1979. Before then, government censorship was common. Television news scripts and films had to be submitted to government censors before they were broadcast. Radio stations had to notify the government two days before carrying a political speech. Those years of censorship were known as "la volencia." Ironically, today, without government censorship, violence reigns in Colombian journalism despite the willingness of officials to provide information, and confirm stories secured elsewhere. The press frequently criticizes government leaders with vigor, though often with partisan reportage. All newspapers are privately owned and have some orientation to a particular political party. A new paper, *La Prensa*, appeared in 1988. It is published by Misael Pastrana Borrero, former Colombian president, presumably to advance his interests in the opposition party. The state controls television and leases time to private operators. The government, however, sets guidelines to assure equal time for political candidates. Starting in 1987, all parties, by law, have access to national television.

Since 1975, the Law of the Journalist has, in effect, licensed the profession. Passed under a Liberal government, the law requires that to be accredited a journalist must hold a journalism degree from a school approved by the government or pass an examination on Colombian culture and journalistic practices. A journalist working without a license, and his employer, can be fined. The law also protects journalists from revealing their sources, assures access to public information but provides for cancellation of the license for breaking laws relating to journalism or damaging third parties.

This law can do nothing to protect the journalist from the daily threats of the narcomafia and the guerrillas. These threats, kidnappings and murders not only influence the content of the print and broadcast media—despite the courageous, often heroic, work of reporters—but inevitably reduce the credibility of the media in the eyes of the public. They know that crucial information may be withheld out of well-placed fear. Other reports may be colored by the affinity of a newspaper with a political party. This notwithstanding, our correspondent lists these journals with high credibility, in this order: *El Tiempo, El Colombiano, El Pais* and *El Espectador.*

Impediments to the free flow of information in Colombia are, primarily, the extraordinary difficulties of covering guerrilla-controlled areas and the narcotics system; secondarily, some restrictions in 1988 on access to information from government sources concerning narcotrafficking and the guerrilla war.

Costa Rica
[IIIIIIIIIIIIIIIIIIIIIIIIIIIIIIIIIIIIIIIIIIIIIIIIIIIIIIIIIIIIIIIIIIIIIIIIIIIIIIIIIIIIIIIIIIIIII___MoF

This country is governed by the oldest uninterrupted democracy in Latin America, and freedom of expression and the press are guaranteed by the constitution. Libel legislation provides for strong penalties, however, and some journalists believe that repeated resort to libel action by elected officials may eventually restrain aggressive investigative reporting.

The country has no army. Order is maintained by urban and rural police forces. Both generally cooperate with the press. There are no military threats or pressures on the press (as in other Latin American countries). The two police forces are separately directed, one by the Ministry of Interior (rural) and the other by the Ministry of Public Security (urban). At times, police officials arbitrarily withhold information after declaring it "secret," creating obstacles to news investigations.

Because it is a small country (2.5 million population, geographically about the size of West Virginia), all major media—newspapers, television, radio and magazines—operate on a national scale. All are edited and produced in San Jose, the capital. All major media enjoy reasonable credibility with the public. Most media, especially newspapers, experience attacks from politicians and political parties, although politicians constantly curry favor from the press.

One deterrent to the free practice of journalism stems from the government-authorized Colegio de Periodistas (journalists' association), established in 1969 under a law approved by the legislative assembly and signed by then-President Jose J. Trejos. Under the law, only members of the association may "practice the profession" in Costa Rica, and new membership is limited to journalism graduates of the University of Costa Rica. Approved foreign journalists may qualify for membership after five years residence in the country.

Through the years, the colegio has obtained control of all official information offices, and openly seeks to limit access to press conferences and other public functions to journalists licensed by the colegio. This attitude, especially noted at international meetings, has been protested repeatedly by the English-language weekly the *Tico Times*, as well as numerous foreign correspondents. Most government officials continue to receive *Tico Times* reporters despite the restrictions, and despite protests from the Colegio.

The colegio law was the basis for a lawsuit by that body against Flavio Vargas, veteran TV sports commentator. The colegio demanded imprisonment for Vargas, as well as prohibition of his further working as a journalist because he refused to apply for colegio membership. The suit was temporarily rejected by a lower court for "lack of merit," and the colegio appealed. Vargas's attorney asked for rejection of the colegio's suit on the basis of the 1985 opinion issued by the Inter-American Court of Human Rights. It ruled unanimously that the colegio law and licensing of journalists are both incompatible with the American Convention on Human Rights, of which Costa Rica is a signatory. Developments in the suit were watched closely, since the Costa Rican law, the first such statute to be officially approved in Latin America, had subsequently been enacted in similar form in twelve other countries of the region.

Vargas was the first Costa Rican to be accused in the eighteen-year history of the Colegio. The colegio charged two U.S. newsmen (1977 and 1980) with "illegal practice of the profession." Both were tried, convicted and given suspended sentences; both subsequently left the country.

In mid-1988, the new minister of information, Guido Fernandez, urged the formation of a government committee to study the legal forms necessary to bring Costa Rica in harmony with the Human Rights Convention. A special commission has been established for this purpose.

In late August, the First Court of Instruction absolved Flavio Vargas of charges of "illegal practice of journalism." The government said it would take the case to a three-member appellate court.

While there are no official impediments to the free flow of news in Costa Rica, our correspondent believes that serious and growing threats to a free and unrestricted press in Costa Rica are posed by the colegio's insistence that only duly licensed journalists may "legally practice the profession," and the growing tendency of political figures to press libel actions.

Cuba
[IIIIII_____MoF

Our correspondent, a distinguished Cuban emigre living in Spain, writes that as the twenty-ninth year of the Castro regime began in 1988, there were no significant changes in the strict control of the press and of the organs of opinion in Cuba. The Cuban government, however, modified certain domestic radio programs in order to compete with Radio Marti, broadcasting news and entertainment from Washington to Cuba. All news and information in Cuba is controlled by the Department of Revolutionary Orientation (DOR), under the direction of an officer of State Security, the Cuban KGB. Through his hands passes all the information that Cubans may or may not receive as well as which names may not be mentioned, books which cannot be read, or music heard. The DOR is the "Ministry of Truth" for Cubans, and is responsible for molding political perceptions and public attitudes. As public discontent increases the hierarchy frequently seeks those responsible. Such a circumstance produced a crisis for the former director of the department, Antonio Perez Herrero.

The president of the Union of Cuban Journalists, Julio Garcia Luis, expectedly provides a diametrically different interpretation for the present "process of transformation" in Cuban journalism. He could not deny that Articles 5 and 52 of the 1976 constitution declare the Communist party the "leading force of society and the state," and the "press, the radio, television, movies and other mass media are state or social property" for "use in the exclusive service" of "society"—the society led by the Party. Reduced to fundamentals, the press is constitutionally at the "exclusive service" of the Communist party.

After twenty-nine years, says Mr. Luis, the "process of rectification" which Cuba is undergoing must include "a definitive change for our press." Cuba, he says, is reflecting "on the role of the press under socialism" and "this has a universal influence." At the latest congress of Cuban journalists, says Luis, "President Fidel Castro outlined a number of guidelines, definitions, and a policy of principles for the press in such a way that today the country's information policy and the policy to be followed is clearly defined." Therefore, the journalists' spokesman adds, "Today our journalism is undergoing a process of very intense debate with these problems being analyzed and resolved collectively." The "problems" refer to applying Castro's "line" to the "different mass media." This is intended to avoid the "old tendency which our press had lived with which was, logically, reflected in a certain professional stagnation of the journalists and in a weakening of their technical work." Finally, the journalists' union leader said that his organization "is involved in very energetic politico-ideological work for achieving the high quality journalism to which we aspire."

There is no indication, however, that the news media of Cuba under the present regime will have diminished its role as the tool of state power.

During a visit to Cuba, Jacobo Timerman, the Argentine journalist, discussed the Cuban press with Castro's vice-president, who said he wanted to improve Cuban journalism. Timerman suggested that *Granma*, the official paper, print "what was happening to [Cuban] troops in Angola." The vice-president said this was not permitted. The trial of a Castro protégé was also being ignored. Timerman, in another conversation, explained to the editor of *Granma* that, "under a totalitarian regime, communist or fascist, there cannot be a press. Not that there cannot be a free press. There cannot be a press at all. The concept of news is suppressed." Timerman told the editor to "forget about the press." But the editor replied, "No, we want to have an interesting press which publishes nothing."[2]

Czechoslovakia
[IIIIIIIIIIIII_____MoF

Our correspondent writes: The constitution spells out the greatest impediment to the free flow of information in Czechoslovakia, as in all of Eastern Europe: By law, all news and information systems must serve "the leading role of the Communist party." This represents a monopoly of power which determines the normative role of the news media. Its purpose within the tight framework of this power-monopoly is reduced to "explaining and defending the policies of the party, and persuading people of their correctness." Also conforming to this pattern are newspapers and journals which are nominally published by other political parties (ed: four small parties used as transmission belts to those citizens who reject the Communist party). For the public, the degree of credibility of the news media is not a reflection of how the policies of the party are explained or defended, but just the opposite: The media are judged by how successful they are in showing what the public is really experiencing and what its beliefs are, by making use of the large or small loopholes that exist in the system. Thus the credibility of the news media increases or decreases according to the current degree of political rigidity in a country ruled by a single political party.

It follows, therefore, that information considered to contradict the current Party "line" is fully exposed to distortion or deletion within the established media. Every fact is explained from the point of view of Party policy, and even straight news items are presented in a propagandist manner. Facts at odds with current policy are suppressed. This is done by overt pressure from Party bodies, and covertly, by the implementation of the "leading role of the Party" deep within every institution, including the media. It is unthinkable that any leading position in such an organization can be held by someone who is not a Party member, or that any journalist could publish or broadcast his own opinion.

The year 1987 in Czechoslovakia was somewhat less rigid in comparison to preceding years. This was the direct result of changes taking place in the Soviet Union. But this process has been ambivalent. On the one hand, it is the outcome of Soviet influence and the pressure of events taking place there—the media are forced to publish some of this, and pay lip service to the current Soviet "line." On the other hand, in domestic policy the media have begun to stress the specific nature of the Czech condition, and the differences which exist in comparison to those in the Soviet Union. Thus, in fact, they contribute to maintaining the rigidity typical of past years.

The outcome is a degree of schizophrenia that further decreases the credibility not only of the media, but of politics in general. The volume of objective information found in the media has increased to a very small extent. Despite the fact that the media continue to serve the powers that be, they employ a number of journalists who are capable and willing to make use of even the slightest political thaw to improve the quality of the media.

* * *

THE LIVELY UNDERGROUND samizdat (self-published or private) press continues to publish alternative views in very small editions. All printing and photocopying equipment is controlled by the Interior Ministry, and even typewriters are restricted. Nevertheless, some Charter 77 signatories publish regularly *Informace o Charte 77* (Information about the Charter). This carries reports of human rights violations and other statements by Charter and other dissenters. Some Czech samizdat, sent abroad, is published by emigres and brought back to Czechoslovakia for avid distribution. Several samizdat writers and publishers received from one to three year sentences in 1987 for engaging in unauthorized publishing.

Citing precedents in the Soviet Union, early in 1988, a group of

Czechoslovak journalists negotiated with the government in order to register an independent newspaper. The paper, *Lidowe Nowiny* (People's News) had already appeared twice. It carried uncensored political, cultural, economic and foreign news. It discussed the war in Afghanistan and reports on Poland, as well as Soviet-American disarmament. The newspaper was produced by journalists who enjoyed some freedom during the Prague Spring of 1968 but were barred from the profession after the Soviets invaded their country.

Jiri Dienstbier, imprisoned and banned for anti-government writing, is a member of the editorial board of Lidowe Nowiny. He said, "I am fifty-one. We are the last people who in their 20s and 30s worked on uncensored newspapers. We are completely without a middle generation. The problem is how to teach a new generation."

He hoped that when Milos Jakes, the new Czech Party leader fully controls the apparatus, he will approve the publication of the independent newspaper.

Denmark

Since 1849 freedom of the press has been guaranteed by law in Denmark —after nearly two centuries of press censorship. Newspapers here are among the freest in the world. Because the citizens are guaranteed the right to form their opinions, they must expect a press that is not only free but independent of government. Newspapers and journalists are not required to register or become licensed, nor is press or broadcast equipment subject to import licensing. Journalists are not required to preserve the secrecy of their sources, but Danish journalists do not reveal them and the courts respect this tradition. Government offices readily provide information to journalists, and criticism of the government or its policies is neither punishable nor censorable.

The question of government subsidies to newspapers has been discussed since the mid-1970s. Some regard this as a loophole in governmental intervention in the press. Subsidies are defended as necessary to support financially troubled newspapers, particularly since the press is regarded in Denmark as an essential element in the democratic system. The government provides low interest loans to publishers, concessions for communications, relief from some taxes and paid advertising to inform readers of governmental activities.

Our correspondent writes: All Danish newspapers are privately owned either by limited companies or private funds. Many limited companies include a large number of shareholders with limited voting rights in order to assure editorial independence and balance the political, philosophical or religious support for the paper. A political party can own a newspaper through a limited company, as for example the Communist party. Other papers are owned by private funds established during the past ten to fifteen years by families. They seek to assure the editorial independence of the paper (ed: and generally support a political tendency, if not party).

Radio and television have been a state monopoly of Denmark Radio, but the field will be liberalized in 1988. A second national television channel, TV2, was scheduled to begin late in the year. Though state-owned, TV2 will be partly financed by advertising. A number of privately-owned radio and television stations began as an experiment in 1987 and have become permanent. They, however, are not allowed to take paid advertising.

From the point of view of the Danish Newspaper Association, Denmark is a very good country for publishing. It has almost ideal relations between the mass media and the authorities, and good working conditions for journalists.

Dominica
[|||||||||||||||||||||||||||||||||||||||||||||||||||||||||||||||| MoF

Radio is the major means of information on this island of 290.5 square miles. The government-owned DBS—Dominica Broadcasting Service—is popular and respected. Video-I is a small, government-owned TV station using a satellite from Barbados for a limited number of local receivers. Marpin TV provides a privately-owned cable service.

The major impediment to the flow of information lies in the inadequacy of local newspapers. There is no daily paper, but there are two weeklies and an occasional paper produced by the governing Freedom party. The *New Chronicle*, eighty years old, is owned 51 percent by the Catholic church and 49 percent by private interests. The *Labourite* is the weekly organ of the Dominica Labor Party.

Dominican Republic
[||||||||||||||||||||||||||||||||||||||||||||||||||||||||||||||||||||| MoF

Dominican journalism has developed more since the demise of the Trujillo dictatorship in 1961 than has the country's economic life. Indeed, the print and broadcast media display professionalism not generally seen in a developing country. Eight privately-owned newspapers freely reflect varied points of view and criticism of the government. None is openly affiliated with a political party, though their editorial views range from conservative to Communist.

There is no government censorship, though in an earlier administration of the present president, Joaquin Balaguer (1966-78), radio stations were frequently repressed. The 138 privately-owned radio stations are a popular platform for opposition politicians.

In March 1987, marshals and police seized equipment and damaged a television station owned by a friend of the previous government. The officers had misinterpreted a court order, and the district attorney halted the operation soon after it began. President Balaguer ordered an investigation, and repeated the government's commitment to basic civil and political freedoms.

The Appeals Court of Santo Domingo in March 1988 declared unconstitutional the colegio of journalists. An appeal was taken to the Supreme Court which at this writing had not ruled.

Egypt
MoF

Since the death of Anwar Sadat in 1981 the press in Egypt has been the freest since the 1952 revolution. Several opposition weekly newspapers, magazines and other publications carry sensationalist stories and strong anti-government criticism. Widely differing news and opinion appears in the major government-owned newspapers.

The press today in Egypt is the freest in the Arab world. However, the four large Arabic daily papers are owned and operated by the Higher Press Council (Shura) established in 1980 by Sadat. Through the Shura, a quasi-governmental, appointed institution of Egyptian notables, the government controls these dailies, and owns 51 percent of their shares. These papers, consequently, generally reflect the government position on issues of consequence. There are also two French- and one English-language daily newspapers controlled by the government, that print mostly official handouts, and are intended for the foreign resident population.

The major opposition parties and groups such as Wafd, Muslim Brothers, Nasserite party, and the left-socialist faction (Tagama) have their own weekly newspapers which are frequently critical of the government. They are uncensored, although it is known that criticism of the president of the republic is out of bounds. There is, however, frequent criticism of other officials and sometimes unbridled attacks on the government. In December 1987 the opposing press launched a virulent campaign against Egypt's interior minister. He had frequently used emergency laws to detain political opponents. The Wafd newspaper's editor acknowledged that "we have a freedom we didn't have for thirty-five years," and said, "we are trying to make that window of freedom larger. We believe it is our duty."

The government does not "guide" the opposition press, though it does occasionally advise journalists working for government-owned newspapers about coverage of certain issues. Such guidance is usually given informally to senior editors by the Information Ministry and is rarely needed because the editors already know the limits of policies they report or discuss.

A crossing of lines developed early in 1988, however. Reporters for the government-owned papers began "moonlighting" as journalists in the opposition press. There they could express their views more freely, earn additional income and gain further professional experience. The administrative council of the state ruled, however, that journalists on government papers could never work for the opposition press, whereupon four opposition political parties condemned the move as an "attack on freedom of expression." The government's response may have been triggered by the earlier attack on the interior minister, which included a sit-in of 200 journalists.

In addition to the press there is a flourishing publication of journals and magazines that often reflect critical perspectives of government policy. Since Sadat's crackdown on opposition or hostile journalists in 1980, there has been no interference with press criticism or journalists. In 1986, several Nasserite journalists were arrested and accused of involvement in terrorist incidents in Cairo, including attacks on two American diplomats.

The press is widely read in Egypt, with the Friday edition of *Al-Ahram*, the largest government-controlled paper, having a circulation of some 1,250,000. The press generally has high credibility in the eyes of the public. Since Sadat's demise, there are relatively few impediments for the print media. The domestic news agency is the government-, or Shura-controlled Middle East News Agency. It is considered fairly reliable by foreign journalists. Numerous foreign news services also operate in Cairo.

All radio and television is government-owned and operated through the Egyptian Broadcasting Corporation. It also owns numerous radio stations and broadcasts in several languages, and widely in Arabic, to the rest of the Arab world. Commercial radio service is offered by Middle East radio. Some three dozen government television stations transmit over two channels.

Ethiopia
[IIIIII_____MoF

Our correspondent writes: Censorship directly or indirectly contributed to the deaths of tens of thousands of Ethiopians during 1982-1985. The drought and famine which made headlines in the rest of the world by 1985 and brought an outpouring of assistance, mainly from the West, went unreported in the country for at least two years. Actual starvation began to ravage

Power, the Press and the Technology of Freedom

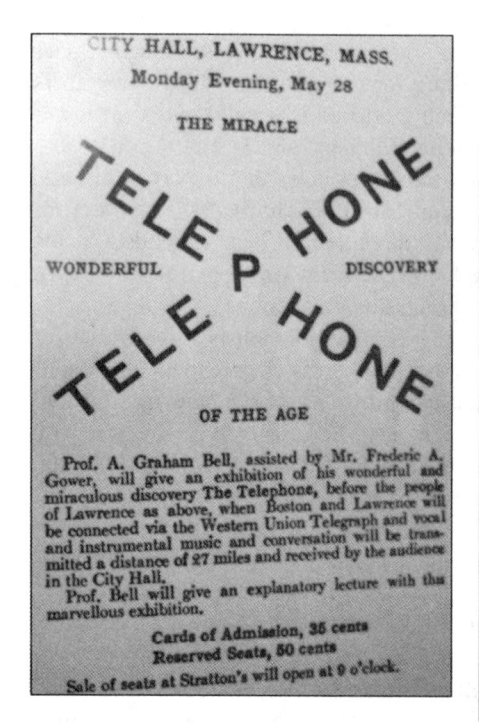

Poster for the meeting on 28 May 1877 at Lawrence, Ma., city hall where Alexander Graham Bell, inventor of the telephone, would demonstrate his "miraculous discovery." *Top r.* Bell's Letters Patent for his "improvement on telegraphy"—the telephone. *Bottom* The author (center right) addresses a seminar in Saigon, South Vietnam, just before the fall, on the values and procedures of a free press. Nearly 100 Vietnamese journalists participated.

Leonard Sussman, director ejecutivo de la Freedom House (Casa de la Libertad), de los EE.UU., momento en que es requerido por la policía sobre sus documentos, aún así no pudo ingresar a la radioemisora.

Top Paraguayan military police briefly detain the author, 10 November 1987, after failing to block him from entering Radio Nanduti in Ascuncion. He had been scheduled to address a meeting at the radio. The meeting had been "permitted," by officials, but "no one could enter"—so read the banner inside the auditorium (center photo) where no audience appeared. The author later held a press conference at a hotel. *Bottom* Humberto Rubin, director of Radio Nanduti.

Power, the Press and the Technology of Freedom

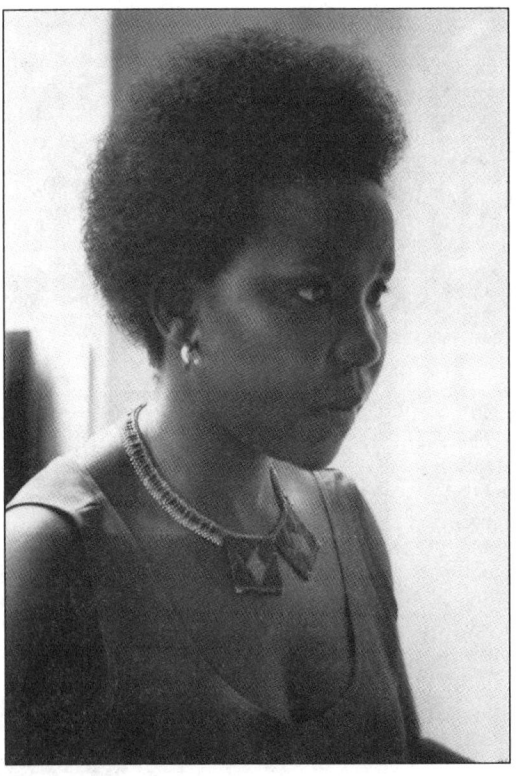

Top Sergei Grigoryants, editor and publisher of *Glasnost* magazine in the Soviet Union. Shown in his apartment in Moscow four months after being released from the second long prison term for writing unauthorized publications. Since glasnost, he has been imprisoned three times for short periods.

Bottom Nomavenda Mathiane, the courageous young essayist-reporter who writes with equal integrity and vigor of black and white racism in South Africa.

The Information-Power Struggle

When the prime minister of Malaysia proposed toughening the Official Secrets Act in 1986 the Malaysian union of journalists circulated posters (top), distributed lapel buttons and campaigned against new restrictions on journalists. The law was strengthened providing prison terms for publishing ill-defined "secrets." *Below* Some of the twenty-three journalists beaten by Chilean special forces the night before shown confronting an official with photos of bloodied newsmen on the table. Assaults came soon after General Pinochet lost a historic plebiscite thereby limiting the dictator's term of office.

Ethiopia as early as 1983. Relief agencies knew it. Many Ethiopian officials knew it, but the government banned any reporting of the oncoming catastrophe in its own media until late 1984. Even when the ban was lifted, coverage had to focus on government "relief and rehabilitation" efforts, not on the famine itself.

Although the Ethiopian government did allow in foreign reporters so that the outside world could come to its assistance, it deliberately hid the dimensions of starvation from its own public. Ethiopia's top relief official, in fact, was not authorized to brief Ethiopian media on the famine; he could brief only foreign media. As explained by the head of the relief agency, Dawit Wolde Giorgis, who subsequently sought asylum in the United States, the government considered the famine "an embarrassment and humiliation to the revolution." It considered famine contrary to the image of progress and accomplishment it wanted to portray on the occasion of the tenth anniversary of the revolution, celebrated 12 September 1984. In the summer and autumn that year, triumphal arches were built, statues erected, meeting halls inaugurated and portraits of chairman Mengistu Hailie Mariam hung to celebrate ten years of revolutionary achievement. The fact that hundreds of thousands of people were actually starving outside the capital was considered an ill-timed inconvenience. In Addis Ababa, bread lines moved from the main streets to less visible parts of the city, away from the view of foreign journalists and Communist guests from abroad who had come to participate in the festivities.

Keeping knowledge of the famine from the Ethiopian public also served to prohibit discussion of its causes. The cover-up also fooled Western journalists who gave almost no coverage to the famine in reporting the anniversary celebrations. The official line was that drought had caused the problem. This was a smokescreen designed to prevent discussion of government agricultural policies, in particular investment in unproductive state farms, discouragement of independent farmers from producing surplus food, lack of attention to the road and distribution system, disruption of agricultural production by military campaigns against secessionists, and the government's massive expenditures on arms.

Ethiopian journalists and film producers were prohibited from reporting on the forcible eviction and resettlement of Ethiopians from northern famine areas to the more fertile south, during which thousands were known to have died. No Ethiopian journalist could write about how people were rounded-up at gunpoint, families separated, the very young, old and infirm left behind, or about the numbers who died along the way after being packed into open trucks without sufficient food, water or protection from the sun, and then exposed to malaria and other diseases in the south.

A cholera epidemic in 1985 that claimed numerous lives was similarly covered up by the Ethiopian media. The government prohibited any acknowledgement of the spread of this infectious disease throughout relief centers evidently because it considered cholera to be a blot on Ethiopia's image. Officials were allowed to admit only to the existence of "acute diarrhea." The refusal to acknowledge the disease impeded the ordering of medicines and the taking of proper measures to prevent its spread.

Increasingly since 1984, with the founding of the Workers party, censorship became stricter, and the ministry of information came more and more under the thumb of the ideology department of the Party. The East Germans secured a coveted role in the training of Ethiopian journalists, and extensive studies began to be made on the reorganization of the ministry of information along Soviet lines.

The most striking feature about censorship in Ethiopia is that it has made the public unwilling to believe most of what is reported by government media. Ideologues at the top of government and those at mid-levels who have a vested interest in the system, believe what they preach. But immense numbers of Ethiopians throughout the country are skeptical. Many regularly listen to the BBC and Voice of America to hear another version. Jokes about political indoctrination abound, expressing popular dissatisfaction with an ideology foreign to Ethiopian history, tradition and culture. The very officials responsible for censorship and propaganda often deride it privately.

Censorship in Ethiopia extracts a cost high in human life. It corrupts many of the officials held under its sway and often destroys those with integrity. It keeps the Ethiopian people in the dark about events in their country as pressing as war and peace, the true causes of starvation and famine, the spread of disease, and the nature of their political system.

Fiji

[IIIIIIIIIIIIIIIIIIIIIIIIIIIIIIIIIIIIIIIIIIIIIIIIIIIIIIIIIIIIIIIIIII_____MoF

Two military coups in 1987 left political liberty and press freedom in shambles on these formerly staunchly democratic islands in the South Pacific.

The turmoil began with the general election in April when an opposition alliance (mainly of ethnic Indians) defeated the ruling party (of largely Melanesian origin). The new government was removed from power 14 May by the armed forces (over 95 percent of whom are Melanesian). The day after reporting the coup, the islands' two daily newspapers were shut down indefinitely. Soldiers were placed in the newsrooms and the newspaper offices were surrounded for a week.

The closure of the papers caused widespread fear and bewilderment across Fiji. When the newspapers failed to appear without explanation, disturbing rumors and leafleting campaigns spread like wildfire. The ministry of information convened a meeting of editors and publishers to discuss guidelines for restoring publication, but both newspapers rejected censorship, and the blackout continued. On 20 May all bans on the newspapers were lifted and editors were given "good advice" on the delicacy of the racial situation, with a plea not to incite ill feelings. Both newspapers returned to the streets on 21 May with strong condemnation of the blackout. Editorials have not spared criticism of the government though information from the military has been difficult to secure.

During the May blackout foreign journalists were evicted from their offices, and their papers and credentials taken by soldiers. The military also pulled photographers from cars, and confiscated their cameras and film. Troops in the central telephone and telex exchange screened articles before transmission abroad. Two foreign journalists were detained, and they as well as domestic reporters were subjected to frequent physical harassment.

On 25 September before another caretaker government could be installed, the military led a second coup during which the newspapers were again taken over at gunpoint. The commercial radio station was also shut down, as it had been in May. Newspaper publishers were detained and journalists physically harassed, despite the fact that the authorities had told the news media they were free to report responsibly.

By year-end, Fiji seemed to have stability restored. The newspaper presses were running again. Political detainees were freed, roadblocks removed, and curfew eased. But Fiji was no longer "the way the world should be," and a new democratic consensus—including the freedom of a responsible press—had not yet been constructed. It was mid-November 1988 before the Internal Security Decree was lifted, and press restrictions ended. Yet even then troops were deployed outside the two radio stations as a new warning.

Finland

Since the first Finnish-language newspaper was printed in 1775, the history of the press here has been marked by a struggle against censorship. Under Russian occupation, a tsarist decree of 1850 banned the publication in Finnish of all except religious and economic tracts. Censorship was relaxed and renewed intermittently until Finland gained its independence in 1917. Today, in covering international affairs, the Finnish press is regarded as somewhat wimpy, though quite credible. Recently, that credibility has been reduced by new and differing journalistic models in the growing fields of evening newspapers, free sheets and local commercial radio stations.

The political system, functioning as a democracy, assures freedom of speech and press. There are no instances of official abuse or judicial decisions restricting that freedom. Although the government does not compel it, the press sometimes displays restraint in covering issues believed to threaten the national interest, particularly in the field of foreign affairs and its large neighbor to the East.

The nation of five million persons is served by 100 daily newspapers and more than 250 magazines. There is one national, state-owned television and radio company (Yle) broadcasting over three channels, sixteen privately-owned, commercial radio stations and one owned by a local community. Yle's channels are also used by the privately-owned, commercial television service, MTV. Yle and MTV, together with a private company, co-own a third television channel. There are also twenty-five private cable TV enterprises. A national news agency is privately owned by newspapers.

There was a major increase in the variety of news and information provided in 1987. There has been some growth in morning newspapers, increased circulation in the country's two evening newspapers, and the appearance of sixteen commercial radio stations after a legislative change two years earlier. Nine percent of the broadcast time on local radio is devoted to news, and the national radio company has increased its news broadcasts. In addition, cable TV now carries international news from France and England (ITV), and the American CNN is presently being marketed. Because of increased competition among media groups, the quality of reporting has shifted somewhat from socially significant events to audience-attracting entertainment and trivia. This is a slight deterioration from formerly high-quality news dissemination.

A general trend towards more commercialism and an increased interest in financial matters ("casino economy") has contributed to a greater influence of commercial and economic considerations in 1987. This is evident primarily in the business oriented journals, some of the new commercial radio stations, and the third TV channel which is openly "friendly to business." The state-owned broadcasting company, Yle, in a rare incident, tried sponsored programming of a series on state-owned industries, but the first program was rapidly cancelled after strong protests from the journalist's union and critical newspaper comments. In this way, the press defends effectively against outside influences. Denunciation of journalists or specific media by politicians is rare, and immediately criticized by the press.

Finland traditionally has a large newspaper readership as well as great trust in newspaper reporting. More than 80 percent of respondents rate their newspapers trustworthy. During the 1980s, the credibility of television has grown and that of newspapers and radio decreased (40-45 percent regarding TV as most credible, 30-35 percent newspapers as most trustworthy, and 25-30 percent believing most in radio). In 1987, there was an inexplicable statistical drop in newspaper credibility, while TV increased its rating.

The public is fairly critical of the press, about one-third of one sample agreeing with "authoritative" (president and premier) criticism of the press. Still more respondents found to be exaggerated the social and political criticism appearing in the press.

Since 1971, the press has been supported by special government subsidies. These include exemptions from certain taxes and delivery costs, as well as financial support to party organs through annual budget appropriations. The state also allocates funds for telecommunications costs incurred by news and press agencies, and their client publications. Most of the newspapers that have disappeared were affiliated with political parties. The center party has best maintained its press, while the right suffered biggest losses, and the socialist papers somewhat less.

The Council for the Mass Media, established in 1968 by journalists and publishers, monitors journalistic ethics. The council, which includes public members, observes broadcasting as well as the print press.

France
[||||||||||||||||||||||||||||||||||||||||||||||||||||||||||||||||||||||||||||||||||||____MoF

In a nation of resolutely independent newspapers, the most significant improvement in France's freedom of expression in 1987 came with the denationalization of its main television network (TF-1). This is by far the most watched, most influential TV channel. A private TV network, Channel 5, was also created. So, for the first time since World War II, the state-owned monopolistic system of television has broken down. There are now two private and two state-owned networks. As may have been anticipated, competition between the private and public sectors made the public sector much more independent in its treatment of news. One negative aspect: The main owner of Channel 5, Robert Hersant, also owns the most widely read conservative newspaper, *Le Figaro*, and many local papers. Hersant controls 38 percent of France's national press and 29 percent of its regional papers. His papers are all highly partisan.

France has more than thirty local radio stations, about half in Paris. Some 200 small "minority" radio stations operating on the FM band were closed down early in the year when the Paris airwaves were deregulated. Ninety-six small stations could be licensed. There are four significant national radio systems: two state-owned (Radio France, with many stations, and Radio Monte Carlo), and two private, politically very influential in their treatment of news and public debate (Europe-1 and Radio Luxembourg).

The single news agency, Agence France-Presse, is semi-official. All

French newspapers and broadcasters subscribe to foreign news agencies as well. In 1987 AFP suffered its worst financial crisis in history. Its deficit topped $34,000,000, though the French government buys some 55 percent of the agency's subscriptions to use in offices around the world. By law, French press representatives have a majority vote on AFP's executive board. Journalists fear, however, that if more government funds are received, editorial independence may suffer. Gerald Long, former head of Reuters, says, "The influence of the French government (on AFP) is very small." But he adds, "AFP journalists have to keep quiet when people claim it's the Voice of France."

Germany (Federal Republic)
[IIIIIIIIIIIIIIIIIIIIIIIIIIIIIIIIIIIIIIIIIIIIIIIIIIIIIIIIIIIIIIIIIIIIIIIIIIIIIIIIIIIIIIIIIIIIIIIIIIIIIIIII|__MoF

Judgments of journalists, not government officials or agents, determine the content of West German print and broadcast media. Commercial interests sometimes influence news coverage, and officials sometimes favor friendly journalists by sharing information. More than 145 privately-owned newspapers and 120 magazines otherwise enjoy full editorial independence. Criticism of the government is unrestricted and uncensored.

Radio and television networks and stations operate generally as corporations under special laws. Broadcast media are managed by independent boards composed of representatives of political parties, churches and other public organizations. Several private television cable channels and local cable TV networks provide programming on an experimental basis. It is likely that during 1989 regular programming will become available over private cable TV channels, with satellite and other "feeds."

Ghana
[IIIIIIIIIIIIIIIIIIIII_____MoF

The Ghana press today is a sad shadow of the original, born in crisis—and hope—when the nation achieved independence just after World War II. The Ghana press was mandated then to "serve its essential role as the watchdog of the peoples' rights and liberties against encroachment by government and its agencies." This was stated in the belief that without a free, courageous press, abusive power, arbitrary and corrupt practices, incompetence and wasteful administrative procedures are likely to go unnoticed and hence unchecked.

Journalism provided Ghana from 1931 on with political education in preparation for independence. In the '30s, more than eighteen newspapers of small circulation reached the public from a number of private press houses. Editors and proprietors of newspapers in the '30s and '40s were "colorful and fearless nationalists." They wrote with conviction and pride in their craft, consistently arguing the case for freedom, justice, human dignity and the need to cast off the colonial yoke. Most editors then were lawyers. Their writings awakened the masses. Other papers established by political parties spearheaded the campaign for self-government. In 1949, Dr. Kwame Nkrumah founded the Convention People's Party (CPP) and established the *Evening News,* adding new impetus to the fight for independence, and also interest in newspapers. Dr. Nkrumah became the first Ghanian prime minister and later president, and himself wrote many interesting articles.

Such was the standing and public image of journalists during the postindependence era that the British Governor released the text of the 1951 constitution to the press two days before official publication. Yet independence and freedom of the press for which editors and proprietors fought during the colonial struggle were quickly curtailed soon after independence in 1957. The process began with the deportation of a Sierra Leonean editor who criticized Dr. Nkrumah. Next, came the enactment in 1958 of the notorious Preventive Detention Act. Not only members of opposition parties, but also outspoken press columnists and editors were detained without trial. The death knell of a free and independent mass media was tolled by the acquisition of the *Daily Graphic* in 1962 and the *Catholic Standard*—the only remaining independent weekly—thereafter carrying hardly any news or commentary of political significance. Government ministers, moreover, made sure that opposition views were rarely published either by the Ghana News Agency (GNA), the radio or any newspapers operating under the Ministry of Information.

Despite efforts in the 1978 constitution to produce conditions in which a free, informed and responsible press could flourish, the 1981 military takeover ended that hope. The press in Ghana continues to be at the mercy of government and its agencies. In the eyes of the public, the media's credibility is nil. Enlightened people, therefore, get news and information especially from the BBC and the Voice of America. French speakers tune to Paris for their news. For example, the death of the head of state of Bourkina Faso was picked up by Ghanians on the BBC and VOA two days before the local press told of the leader's death. Even then, the Ghanian statement was doctored to suit the government's special relationship with the slain head of state.

The flow of news and information inside the country is substantially altered by the government. Editors are in constant touch with the executive or the ministry of information to be told what to use and what not, or how to handle a particular story. Editorials in the press are written by randomly selected officials in government agencies for editors to publish. The secretary of information or his representative are generally on hand at the newspapers to approve the contents in the two government-owned dailies and their sister weeklies. Private press editors dare not risk publishing anything unpalatable. They may lose their supply of newsprint, which is controlled by the information ministry. The government imports newsprint in bulk and distributes it to both government and private publishing houses.

There was no improvement in the news flow in 1987. Journalists, however, generally fare well, provided they practice self-censorship. It is the formula for survival and well-being, in the public as well as the private press. Journalists have recently enjoyed salary increases and been invited to many seminars explaining how to deal with the needs of a revolutionary government. For a journalist, it is either "taking it or leaving it," as he sees fit. There have been, nevertheless, many resignations recently from the government press houses, and radio and television. The private press continues to deteriorate, mainly due to lack of funds. It is difficult to start new papers for fear of running against revolutionary policies. The trend now favors newspapers devoted only to sports.

Ten years ago, the Essa Committee on Service Conditions of Journalists declared:

> Some journalists, it is sad to observe, do not wait for the government to give them a line. Motivated by opportunism to the exclusion of professional ethics or regard for public interest, they launch witch-hunts against political dissenters and carry out distortion of facts, if they surmise that such activities while not ordered by government will not unduly displease it. As a result of such activities, journalists in the public service have largely become objects of hatred and contempt in the society that normally expects so much of them. Instead of being regarded as leaders of opinion, they are almost generally regarded as stooges, ignorant dealers in misinformation or preventers of truth.
>
> The effect on the morale of those journalists who retain a modicum of the decent standards taught them in training schools or on the job has been catastrophic. They hold their heads low in society because they know what they are part of, but cannot change. Nor can they resign easily because they see few other avenues open to them to practice their trade.

Those words from 1978 are no less applicable in 1988.

Greece
[IIIIIIIIIIIIIIIIIIIIIIIIIIIIIIIIIIIIIIIIIIIIIIIIIIIIIIIIIIIIIIIIIIIIIIIIIIIIIIIIIIIII_____MoF

No country in the West approaches Greece for having as many newspapers and other publications relative to low population and readership statistics. Also, the excessive influence on political affairs which the press wields, considering the low readership, is unique to this Western country.

There are seventeen national dailies here, with a total daily circulation of approximately 950,000 copies. This is an extremely low per capita readership, and accounts for the fact that only three newspapers actually make a profit from the standard channels of sales and advertising. There can be no explanation for how the rest survive other than by obtaining "outside" support either through government loans on which they default, or from private business, party interests or from abroad. For those few publishers who don't resort to such means of financing, their newspapers become a valuable instrument of power. They use this power to exercise political and business pressure to extract benefits in other areas. This last possibility explains the fact that despite the glut, newspapers and media outlets continue to appear on the scene in Greece. Three more dailies were about to be launched in 1988, as were a series of private or municipal radio stations, the result of the government's recent liberalizing measures concerning the operation of broadcast facilities.

The papers with the highest circulation in Greece are *Ethnos* and *Ta Nea,* both progovernment and anti-Western in orientation, and with daily sales of about 160,000 copies. They are followed by *Eleftheros Typos* (conservative), *Eleftherotypia* (pro-Socialist government) and *Apogevmatini* (conservative). On the other end of the circulation scale are *Estia* and *Eleftheri Ora,* both extreme rightist and having circulations of only a few thousand.

On average, the combined circulation of the afternoon papers is twice as high as that for morning papers. Of the latter group, the newspaper with by far the greatest circulation (50,000 copies) is *Rizospastis,* the official organ of the pro-Moscow, Greek Communist Party (KKE). (It has conclusively been proved that the KKE's extensive publishing network in Greece was financed through a front company set up in Luxembourg by the East German regime.)

In addition to the national dailies, every major Greek town has one or two regional newspapers. Their impact is limited to local politics and news, the exception being the three papers published in Salonika, capital of northern Greece, which have a somewhat broader appeal. The media blitz is completed (almost) by a large number of weeklies, fortnightlies and monthlies. The most prominent of these, as far as their political impact is concerned, are weeklies *Pontiki* (center-left), *Tachydromos* (centrist), *ENA* (moderate-conservative), and fortnightly *ANTI* (moderate-left).

The real explosion, however, is not taking place in the sphere of the printed word but in broadcasting. The government recently ended one of the most undemocratic features of public life in Greece—the absolute state monopoly over radio and television. Both media were used by all preceding conservative governments, and by the present socialist one as instruments of mass propaganda.

Now, municipal radio stations have begun broadcasting in the major cities of Athens, Piraeus and Salonika and are proving enormously popular. For the present, they are largely in the hands of conservative political forces, as the conservatives won the 1986 municipal elections in all three cities. More important, they have added a refreshing dose of democratic pluralism through the airwaves, something taken for granted in every other parliamentary democracy in the West, but not in Greece. Now the major publishers, such as the Ethnos and the Lambrakis groups, are also preparing to launch radio stations, as are independent groups of journalists.

These liberalizing measures have yet to be extended to television, the most powerful media instrument today. The country's two networks are government-controlled and are exploited accordingly. Cable and private television networks remain a distant dream, the main reason being the financial costs involved. The only comparable exception is a novel enterprise calling itself "TV Press," which is composed of a group of dynamic young journalists who distribute news features on video cassettes to several thousand subscribers throughout the country. Again, this venture is proving enormously popular because viewers are able to "watch" events, see personalities and listen to views which the state networks would never screen.

Amidst the current media explosion in Greece, one sore point remains: Greece has no school of journalism or any university-level course for the study of communications. Greek journalists usually start from scratch, with little training or particular knowledge of their profession. If one adds to this the fiercely partisan nature of the Greek press and the fact that every publication is totally given over to some particular political entity, then it should be no surprise that the standards of Greek journalism remain low. This, in turn, has an adverse effect on the quality of news and information provided for the Greek public which is, for the most part, poorly informed or confused over public affairs. Fortunately for Greek newspapers, the country's libel laws in practice are very lenient.

Successive governments, and the narrow interests of the publishing entrepreneurs, are largely to blame for this state of affairs. A newspaper in Greece today—or a radio network tomorrow—are often used more as instruments of power, of exchange, of political influence, than as conduits of impartial information and as defenders of the public's "right to know."

Finally, there is now undeniable evidence of Soviet interference in the Greek press as a means of spreading disinformation and cultivating anti-Western sentiments. There is court evidence provided under oath, from former professional Soviet propaganda and disinformation experts, as to how the USSR decided to loosen Greece's ties with the West through the process of "active measures" (political warfare), and disinformation. The main technique used, they explained, was the injection of pro-Soviet and anti-Western material into the mainstream press.

Overall, the picture provided by the Greek press today is unsettling.

Grenada
[IIIIIIIIIIIIIIIIIIIIIIIIIIIIIIIIIIIIIIIIIIIIIIIIIIIIIIIIIIIIIIIIIIIIIIIIIIIIIIIIIIII_____MoF

In October 1983, United States armed forces, supported by a Caribbean peacekeeping force, intervened militarily in Grenada. That marked a watershed for the island's news media.

For nearly four years earlier, Grenada had been dominated by the Peoples Revolutionary Government (PRG) which seized power in March 1979. On 19 October 1983, in a bloody coup, the PRG was overthrown by a faction of the Peoples Revolutionary Army. Until completion of the successful military intervention six days later, all Grenadians (including the media and its representatives) were under complete subjugation.

Under the PRG, press freedom was severely curtailed. The *Torchlight* newspaper had been shut down, the *Catholic Focus*, an organ of the Roman Catholic Church, was banned and retroactive legislation was passed banning production of newspapers. The law said that "No newspaper or other paper, pamphlet or publication containing any public news, intelligence or report of any occurrence or any remarks or observations thereon or upon any political matter...shall be produced, printed, published or distributed in Grenada..."

As a result, the assets of the *Grenadian Voice* (published before the law was enacted but subject to it retroactively) were seized by the PRG. The editor and two of the newspaper's legal advisers were imprisoned. They were to spend more than two years in jail because no charges were brought against them, and they had no opportunity to defend themselves in court. Others associated with the paper were harassed, had their homes placed under surveillance by the Peoples Revolutionary Army, were publicly vilified, and prevented from leaving the island.

Following the October intervention, and for a period of some twelve months, the government was under the control of an interim administration appointed by the governor-general. During that period, there was an immediate and dramatic improvement in press freedom with all restrictive legislation removed.

There were, however, suspicious circumstances suggesting the victimization of journalists who had been employed by the PRG at the state-owned Radio Free Grenada. The staff at that time numbered about sixty and, following the intervention during which the radio station was destroyed, thirty-five were not reemployed. About half of these were nonappointed trainees who could have been dismissed without state reasons. An analysis of appointed staff members who lost their jobs at that time, however, suggests that these were the ones thought to have strong leanings toward the PRG, and so posed a security risk. An elected government took office in December 1984 and, since then, there are no known incidents of this kind. Not only are the organs of the media free, but there is no victimization of journalists or applied pressure, political or otherwise, to influence the flow of news.

Of the seven newspapers published in the island, four are put out by political parties. Their credibility is tempered by public knowledge that they present a partisan point of view. Two of the other three papers enjoy good credibility but the third sometimes suffers because of its efforts to be sensational.

The only radio station, Radio Grenada, is state-owned. Its newsroom relies to a considerable extent on releases from the government information service. News presentation, while never unfavorable to the government, is not blatantly slanted in favor of the party in power. The station enjoys a fair degree of credibility.

Grenada has one television channel. It was established after the 1983 intervention, and the basis of its operation is unclear. According to press releases, the station was donated to Grenada by a United States nonprofit organization, and has the approval of the president of the United States in its efforts "to help encourage free enterprise..." The station enjoys fair credibility, but large segments carry advertising geared to United States viewers. Inquiries made of its foreign station management drew the reply that the supply of free programs depends on their being aired without advertising cuts. This opens speculation as to what the station is attempting to achieve, and how much its broadcast content is influenced by outside factors.

Influences on the Grenadian public cannot be confined to those organs based in the island. From the Eastern Caribbean, Grenadians can choose radio broadcasts from a half dozen other nearby countries. These programs carry news and information of and about Grenada, the region and the world. There is also limited reception of television programs from three islands, and some flow of information through newspapers published in other islands and on sale locally. As a result, Grenadians have free access to a variety of opinions, and there is no impediment to the flow of news and information.

Press freedom flourishes in Grenada now. Following the oppressive years of the revolution, conditions have improved steadily, and there are no indications that this trend will not continue.

Guatemala
[IIIIIIIIIIIIIIIIIIIIIIIIIIIIIIIIIIIIIIIIIIIIIIIIII]_____ MoF

With the arrival in January 1986 of the first civilian president in sixteen years, violations of human rights of Guatemalans has steadily declined. The new democratic system, following elections in 1985, made advances observable in the freer flow of news and information. Several thousand Communist guerrillas continue to fight in inaccessible areas. Their use of landmines and murder of presumed opponents continue to cast fear in the population. Two journalists were abducted, tortured and killed in 1987. Rightist counterparts in 1988 bombed the Soviet news agency Tass and threatened the Cuban press service Prensa Latina.

While there is no formal press censorship, the news media in this climate exercise some self-censorship on highly sensitive subjects. The government was criticized in 1987 for prohibiting at first the showing of some television commercials which criticized tax policy. The government maintained the commercials included false information. But the advertisers secured a court injunction, and the ads were televised. Most advertisements attacking the government, including declarations by the insurgents, are carried without government interference over the officially supervised radio and television system.

In mid-1987, under a nine-year-old law, the government began requiring broadcasters to carry a daily government program. Previously, such requirements were rarely made. Journalists regarded this new requirement an abuse of official authority. Of some ninety radio stations, five are government-operated. There are four commercial TV channels.

Among numerous questions, a public opinion poll, made several times in 1987 by the Chamber of Free Enterprise, tested the credibility of the news media in the eyes of the public. The respondents, men and women of different socio-economic strata, resident in the city of Guatemala, expressed increasingly favorable attitudes from January through September with respect to news media performance. The positive replies ranged from 62 to 67 percent and the negatives dropped from 11 to 9 percent. Those giving medium scores went from 27 to 22 percent. On the question of violations of human rights, the replies in September showed 42 percent believing the situation had improved, 40 percent worsened, 16 percent unchanged and 3 percent didn't know.

Our correspondent summarizes: The news media had freedom to work in 1987. Compared to years past, this was the best year. In March 1988 the leading daily, *Prensa Libre*, took President Vinicio Cerezo to task for an extramarital affair. He, in turn, commented that "there is positive news and negative news, and newspapers publish the negative news to sell more copies." Yet *Prensa Libre* found space to cover the murders of political activists. It also attacked a tax bill President Cerezo supported. In all, this was a robust demonstration of a newspaper tussling with presidential authority, and receiving political brickbats as hard as those it tossed. It is likely that the crux of the issue was the power of the military—not that of the presidency or the press—the military demanding that the businessmen, finally, pay for the war against the insurgency. The press will be less likely to challenge the military directly. Yet opposing the tax bill is a step in that direction.

A coup attempt by lower-level military men in May 1988 failed. Yet the settlement further reduced President Cerezo's actual power to rule. In June, the liberal daily *La Epoca* was firebombed to ashes. Byron Barrera, the director, told us that the arsonists are known but will not be prosecuted. He fled the country for several months and then returned to publish *La Epoca*—this time, he hoped, with some "guarantees" that "freedom of expression can be assured in a country that calls itself democratic."

Guyana
[IIIIIIIIIIIIIII_____MoF

Since the *Guyana Graphic* was taken over by the government in 1974, the only daily newspaper has been the *Chronicle,* which is owned by the government. It is little more than an outlet for official news; opposition views are barely covered. Unfavorable news is often ignored or reported some time after the event, accompanied by a rationalization. Newsreporting on the whole is tendentious.

Newsprint is used as a news-control mechanism. It can be imported only with a government-issued license. None was given to the opposition weeklies, the *Mirror* and the *Catholic Standard*, even for gifts of newsprint. They had to purchase newsprint "ends" from the *Chronicle* to survive.

Since the death of President Forbes Burnham in 1985, there has been some liberalization under his successor Desmond Hoyte. The most significant development in the media field was the president's agreement to permit the publishing late in 1986 of a new independent newspaper, the *Stabroek News*. The president refused, however, to make foreign exchange available to the newspaper. A start-up grant was secured from the National Endowment for Democracy. (NED has a bipartisan board, mainly of private citizens, who dispense funds allocated by the U.S. Congress.) The paper was printed at first on the presses of the *Express* in Trinidad, but has since been published on a second-hand press in Guyana. Some 20,000 copies are sold. Circulation would be greater, the publisher believes, if there were no newsprint constraints. The paper has filled a need for the free and open discussion of important issues. In its editorial column, the *Stabroek* calls for open government, and free and independent media. It supports the rule of law and other human rights issues. The *Stabroek* initially encountered resistance to interviews by government officials. There had been full control of news for so long that officials seemed afraid to discuss problems. After government ministers, heads of government corporations and the vice-chancellor of the university were interviewed, public discussion has widened. Although the paper is commercially viable—it sells all the copies it can print—foreign exchange restrictions threaten the *Stabroek*'s existence.

The government-owned *Chronicle* has changed its editor and raised the quality of the daily newspaper. The independent weekly, the *Catholic Standard*, edited by Jesuit Father Andrew Morrison, suffered a serious setback in 1988. It finally lost two libel suits brought against it in 1982 by President Hoyte and two others. In one suit, the *Standard* published a release from the opposition accusing the ruling party of deceiving the U.N. Development Program when applying for funds. Hoyte, though unnamed in the story, was then head of the ministry which applied for the funds. In the second case, the paper quoted an insurance official claiming the government lied to the World Bank and International Monetary Fund. Hoyte was also responsible for relations with the bank. Hoyte produced men who testified they believed that, though not mentioned, he was the subject of the articles. The appeals court ruled that a paper could be guilty even if it did not intend to libel the individual or even if the paper did not know the libelled person exists, as long as reasonable people reading the article could make the association. The *Standard*'s lawyer said that hereafter journalists would not be able to attack the government without fear of being sued for libel by those who implement policy. After appeals at Catholic masses, the *Standard* paid nearly $15,550 in damages and court costs.

The only radio station in Guyana is government controlled. News coverage is progovernment. There is one radio receiver for every two citizens. Many capable announcers have resigned over the years, and emigrated. There is no television system.

There was hope that conditions would change for the better under President Hoyte. Indeed, there is less victimization, fear and oppression than under the late President Burnham. But the economic situation has not improved, and that provides an argument for some remaining restrictions.

Haiti
[IIIIIIII_____MoF

In mid-1986, the government that followed the three-decade-long Duvalier dictatorship in Haiti guaranteed "freedom of expression" but prohibited publications damaging to "morality or public order." In the several privately-owned daily and weekly newspapers and radio stations, including broadcasts from the influential Protestant and Catholic outlets, vigorous discussions of public issues followed Duvalier's departure. The newspapers and radios, including those owned by the government, were highly critical of officials prior to the November 1987 elections.

Violence by the armed forces and civilians, and disruption of the election, cast a pall over the country, including the news media. One television cameraman was killed, two others wounded, other photographers harassed and foreign journalists threatened. The capital's seven largest radio stations were systematically sprayed with automatic weaponfire. Transmitters of four radio stations were firebombed.

If the election was a fiasco, the military attacks on journalists served to dampen significantly the spirit of freedom promised in the wake of the Duvalier era. The resort to self-censorship was increased by the Ministry of Information's denunciation of the press in March 1988. It referred to "cynical insinuations" by the media, and said such abuses of freedom of speech could be punished by the Penal Code.

Hungary
[||||||||||||||||||||||_____MoF

Despite government ownership and control of all official means of mass communication, Hungary permits greater latitude to unofficial (samizdat) writers and publishers than other communist countries. (Polish samizdat writers may publish more than Hungarian counterparts, but Polish dissent is less tolerated.) The Hungarian prime minister said in September 1987 that his regime will permit well-intentioned opposition provided it remains under strict surveillance and control. Hungary's first independent weekly news magazine, *Reform,* came out for the first time in September 1988. A biweekly journal, *Credit,* was scheduled to debut as the spokesman of the Hungarian Democratic Forum, a new independent political movement. By December, the central committee abolished its propaganda section, and a new law permitted private individuals to start their own newspaper.

The government continues to harass political opponents, and occasionally to expel them or encourage their emigration to the West. In early February 1987, police seized copies of a new samizdat magazine; in March, they took copies of other samizdat from the homes of three editors; in March 1988 Budapest police detained five samizdat journalists including the co-editors of *Beszelo.*

This publication, in a special edition of 60 pages, circulated widely under the title, "The Social Contract: Prerequisites for Resolving the Political Crisis." The document demanded constitutional checks on party rule, a law governing the Communist party's role, many economic and social reforms, freedom of information and press freedom governed by law with nondiscretionary licensing of the media. Licensing of the media is often attacked in the West as a step toward state control of independent news media; here, licensing is proposed by samizdat writers as a check on governmental power. This suggests that methods, institutions, even constitutions are sometimes less important than pragmatic policy and human intent and objectives.

The first page of the samizdat document stated: "Kadar must go." By mid-1988, Kadar was replaced as general secretary of the Party. *Beszelo* called for a law defining the government's obligation to provide information to the public on a great variety of political and economic subjects. The law would limit the classification of state secrets, and judicial review would be permitted to determine whether classification is lawful. Unlawful withholding of information by the government would be considered a felony. The law would also specify the frequency and manner in which information would be released to the public, including formal announcements and press conferences. The samizdat publication urged that press freedom be mandated in law because the established order of communication is an area of tension "between the power structure and the party still treats the press, radio and television as direct tools of propaganda." The central committee's agitprop department "determines which news media, when and in what tone, may discuss topics considered to be of political importance." Further, "It prevents making public the debates preceding a decision, launches press campaigns for the acceptance of [these] decisions as accomplished facts, and permits the reporting of contrasting opinions, at most, as indications of the public mood. The party does not tolerate criticism of official policy, not even in retrospect; and when the problem is already undeniable, it has to be attributed to mistakes in implementation."

Another statement signed by twenty-two journalists on 30 March 1987 also addressed media reform. They called it, "Turning Point and Reform —Suggestions for Media Reform." The Hungarian word "nyilvanossag" could also be translated "glasnost," or "publicity," as well as "media." The document described how the public's misconceptions about Hungary's economic problems had arisen as a direct result of distortions in the mass media. The report described the media as a "collective agitator" that helped the authorities retain power. It urged the government to expand the 1986 press law that would guarantee the rights of journalists and the limitations on state secrecy.

It concluded that substantive changes in the media must be accompanied by political reform, but that, in turn, could not be achieved without a free press. The report, created for the Patriotic People's Front, not samizdat, described without saying how censorship works in Hungary. [For further coverage of Hungary see Chapter VI.]

Iceland
[IIIIIIIIIIIIIIIIIIIIIIIIIIIIIIIIIIIIIIIIIIIIIIIIIIIIIIIIIIIIIIIIIIIIIIIIIIIIIIIIIIIIIIIIIIIIIIIIIIIII____MoF

Icelandic news media covered their greatest domestic story in 1986 when President Reagan and Soviet leader Gorbachev chose Reykjavik for their summit meeting. World politics, rather than fishing or tourism, became the top story.

The Icelandic newspapers (three dailies leaning toward local parties, one independent) and the State Broadcasting Service all operated, as usual, completely uncensored and with full right to criticize this or any other government in this highly pluralistic society.

India
[IIIIIIIIIIIIIIIIIIIIIIIIIIIIIIIIIIIIIIIIIIIIIIIIIIIIIIIIIIIIIIIIIIIIIIIIIIIIIIIIIIII_____MoF

The world's largest democracy presents many stark contrasts, perhaps none as puzzling as the promise of full press freedom and governmental practices which often belie the constitutional guarantee.

The most threatening since Indira Gandhi's "emergency" crackdowns on the press in 1975, was her son's, Prime Minister Rajiv Gandhi's attempt in 1988 to limit the mounting criticism of his policies. Thousands of Indian journalists held a one-day strike 6 September 1988 to oppose the Defamation Bill. They called the measure, pushed through the lower house, "Draconian" and "a gigantic fraud on the Indian public." The journalists' statement said the bill was the act of a "desperate government weighed under its own guilt," which needed the protective cover of secrecy.

The press argued that existing civil and criminal law is sufficient to defend the legitimate interests of the government or anyone else. The bill would have shifted the burden of proof from the complainant to the defendant, prevented the courts from dispensing with personal appearance for newspaper editors, provided heavy minimum penalties of fine and minister at first deferred the bill, and then withdrew it. The press, including papers that generally support the government, applauded the withdrawal as "a victory for journalists." The prime minister pledged to "uphold the legacy" of a free press.

Radio and television are monopolies in the hands of the state. Starting in 1988, 55 percent of the country was enabled to tune in to a single program broadcast simultaneously for six hours over All India Radio. This will make India one of the world's most inclusive domestic broadcasters. There are more than fifty million radio and five million television receivers—all tuned to programs controlled solely by the central government. The television network went through a brief reform in 1988, but soon reverted to its old role as a propaganda machine for the government. India-TV is watched by 120 million viewers. All India Radio (AIR) was putting into operation in 1989 Asia's largest shortwave transmitter. It will make India's voice heard around the world (except perhaps in North America).

State governments under the control of political parties in opposition to the ruling Congress (I) suffer as much as the ordinary citizen by news management, dissemination of false information, and blatant misuse of power. By contrast, there are more than 20,000 newspapers and publications in the country, most of them privately owned. They should enjoy a much higher credibility than the official media. But over the past two decades, government pressures and political and economic inducements offered for being compliant have eroded some of this credibility and goodwill.

There are two news agencies in India. They are owned by the newspapers themselves, and they compete actively. However, because of financial constraints which are aggravated by perhaps the highest communications tariffs imposed by governments anywhere, their dependence on state assistance has increased—first by the level of subscriptions negotiated on behalf of government radio and television and second, by block subscriptions by various government departments and embassies abroad. Incoming news copy has to be channeled through either of these two news agencies, but outgoing copy can, in theory, be freely sent. A widespread system of eavesdropping and unauthorized copies taken at the central dispatch stations under government control, all in the name of national security, water down the sanctity of professional news dissemination.

In 1987 there was a substantial increase in the level of intolerance of dissent by the government. The worst incident took place in September when the *Indian Express* newspaper chain was raided by hundreds of intelligence and special branch officials supported by paramilitary personnel. This brazen attempt at intimidation and harassment earned almost unanimous protest from the Indian press, although the level of criticism was tempered in cases where the government was able to continue to exercise pressures. The only exceptions were a handful of newspapers owned by the ruling party and their camp followers in the press.

Thanks also to the chorus of international protests, the interference with the newspaper's operations was given up on the same day, but the harassment continued. Ten cases had been filed against the newspaper and twenty-four show-cause notices issued. The charges sought to be leveled are those of which the government's friends in the media are equally, if not more, guilty. No action has been taken or is likely to be taken against them. The government is clearly guilty of highly selective harassment.

For journalists themselves, the profession could still be dangerous. Several were kidnapped in 1987. Five were murdered in 1988. A reporter for the Hindi-language daily *Janpath Samachar* was kidnapped from his home one day and his severed head found the day after.

Yet when two Indian states, Tamil Nadu and Bihar, attempted to introduce bills in their legislatures to restrain press freedom, a Bombay high court struck down the effort.

Indonesia
[IIIIIIIIIIIIIIIIIIIIIIIIIIIIIIIIIIIII_____MoF

The concept of guided democracy, applied to the Indonesian press, was expanded in the Press Law of 1982, based upon the 1945 constitution. Press guidance provides not only occasional oral demands from government that some events be covered or ignored, but a general requirement for coverage without specific guidelines. This generates a high degree of self-censorship in the news and information systems.

The Press Law stipulates that the mass media should support government strategy. Slogans of the regime's New Order to be advanced by the press make it an "instrument of the national struggle," "instrument of the driving force for national development," and "guard of Pancasila ideology."

The duties and obligations of the press are now defined along the lines of the New Order slogans. A further obligation of the news media is to "struggle for the realization of a new international order in the field of information and communication." This commitment, raised first by the Nonaligned Movement, became the driving force of Third World proponents for a decade at Unesco.

The 1982 law also strengthened the government's position in controlling the press council, and thereby the press itself. The council is chaired by the minister of information. The new law also provides licensing for the press. Licensing is not new but this version of the statute raised new fear among the news media that licensing would be used for political ends. The information minister in 1983 had hinted at this possibility. He said there would be a ban "if a paper is clearly spreading...Marxist/Communist ideology—in short, when the press is not in line with the philosophy of the nation and the state."

Despite the constitutional stipulation that "the national press will not be subject to censorship or muzzling," the independent newspaper *Prioritas*, a leading Jakarta daily, was banned in late June 1987. The paper was accused of publishing untrue, cynical, insinuating and "tendentious" stories. The Information Ministry had warned *Prioritas* in January that it had used "misleading headlines, sensational reporting and contents inconsistent with guidelines laid down in its government-issue publishing permit." The minister said that the government would prefer to use "persuasive action" to guide the Indonesian press but it would take repressive steps if necessary. He said that Indonesia's press freedom should be used by the media to "exercise constructive control for the good of national development." The banning came under the licensing system set forth in the 1982 Press Law.

As though to demonstrate this, the government in mid-1987 strongly warned *Prisma*, a publication of an economic and social research institute. The warning discussed four specific authors and the content of articles. Around the same time, Indonesia expelled the correspondent of the *Far Eastern Economic Review*. The government earlier in the year had set forth restrictions on foreign journalists covering the national election campaign in Indonesia. Some correspondents were denied visas, and permission to others was delayed. All foreign writers were kept out of villages where most of the voting took place.

Restrictions on the size and advertising content of newspapers are a major source of government control since most of the papers are privately owned. Officials said that these regulations of the Press Council were intended to help small papers with financial difficulties, but the controls were generally believed to be aimed at curbing the influence of mass-circulation, independent dailies such as *Kompas* and *Simar Harapan*. In 1986, the government seemed to substantiate this fear by banning *Simar Harapan*. In 1987, however, the government licensed *Suara Pembaruan* to publish with many of the same journalists as the banned daily.

It has been fourteen years since the wholesale banning of the press. Eleven newspapers and one magazine were shut down in the wake of student demonstrations following the visit to Jakarta of Japanese Prime Minister Tanaka. Some of the most influential newspapers were shut and have not yet reappeared. This banning included the detention of journalists. The memory of those bannings continues to strengthen the power of governmental guidance and its inevitable response, self-censorship.

Israel

[IIIIIIIIIIIIIIIIIIIIIIIIIIIIIIIIIIIIIIIIIIIIIIIIIIIIIIIIIIIIIIIIIIIIIIIII_____MoF

Democratic countries in times of emergency apply varying degrees of censorship. Occupying authorities, particularly during an insurgency, control the flow of news and information with varying degrees of oppression. Israel in 1987-88 acted—mainly in Gaza and the West Bank but also throughout Israel—as a democracy applying censorship to enhance security at home and protect its democratic image abroad.

The change in Israeli handling of the press is illustrated by Joel Greenberg who covers the West Bank for the *Jerusalem Post*.[3] In mid-January (1988), Israeli soldiers prepared to charge Palestinian boys pelting them with stones at the al-Amari refugee camp in the West Bank. The soldiers strapped on their helmets in full view of a battery of television cameras lined up behind them. Before charging the boys, a soldier turned to the camera crews and asked casually, "You guys have enough light? Can we get started?" The cameramen nodded, says Greenberg, "and the confrontation began." Four months later, however, when Greenberg filed a story about the latest leaflet published by the uprising's leaders the censor killed the story. These incidents, says Greenberg, "illustrate the transformation that has taken place in the attitude of the Israeli government." Tolerance and openness, he says, "have been replaced by hostility and increasing press restrictions."

Despite restrictions, and sometimes in defiance of them, the privately-owned Israeli press and foreign journalists provided vigorous coverage of the insurgency, dubbed "the uprising" of Palestinians. Television coverage in the West showed stone-throwing Palestine youths held at bay by young Israeli soldiers who first used live ammunition, killing some stone throwers and arousing foreign indignation at the "overkill." These photos were kept off government-controlled Israeli television, though the domestic press reported the incidents and aroused concern among Israelis over government policy and their country's image abroad. Israeli Defense Force tactics changed. Beatings and rubber bullets replaced live ammunition for the most part, but international press reports pictured the IDF as bloodily repressive. The IDF restricted the press from areas of combat—most of the occupied territories.

Control of the press is maintained by licensing of the media, censorship of those operating, and confiscation of those regarded as threatening Israel's policies. In 1987, the government banned an alternative Jewish newsletter and information center, withdrew the permit of the Palestinian daily *al-Shaab* to distribute in the West Bank, and ordered the closure of an Arab-owned news office in the West Bank. Many Palestinian journalists were arrested and imprisoned for extensive periods, several placed under house arrest, and others forbidden to travel abroad. Several foreign journalists were detained and released. Police broke into the offices of the magazine *Al-Sirat*, confiscated several issues and restricted the travel of its editor. On one occasion, the Israeli army arrested its own radio newsman and detained him for several hours.

In 1988, the uprising became more intense. The government shut down the Palestine Press Service, and banned a pro-Arab Hebrew publication as well as a Communist party daily. At least forty Palestinian journalists were arrested and held without charges or trials during the first ten months of the year. Jewish-Israeli photographers have been beaten. The government withdrew the press credentials of American correspondents for NBC News and the *Washington Post*. *Time* and *Newsweek* sought an injunction from the Israeli courts to recover news film the authorities had confiscated.

As in South Africa, the denial of press access to the scenes of intergroup conflict eliminated much television coverage of the beating and killing of protesters, but did not end foreign criticism of such policies. The dilemma—how to maintain a democratic system under fire, while applying censorship in the name of security—remains. The conservative newspaper *Ma'ariv* editorialized in December: "The blame attributed to the media for inciting unrest by reporting what is happening is a pathetic attempt to block off reality."

Italy
[IIIIIIIIIIIIIIIIIIIIIIIIIIIIIIIIIIIIIIIIIIIIIIIIIIIIIIIIIIIIIIIIIIIIIIIIIIIIIIIIIIIII_____MoF

Among Western democracies, Italy has one of the most subtle ways of conditioning its relatively free press—through the ownership of newspapers by a handful of industrial magnates well connected to the political power structure. In 1987, this subtle grip got a little tighter.

One of the first acts of newly-chosen Italian Premier Giovanni Goria was to write what he intended to be a confidential letter to the major newspapers asking them to ignore his personal life. Goria's small gesture suggests that Italy's power elite—descendants of its not-so-remote crown, church, and class systems—still has the traditional notion that it can mold the press into a submissive mouthpiece.

The widespread ownership of Italian newspapers by large industries caught in an internal web of political and financial pressure groups spawns an invisible but foolproof form of influence over journalists. This powerful lobby loosely holds the reins of the Italian press by imposing an ad hoc rule: Don't bite the hand that feeds you. "Never before has industry had such a strong influence on Italian newspapers," says Giuseppe Morello, president of the National Order of Journalists, the watchdog body of professional journalists. "On the other hand, industry has helped newspapers to regain their profits when they were in the red, and to acquire technology," he says "There is no formal limit to the freedom of the press, but the moment a journalist works for a paper owned by certain interests, the limits are defined—he can write anything he wishes, but he may find himself out of a job," Morello adds.

One example of industry's clout occurred in 1987 at the leading daily

La Repubblica, which is partly owned by Carlo De Benedetti and his Olivetti office equipment empire. When De Benedetti was questioned in a judicial case connected to Italy's biggest postwar banking scandal, the paper treated the story—which was given prominent general coverage— in a cursory way.

A law introduced in 1981 and renewed in 1987 required investors backing newspapers to reveal the nature of their interests. But it does not prevent the wealthy few who own Italy's major newspapers from using them as bargaining cards to gain favors, and journalists as dealers to load their special-interests dice. "At least now, with this law, we know publicly what is happening," says Morello.

Turin's daily *La Stampa*, one of the country's top three newspapers, is owned by industrialist Giovanni Agnelli and his Fiat auto company, which controls a total of 25 percent of Italy's daily newspaper market. *La Stampa* regularly lets Italy's industrial "prince," Agnelli, off the hook. Recently, Fiat's 50 percent ownership of the Valsella Meccanotecnica company, a producer of mines allegedly destined for Iran, was all but ignored by the Turin daily. When Fiat bought back Libya's controversial share in the company last year, the automaker provided a cost-free airplane and meals to lure journalists to Turin for a press conference dominated by Agnelli—a regular practice among Italy's major companies. Agnelli has also been known to grant exclusive interviews only on condition that they be printed on page one, and to blacklist news organizations that broke the promise. "Everyone knows the paper is owned by Agnelli—so they don't buy *La Stampa* to read about Fiat, they know what bias to expect," says Sebastisno Sortino, director of the Federation of Italian Newspaper Publishers. "But newspapers should not be objective, they should have an opinion," he says, "and, since newspaper-owning companies compete with each other, the newspapers are competitive."

This business-controlled hand feeds some of Italy's underpaid journalists by subsidizing their travel expenses and luxury hotels, while they dutifully cover events through rose-colored glasses—and assure themselves of the next invitation, along with a self-imposed press muzzle. As the profits of Italian industries soared in the past years, the feasts offered were bigger and better. In September 1987, when the Fiat-owned Alfa Romeo automaker came out with a new luxury model, 600 journalists were wined and dined in Milan for the car's debut, which was later televised live as a "bash" for 1,200 guests. They followed up by giving the new product a free blitz publicity campaign on news pages, complete with photos of Italian President Francesco Cossiga stepping into the four-wheeled wonder, of which gift samples were delivered to the Italian government. "The new Alfa Romeo seems to have all the winning numbers," wrote one journalist. "It is a lady with a great tradition." But, says Morello, "It's a problem when a journalist becomes the conscious or subconscious vehicle of the advertising machine." He adds, "It affects the way journalists operate, their professional dignity and quality."

Milan's conservative *Corriere dela Sera*, another of the top three papers in Italy, is also owned by a company controlled by Fiat. Milan's daily *Il Giorno* is owned by the state-run ENI petroleum conglomerate and Rome's *Messaggero* by Montedison, a chemical giant. *Italia Oggi*, Italy's newest and only independent financial paper, has been bought by Isvim, a private financial holding company. Its rival, *Il Sole 24-Ore*, has Ferdinando Borletti, the chairman of Valsella Meccanoteccnica, on its board of directors. Borletti is also on the board of Fiat. "It would be difficult to find a paper that writes badly about the interests of its owner, but for that reason, it writes twice as badly about its rival newspaper owner," says Sortino. "Italy is an anarchistic system—you have to buy at least two or three papers to get the whole picture of the day's news."

Italy also has one of the lowest per capita newspaper readerships in Europe—six million dailies bought by a population of fifty-seven million—further threatened by rising competition from a growing private TV network. As a result, financial pressures on the papers and an oversupply of journalists often act as natural barriers to press freedom by pushing many journalists into a lobbyist role for their sponsors, at least on some issues. "At each newspaper, the important thing is not to speak badly of the owner's interests—as for other people, you can write whatever you want," says Morello.

The Communist party newspaper *L'Unita* has a weekly satirical insert, "Tango," which often makes vicious fun of everyone, including its own Party members. "There is a very high degree of freedom in the Italian press—no other country's press publishes the kind of brutal or satirical attacks on its leaders, on the pope, on anyone, that you find in Italy's papers, for whom nothing is sacred," says Sortino.

At the same time, Parliament is debating proposals for new laws that would break up the concentration of industrial giants looming over the press. Italy, which a newspaper has called "the Cinderella of the EEC," is the only country in Europe without an antitrust law. According to the proposals, a single industrial group could not own more than 10 percent of the newspaper market. A more direct form of press control comes out of Italian courts, whose inconsistent interpretation of press laws is a chronic hurdle to journalists.

In the eyes of the law, reporters are expected to reveal their sources—but not always. Legal precedents do not establish a framework for future judicial conduct, so if a journalist was prosecuted for not revealing his sources last year, he can take his chances again this year and maybe get away with it. But if he risks using anonymous sources, is then called by a magistrate to testify as a witness to a crime and refuses to do so, he can be charged with aiding and abetting, and can be fined, jailed or suspended from his profession. Judges also decide occasionally to enforce the "segreto istruttorio," the pretrial investigation that is to be kept secret until the trial, but is routinely violated because of Italy's slow justice system. "Usually it is a magistrate who leaks pretrial information to the journalist—and his fellow magistrate who prosecutes the journalist for printing the information," says Sortino. This extremely wide interpretation on the publication ban of information from secret investigations keeps the press and courts at a constant standoff.

Such judicial inconsistencies, coupled with the lack of an official shield law, give the courts great powers of intimidation. But the interpretation of Italian press laws has become increasingly lenient and the penalties for infringing them softer. The restrictive climate of 1985-86, when more than a dozen journalists were suspended, jailed or killed by the Mafia, eased in 1987. This year, no journalists were prosecuted by any court. In addition, some previously filed suits were dropped, as evidenced by the defamation case against reporters who wrote about Licio Gelli, the grand master of the secret Italian Masonic lodge implicated in right-wing terrorism, a plot to destroy Italy's government and involving financial fraud. In September, a Roman court absolved the journalists for articles on the recently-arrested Gelli written before his first imprisonment and escape in 1984 to Switzerland.

In 1988, the most brutal way of silencing the Italian press was still in the form of hushed-up threats of violence as wielded by the Mafia, though this method has been on the decline. In short, the barriers to Italy's press freedom are not ironclad. Italian journalists can in principle write whatever they deem to be true, even if it is unpleasant for the establishment—but some of them may slowly find themselves being shut out of a secure job, and deprived of "la dolce vita."

Jamaica
[|||||||||||||||||||||||||||||||||||||||||||||||||||||||||||||||||||||||||||||||__MoF

Some improvement was recorded in 1987, continuing in 1988, for the free flow of news and information. There have been no impediments to the freedom of the print press—the privately-owned newspapers of the 153-year-old Gleaner Company and several other independent, small papers —but much of the electronic media have been owned by the government. The big change in 1987 was the government's offering to divest Jamaica Broadcasting Corporation (JBC) radio and television, and three regional radio stations. The government also agreed to license a religious television and radio station.

A public media commission, in its first year of operation, monitored the news systems owned by the government, and investigated complaints. The government station, at the suggestion of the commission, was made to apologize to the opposition for inaccuracy.

Generally, press freedom is guaranteed in the constitution and observed in practice within the broad limits of libel laws and the State Secrets Act. The major newspaper, the *Daily Gleaner*, has criticized all Jamaican governments while the JBC is usually accused of favoring government policies whichever party is in power. The privately-owned Radio Jamaica (RJR) is independent and often critical of the government though the government owns some stock in the company.

This improved picture of information freedom was marred somewhat by the precedent set in the passage of the bill to prevent the press from publishing details of evidence in divorce cases, though given in open court. Journalists protested and some amendments were made, but the offensive bill still passed. The press continues to publish with impunity.

Japan

It is equally accurate to state that the Japanese press exerts a strong influence on public policy, and the Japanese government exerts a great influence on the Japanese news media.

The privately-owned newspapers publish more than a dozen main editions a day. They reach all parts of the country almost simultaneously, and per capita circulation figures are among the highest in the world. *Asahi Shimbun* sells 11.2 million copies in the morning and 6.6 million in the evening. *Yomiuri Shimbun* has a circulation of nine million morning, five million evening, and *Mainichi Shimbun* serves 6.8 million morning readers, 3.5 million in the evening. Several other dailies sell about two million copies each. Such coverage is bound to influence a large audience, and therefore public policy.

Japanese journalism, however, is as close to the political establishment as are the industrial and financial systems. The interaction through informal press clubs attached to several Japanese ministries assures the cluing-in of trusted journalists through the formulation of policy and implementation. Veteran journalists and permanent officials are bonded through these informal press clubs.

Yet, clearly, journalists in Japan today enjoy a high degree of political freedom. There is no censorship, formal press code and, in the case of the print media, no explicit regulations. (An exception: In Okinawa prefecture where U.S. air forces and navy occupy much of the territory as their important military bases, the activity of journalists is often restricted.) Despite criticisms that the Japanese mass media engage in defamation and invasion of privacy, they receive relatively high marks from the Japanese public. A 1987 survey by the *Asahi Shimbun* reported that more than 60 percent of the Japanese public is satisfied with what newspapers report. About 30 percent of those polled said that newspapers are not always careful in respecting the individual's privacy and good name. According to the same survey, more than 60 percent of the respondents suspected that the editorial content of newspapers is subject to outside influences.

The great freedom notwithstanding, there is not a great variety of information, comment and opinion in the Japanese press. Indeed the striking feature of the news media is its similarity in editorial content. Ironically, freedom itself may explain the lack of initiative of some journalists in the postwar generation. Protected by the constitution against impediments from authority, some young journalists are not as enthusiastic about the quest for truth as are older journalists. The younger practitioners are likely to be driven by their emotions, and there is a tendency for them to write articles without sufficient facts.

The most violent incident occurred in May 1987 when a gunman from a far right nationalist group forced his way into a local bureau of the *Asahi Shimbun*, fired without warning and killed one reporter, seriously injuring another. The attacker has not been found. Earlier in the year, a time bomb was discovered in the parking lot of the same newspaper in another city.

There are two separate, extensive radio and television broadcasting systems. The Japanese Broadcasting Corporation (NHK), a public system, operates three nationwide radio networks and two nationwide television networks. NHK has a twenty-year lead in developing high-definition television, is the widest seller of direct-broadcast dishes, and has a vast research and development facility. Some 900,000 Japanese households have DBS. In addition, there are more than 130 independent members of the National Association of Commercial Broadcasting who operate more that 6,300 radio and television stations supported entirely from advertising revenues. Some political and economic influences are apparent in the content of radio and television programming, though journalists themselves play a strong role in editorial judgments.

Jordan
[|||||||||||||_____MoF

Although the daily press in Jordan is privately owned, it is strictly controlled by the government, especially by the Ministry of Information. The Ministry is also believed to provide indirect government subsidies to the four Arabic and one English-language daily newspapers. They all stick close to the line put out by the Information Ministry.

In recent years, one or another of the dailies has been suspended for coverage deemed objectionable by the government. The editor of the pro-government *Sawt al-Shaab* was fired from his post in 1987 for running afoul of a government directive on the Palestine Liberation Organization. Five other journalists were detained for breaching information policy. In May 1988, authorities withdrew press credentials of two Jordanian women reporters, one who worked for the *New York Times*, and the other for the *Washington Post* and NBC-News.

The domestic news agency is the Jordan News Agency operated by the Ministry of Information. The several radio and television stations, all controlled by the government, carry programs in Hebrew and English, as well as Arabic.

The principal impediment to the free flow of information in Jordan is the Ministry of Information. The press is perceived as a purveyor of the government line rather than as a source of active information concerning problems of the country or the region. The ministry acts as censor over news it considers objectionable and planter of information which the government desires to convey. Jordanian journalists are conditioned to such guidance and to the requirements for maintaining their status in the country. Since nearly all toe the line there are relatively few incidents involving interference with their activity.

Kenya
[|||||||||||||||||||||||||||||||||||||||||||||_____MoF

There is a strongly guided "free" press in Kenya—with certain generally understood and important limitations—and competition among the daily and weekly newspapers, most of which are privately owned. There is no systematic or formal censorship of the press, but self-censorship is widely practiced. Criticism and commentary are published within broadly understood limits which are not legally set forth. The press, for example, criticizes government policies and particular officials, but never attacks the President. The government, however, may guide editors in handling sensitive issues. Government pressure may be brought on journalists and publications believed to have avoided or undermined the government position.

During 1987, party officials warned they would censor foreign journalists because of their "smear campaign against the country." A *Washington Post* correspondent was expelled in June. Swedish or Norwegian journalists would be barred in the future, and the status of all foreign correspondents reviewed later in the year. Four foreign journalists were beaten by the police during a student riot in November, and President Daniel arap Moi warned that foreign correspondents would be "thinned out" to a "manageable level" for being "irresponsible."

Among domestic journalists in 1987, a correspondent of the *Daily Nation* was sentenced to two years' imprisonment, a Kenyan journalist working for BBC and Reuters was held incommunicado for sixteen days before authorities reported his detention, an editor of the Voice of Kenya (two weeks after he was reported missing) was revealed to have been detained, and four other Kenyan journalists were held for political offence. Two journalists, released in 1988, claimed they had been tortured while imprisoned, one since May 1986, the other since August 1987. The monthly magazine of Kenya's National Christian Council, *Beyond*, was banned in March and its editor detained and released on bond. The editor of *Nairobi Law* magazine was also detained and released on the same day. In August, the editor of a church magazine was sentenced to nine months in jail.

In the face of such events, journalists and editors feel strong political as well as commercial influences directed at the editorial content of the media. There is competition among the three English daily papers, the *Daily Nation* owned by the Aga Khan (Ismailia) group; the *Standard* (established in 1902) owned by the British conglomerate, Rhonro with Tiny Roland of London; and the *Kenya Times* owned by the government through the Kenya African National Union (KANU). Both the *Nation* and the *Times* have daily sister newspapers in the Swahili vernacular—*Taifa Leo* and *Kenya Leo*. All three principal papers have Sunday editions.

Because of the relatively high literacy rate in Kenya, the newspapers find it necessary to tone down blatant propaganda material, or lose readership. The government-owned papers in Swahili have improved their circulation in a short time by eliminating much official propaganda. As a consequence, one finds the same stories in all the dailies though with different leads.

There are also scores of magazines catering to particular interests—law, parents, finance, secretaryship, management, humor, etc.

Though some journalists have been detained and pressed to reveal their sources of sensitive information, the relationship between the government and journalists is stable. The press has generally heeded the official appeal to stress national development though there are more analytical, investigative and educative articles, well researched, than ever before.

The Kenya press has led crusades against corruption in both private and public sectors. As a result of investigative reporting, many formerly untouchable leaders have been hauled into court. The press has provided some constructive criticism. A story appearing in print one day may bring government action later on.

Despite threats to reduce their number, many foreign correspondents from around the world covering East, Central and Southern Africa, make Nairobi, the Kenyan capital, their base.

Kiribati
[IIIIIIIIIIIIIIIIIIIIIIIIIIIIIIIIIIIIIIIIIIIIIIIIIIIIIIIIIIIIIIIIIIIIIIIIIIII_____ MoF

The Kiribati constitution, in the British tradition, provides for freedom of press and speech. The island's sole newspaper and radio station are owned by the government on Tarawa. Catholic and Protestant newsletters are privately published. There is no censorship or notable limitations on press freedom.

Republic of Korea (South)
[IIIIIIIIIIIIIIIIIIIIIIIIIIIIIIIIIIIIIIIIIIIIIIIIIIIIIIIIIIIIIIIIIIIIIIIIIII_____MoF

If 1987 was the year of democratic promise for Korea, 1988 was the time of dramatic movement in democratizing governmental institutions and policies. Major beneficiaries, as well as important players in the process, were Korea's journalists.

The first direct presidential election in sixteen years was held in 1987. Two major opposition candidates divided a majority of the votes, a sign of a generally free and fair election. A peaceful inauguration was followed by an election in April 1988 in which, for the first time since 1950, a ruling party was denied a majority of seats in the National Assembly. The challenge to President Roh Tae Woo's Democratic Justice Party (DJP)—as his minister of information explained to this writer in a private interview—was finding consensual policies to which all parties can agree. "There are now four ruling parties," he said. Democracy demands compromise, consensus.

The press of Korea played a major role in the democratization process. In May 1987, 133 reporters from the *Dong-A Ilbo*, the country's most prestigious newspaper, demanded that the government lift press restrictions and carry out constitutional and democratic reforms. They attacked the daily government press guidelines that ban or restrict news coverage of sensitive issues. Later, 145 reporters from another leading newspaper, the *Hankook Ilbo*, issued a statement demanding that the president reopen discussion of constitutional reforms including press freedom. The major target of the journalists was the Basic Press Law passed in 1980.

The law merged broadcasting networks and newspapers, licensed media organizations, prohibited national newspapers from placing reporters in provincial cities and established a government-owned public television corporation. There were also provisions in the criminal code against spreading "rumors which eventually disturb peace and order" or "defiling the state." Such statutes were used to muzzle the press and punish dissident views. Under the law, the news media censored themselves according to "guidelines" issued regularly by the government. These came in conversations with editors. Journalists who resisted or objected to the guidelines, or who carried critical stories on politically sensitive issues were sometimes harassed or detained by security officers.

By the end of June 1987, the media climate changed dramatically. This was symbolized first by the appearance of a photograph of Kim Dae Jung, the opposition leader, published in the morning newspapers. Newspaper guidelines were regularly ignored. "Top government people are furious about it," said one journalist, but press freedom "is gaining momentum." In a poll conducted by *Dong-A Ilbo*, Koreans put press freedom first on a list of changes they wanted to see made by their government. "For me, democracy means press freedom," said one young man after hearing President Roh's historic speech the day before—now known as the "29 June 1987 Declaration of Democratization" by Roh Tae Woo, then the presidential candidate of the DJP. The speech promised to release political prisoners, restore other human rights, and remove restrictions on the press. The consequences of the election and unremitting press clamor, linked to demonstrations in the streets and the need to set Korea's house in order before the world watched the fall 1988 Olympic Games in Seoul—all of this helped produce significant reforms.

A new constitution, bipartisanly revised in 1987, went into effect in February 1988. The National Assembly, given enlarged powers, became dominated by the opposition in April 1988. The movement toward democracy was most visible in the press. The onerous Basic Press Law was abrogated by the assembly while still under the control of the DJP. The four major daily newspapers now frequently criticize or challenge the government and other parties.

Most significant, the minister of information told this writer in April, the government ceased issuing guidelines to the press immediately following the 29 June 1987 declaration. Before then, the ministry had instructed editors on whether certain coverage was "possible," "impossible," or "absolutely impossible." Three journalists were arrested in late 1986 for publishing the press guidelines issued by the government. Though convicted in 1987, their sentence was suspended. The minister told me in April, "Since last June 29, there is freedom of the press in Korea. We have not made a single phone call to the newspapers." The minister also found encouraging the present unionization of reporters and pressroom staffs on local newspapers. He told us that some twenty applications had been received and approved to start new publications. Among these is a major daily staffed by journalists fired at the demand of earlier Korean governments. The minister made what he regarded as a pertinent observation of the sea change in governmental policy. "I am the first minister of information who is a poet and an academic," he remarked. His predecessors were usually journalists-turned-censors.

He turned the questions to this writer: "What do you have to say about the media serving the society responsibly?" This is the perennial question of transitional governments fearful of freeing the press. "Shouldn't you watch whether the press violates human rights?," he asked. We assured him that responsible journalism is the other side of the coin of press freedom. The individual journalist must be aware of his responsibility to the citizens to provide accurate news and information, and access to diverse views. But the journalist's sense of professional integrity, not the government, should be the enforcer.

The publicly-owned Korean Broadcasting System, a nationwide, noncommercial radio and television network, is significantly fairer to varying points of view in its reporting and commentary. A manager of *Dong-A Ilbo* reports with relief that there have been no untoward government/press clashes such as that in August 1986 when the managing editor, political editor and a political journalist of *Dong-A Ilbo* were detained by the authorities. Another correspondent writes, "Press freedom in Korea remarkably expanded" during 1987.

From interviews in Seoul in mid-1988, it seemed clear that—student street demonstrations notwithstanding—the process of democratization of the party system and the news media has been significantly institutionalized. Short of a military challenge from the north or widespread internal disruptions from a small minority of the population, the suddenly freed press can help sustain the broader opening in the Korean society.

By June 1988 President Roh responded to student demands for talks with North Korea. He promised to allow greater access to newspapers and other materials from North Korea, to help public understanding of the Communist nation. The two Koreas have been divided since 1945.

Perhaps far more significant in the long term than the generally positive international coverage of Korea during the 1988 Olympics, was the domestic criticism during that period of the powerful army intelligence service. After members of the service ordered the knifing of journalists who had criticized the military, the defense minister issued a public apology for the attack and dismissed the chief of army intelligence. [Further coverage of South Korea see Chapter VI.]

Liberia
[IIIIIIIIIIIIIIIIIIIIIIIIIIII_____MoF

A presidential press conference in May 1987 signaled a sharp improvement in the status of the Liberian press. President Samuel K. Doe announced that the press hereafter was free to expose corruption whenever and wherever it occurs. He also gave permission to two newspapers previously shut down by the government to begin publishing again. The papers, *Footprints Today* and *Sun Times*, resumed publication. Both papers had been attacked by the government and closed for varying periods for some years prior to 1987.

Their travail did not end with the president's welcoming speech in May 1987. Eleven months later (April 1988), the government banned both newspapers and arrested five journalists, including the publisher of the *Sun Times*. Now, President Doe warned that his government would take a firm stand against those who wish to destroy "what this government has tried to build." He said some members of the press have been systematically trying to undermine the government's economic programs. Independent newspaper publishers declared a "week of mourning" in April and suspended publication for five days—the first press strike in Liberian history. The Press Union of Liberia filed suit against the government for the banning of the *Sun Times* and the arrests. The information minister ordered government-run corporations to stop advertising in the independent press. Water and electricity lines to various newspaper offices were also cut for alleged payment arrears. The government also banned the retransmission of BBC and Voice of America foreign news broadcasts on the official Liberian radio. A BBC reporter, Isaac Bantu, was arrested and imprisoned twice by the government.

All of this followed by less than a year the president's announcement that press freedom would be enlarged. Soon after that statement, *Footprints Today* exposed one of the greatest governmental scandals since the military takeover in 1980. The press exposures led to the dismissal of the foreign minister and the suspension of the chief of protocol in the ministry. Since the $575,000 scandal other newspapers exposed corruption, causing the president to establish a seven-man commission to probe such allegations.

In the immediate aftermath, journalists called 1987 "The Golden Age of the Liberian Press." Constitutional rule had returned the previous year and now the president was openly inviting the press to play a watchdog role. Journalists assumed, however, that self-censorship would still be in vogue. While the constitution carries provisions for press freedom it also stipulates that persons be held "fully responsible for the abuse" of this right. Decree 88a (1984) stipulates that spreading "rumors, lies and disinformation" is a felony. The decree is generally believed to be unconstitutional, and no individual has ever been convicted under it, but the threat persists. Journalists repeatedly call for the abrogation of the decree.

The government sent another mixed signal in 1987. It reinstituted licensing of journalists, and issued credentials to nearly 400 print and broadcast journalists. The Press Union objected to licensing as a form of "prior restraint," though there was no indication that the government had used mandatory licensing to deny accreditation to any journalist.

Government-owned radio stations and newspapers seldom carry information that does not follow precisely the official viewpoints. These media do not have as high credibility among the public as the independent newspapers and radio stations. Readership of the Information Ministry's newspaper, the *New Liberian*, is low, and it is regarded as a propaganda machine of the government. The ruling National Democratic Party as well as three opposition parties have monthly political publications. The Catholic church and other religious groups operate community radio stations. Their credibility is higher than that of the official channels.

Apart from government, other influential persons directly or indirectly may threaten media practitioners when they disapprove of news published about them. Even if reports are true, they tend to intimidate the journalists involved. Certain sources of information, moreover, refuse to cooperate with journalists. Reporting is made more difficult when stories are published without adequate comments from all relevant sources. The major improvement in the news flow in the latter half of 1987 was followed by further restrictions early in 1988. Two newspapers, *Sun Times* and *Footprints Today*, were closed down in April after reprinting skeptical reports about an alleged coup plot.

Print and broadcast journalists continue to exercise considerable caution. Some practice self-censorship or restraint because they feel there are obvious, unwritten limits beyond which they cannot go.

Luxembourg
[IIIIIIIIIIIIIIIIIIIIIIIIIIIIIIIIIIIIIIIIIIIIIIIIIIIIIIIIIIIIIIIIIIIIIIIIIIIIIIIIIII___MoF

Strong political influences—but not governmental pressures—are reflected in newspaper coverage in Luxembourg. But there is no censorship (except for bans on pornography). Opposition to the government is freely published and broadcast, and a variety of views are widely circulated. Newspapers are privately owned and, except for 4 percent indirectly controlled by the French government, the broadcast media are also independently owned. There were no claims reported in 1987 or 1988 of the government or other nonjournalists trying to inhibit the news media.

Malaysia
[IIIIIIIIIIIIIIIIIII_____MoF

The press/government confrontation has been growing since Prime Minister Mahathir Mohamad told the Malaysian journalists union late in 1986 that its members must behave responsibly or else the government would crack down. Dr. Mahathir justified this on the ground that he was elected by the people, and is therefore empowered to determine what the mass media should or should not convey.

Early in 1987, the prime minister foresaw threats to Malaysian stability from the growing political power of the ethnic Chinese. More than half the population is ethnic Malay (as is the ruling party), slightly more than one-third is Chinese and about 10 percent is Indian. In October, to secure virtually absolute control of the news media, the prime minister revoked the licenses and thereby closed four newspapers—the *Star,* the *Sunday Star, Sin Chew Jit Poh,* and the *Watan*—and arrested more than ninety people under the Internal Security Act. The step was said to be taken to reduce racial tension, but it may also have been aimed at suppressing political opposition. Coordinated with the shutdown was the introduction in the Parliament of strong, new limitations on the media to be imposed by amending the Official Secrets Act.

Under the amendments, the definition of "official secret" is so broad that, in the words of the women lawyers' association, "a wall of secrecy will surround almost any official information." The penalties of fine were removed and anyone prosecuted under the act would face a minimum of one year's mandatory jail sentence. The amendments were approved in December.

Also in December, Parliament amended the Printing Presses and Publications Act of 1984 to further tighten government restrictions on the news media. It is now a punishable offense to publish "malicious" news. Malice is "presumed" in the absence of evidence "showing that the accused took reasonable measures to verify the truth of the news." Further, the minister of home affairs is authorized to ban any domestic or foreign publication "which is likely to alarm public opinion." Court challenges to the minister's decision to suspend or revoke licenses to publish, are now explicitly prohibited.

Near the end of March 1988, the prime minister restored the licenses of the four newspapers he had banned five months earlier. The tension between the press and the government subsided, but the love-hate relationship continues. The government clearly has increased power to crack down on the newspapers, and it already controls broadcasting through Radio Television Malaysia. All independent journalism in Malaysia must be influenced—stridently forced to self-censorship—by the events starting late in 1986. [For additional coverage of Malaysia see Chapter II.]

Malta
[IIIIIIIIIIIIIIIIIIIIIIIIIIIIIIIIIIIIIIIIIIIIIIIIIIIIIIIIIIIIIIIIIIIIIIIIIIIIII_____MoF

The general elections of 1987 produced a dramatic change in government, and a favorable climate for Malta's news media.

The new Nationalist government favors broader dialogue for citizens and the press, as well as membership in the European Community. While practicing a policy of neutrality, it leans toward the West. Prior to the May elections, newspapers that did not support the Malta Labor government (in power for sixteen years, though for much of that time the Nationalists won more than 51 percent of the popular vote) were frequently harassed.

Malta has a strict libel law. Though fines are small, editors can be charged for trivial matters and must prove their innocence before the court. There is also an incitement law under which a journalist can be charged for creating a riot. Malta also has a "breach of privilege" under which editors can be charged and judged by Parliament for publishing privileged information. The new government permitted a challenge to the legality of the "breach of privilege" in the European Court of Justice. This challenge was not permitted under the previous government.

In that administration, a leading editor was arrested for a long period for interviewing the leader of the opposition (now the prime minister) for broadcast on radio. The building and press of an independent newspaper which had criticized the socialists was burned to the ground while police failed to intervene or arrest those responsible. An opposition press was bombed on three occasions. Reporters and photographers were repeatedly impeded and harassed.

After 9 May 1987, the number of libel suits was greatly reduced. The news media are free to comment and publish without restrictions. There have been no arrests of journalists, and they are permitted to engage in investigative questioning and writing. The Foreign Interference Act, which hampered foreign journalists in visiting Malta, has been lifted. No special permission is required for journalists to enter and report, and Maltese journalists are not restricted in traveling abroad.

Malta is a small country with an almost even division along political lines. Newspapers published by the parties tend to be biased in favor of the owners. The independent papers are more balanced and the broadcast media have moved significantly toward more balanced programming. There is no overt political interference or pressure on the media, though some self-control is exercised by the journalists and editors themselves.

Nauru
[||||||||||||||||||||||||||||||||||||||||||||||||||||_____MoF

On this tiny island republic in the west-central Pacific, press freedom prevails. There is only a weekly bulletin and a radio, both owned by the government. Since the creation in 1987 of an opposition party, some political influences affect the contents of publications and broadcasts.

Nepal
MoF

The press in Nepal is controlled by the state, despite the fact that there are only two government-owned newspapers out of some 472 privately-owned papers. The government papers are by far the largest, best financed and supported by advertising (often denied the independent papers by government action). The Press and Publication Act (1975) substantially abrogated traditional freedom of the press in Nepal.

The act governs all important aspects of Nepalese journalism. Licenses are required for publishing or for work as a journalist, whether one is a Nepali citizen or a foreigner. Licenses may be cancelled at any time by the government for the slightest violation of the press law. In 1987, the government closed nineteen papers for publishing objectionable material, and arrested nine editors and reporters. The courts ruled, however, that the government could not suspend publication of newspapers, but only cease supporting them. The Supreme Court left unchanged, however, the government's power to withdraw a newspaper's license. Consequently, despite the court action, the government still is empowered to close down the press.

The two government dailies carry some coverage of opposition activities, though there are no political parties in Nepal. There have been efforts to introduce the party system. There are, in fact, eight different Communist groupings leaning toward either Moscow or Beijing, and a group which calls itself the Nepal Congress Party. The country is ruled by a National Assembly under King Birendra.

Under the constitution, the press may not support a political party or write ill of the monarch or his family. Criticism of the government, however, is permitted. The distinction between the monarchy and the government leads to the uncertainties which limit the practice of journalism in Nepal. The independent press, though, criticizes government policies vigorously. Although the press is often accused of turning sensational, there are few libel cases because most papers cannot afford to pay damages. Foreign publications are confiscated or banned for carrying reports unfavorable to either the government or the monarchy.

Journalists must take into account not only the constitution and the press law, but the Secrecy of Confidential Documents Act, Freedom of Speech and Publication Act, Civil Liberty Act, and the Libel Act. It may be considered libelous, for example, to publish articles deemed to promote disharmony, enmity, or hatred between religious communities; or which cast aspersions on the dignity of the king; or have an adverse effect on the country's integrity; or which encourage party politics or ideology.

There was a heated debate in the national legislature in 1987 over the arrest of the editor of the *Janatyoti*, and the banning of the publication. Replying to questions from legislators, the communication minister said that while press freedom is guaranteed by the constitution, no one may dishonor the law and damage the national dignity on the pretext of invoking that freedom. Yet in a subsequent editorial the English-language daily *Motherland* called the arrest an insult to the press. Critics of the arrest agreed that there are limits to freedom, but they should not be arbitrarily set. One such limitation is the government's insistence that newspapers be published either in Nepali or in English, though a major segment of the population is literate only in other vernaculars.

If 1987 was a comparatively better year for Nepalese journalism—the number of bannings and arrests of journalists, though high, was reduced from previous years—the first half of 1988 was still better in this regard. The same uncertainties facing the journalist on his daily rounds, nevertheless, prevail.

Nicaragua

Overwhelmingly, the flow of news and information inside Nicaragua is controlled by the Sandinista government. This control is maintained by the publication of one Party daily newspaper, *Barricada*, and another pro-Sandinista paper, *El Nuevo Diario*; government-owned radio and television; the official Nicaraguan News Agency; and scores of other publications. The single opposition press voice is the veteran daily *La Prensa* which was among the few vocal opponents of the Somoza dictatorship. Occasional critical views may be heard on Radio Catolica, operated by the Catholic church. The highly limited freedom granted *La Prensa* and Radio Catolica is intermittently threatened, completely snuffed out, or haltingly restored—all depending upon the momentary whim or political necessity of the Sandinista officials.

The government leaders state that strict controls are necessary because of the insurgency and the state of emergency. Such power retention, however, has been part of the fundamental Sandinista objective since before it came to rule. The 1977 platform of the FSLN (Sandinista National Liberation Front) declared that once its revolution succeeded, "we will be able to develop openly along progressive Marxist-Leninist lines. We will be a party of iron, forged and tempered in the same process to enable us to fully organize and mobilize the masses." That long preceded the contra insurgency. Decrees 511 and 512 strictly regulate the publication of information about internal security, national defense and even economic issues. The minister of the interior may apply prior censorship at any time.

For several years before *La Prensa* was shut down by the Sandinistas in 1986, the newspaper was thoroughly censored by a Sandinista representative every day before publication. On some days, major sections were eliminated. *La Prensa* resumed publication 1 October 1987, after fifteen months of closure, because the Sandinistas had agreed to the five-nation pact to democratize the Central American region, and cease hostilities.

In permitting the reopening of the paper, President Daniel Ortega co-signed a remarkable communique with Violeta Chamorro, publisher of *La Prensa*. The president at first demanded that prior censorship be made a condition for resuming publication. Mrs. Chamorro refused, and President Ortega agreed, in writing, that *La Prensa* could appear "without any restrictions except those imposed by responsible journalism." The definition of "responsibility" remains with the police power of the state and not the professional integrity of the publisher.

In mid-April 1988, the Sandinistas managed another shutdown of *La Prensa*, this time using the government's control of newsprint. *La Prensa* was denied access to paper, but when a gift of newsprint was offered from the United States, the government relented and publication resumed. Newsprint control places *La Prensa* at the mercy of the state.

When the government permitted Radio Catolica to resume broadcasting, the station was barred from carrying news programs. Later, some broadcast restrictions were removed but in May 1988 several of the radio's news programs were again banned. Another twenty-two radio stations shut down earlier by the Sandinistas have not been permitted to resume broadcasting; nor are a number of trade union and other publications allowed to publish again. There is no free access to television in Nicaragua. The government has a complete monopoly of ownership, and strictly prohibits access to the medium by independent citizens or opposition groups. Independent news media, moreover, are regularly attacked and threatened over the official television channels.

In May 1988, as negotiations between the Sandinistas and the contras loomed, the government used *La Prensa* once again as a pawn in the power struggle. New restrictions on the opposition press were imposed. These restrictions applied to both radio and the written press, though the cease-fire agreement signed with the contras in March provided for "unrestricted freedom of expression." Said the news director of Radio Catolica, "We are going backward."

The manner in which progovernment media are controlled was revealed in the publication of a 1981 memorandum leaked to *La Prensa*. The head of the FSLN department of mass media was concerned that Poland's put-down of the Solidarity labor union might suggest a parallel in Nicaragua. He addressed the managers and editors of state-controlled media in an official memo:

Reflect the difficult situation that faces the Polish revolutionary movement from an objective viewpoint, reporting only those facts that have been confirmed by Tass and Prensa Latina (Cuban) news agencies...Stress that the emergency measures taken by the party and government in Poland are aimed at rescuing the country from the crisis created in Poland by the violation of socialist principles on the part of some ex-leaders of the (Solidarity) union, today under arrest, and the manipulation of these errors by the counterrevolutionaries directed by imperialism...They are not repressing the workers but only the counterrevolutionary elements that wish to hand over Poland to imperialism...It shouldn't be stressed that strikes are forbidden...Although it is not possible to neutralize the reactionaries' anti-Sovietism in reference to Poland, we can neutralize the possible analogies that the reactionaries may make between Nicaragua and Poland...

The narrow, and narrowing, window of press freedom in Nicaragua was further clouded through mid-1988 by the power needs, changing from day to day, of the Sandinista leadership. *La Prensa* was shut down for fifteen days and Radio Catolica indefinitely for allegedly supporting a large public rally for democratization, and against the Sandinistas. Jaime Chamorro, on *La Prensa's* board of trustees, doesn't deny his paper's strong anti-Sandinista position. He says: "We can't be objective. This is a war—not of arms but of ideology. We're engaged in a life-or-death struggle. If the Sandinistas consolidate their system, *La Prensa* dies."

Subtle pressures were seldom used to restrict or ban the small opposition voices. Yet those voices, especially when muzzled or silenced, were heard outside Nicaragua as portents of still more pervasive political controls to come. Perhaps the clearest symbol of those portents came in May 1988 when Interior Minister Tomas Borge punched in the mouth the director of Radio Corporacion, a small independent broadcaster, for allegedly reporting false information about a hunger strike.

The struggle over *La Prensa* is, indeed, a vital battle for power. The power of the independent press, which for decades played a major role in Nicaraguan life, even as a critic of Somoza, has been severely limited. The claim that the Sandinistas seek a "broad-based popular hegemony" is a catchphrase for absolute government monopoly of information power, oriented, in the Nicaraguan/ Sandinista case, to Marxist ideology.

It is misleading to say that "Nicaragua has the freest print media in Central America, because there are now two powerful sides spewing contempt at each other."[4] First, *La Prensa* is no longer powerful except in the physical perseverance and editorial independence it demonstrates; realistically, the newspaper is held hostage to the daily whims of the government. Toward the end of 1988, *La Prensa* signaled that new restrictive laws forbidding U.S. funds to enter Nicaragua for private purposes could put the opposition newspaper out of business in January 1989. [See Appendix C: Mrs. Chamorro's letter to President Ortega.] A small weekly, *La Cronica*, appearing for the first time in November as "centrist," still had to face Sandinista responses. The very opposition of *La Prensa* is a factor exploited by the Sandinistas to smoke out domestic opponents, woo foreign governments, particularly European financial supporters, and satisfy American congressmen opposed to aiding the contras. With all the power in the hands of the government, the daily headlines of *La Prensa* are little more than a charade in a larger game—hardly a sign of "free print media."

Nigeria
[IIIIIIIIIIIIIIIIIIIIIIIIIIIIIIIIIIIIIIIIIIIIIIIIIIIIIIIIIIIIIIIIIIIIIIIII_____MoF

Newspapers here are still among the most vigorous and varied in Africa. There are seven privately owned national dailies, and one daily owned by the federal government, which also holds a majority share in another newspaper. Private papers compete with state-owned publications. All social and political issues are discussed with fervor in the press. But the constitution permits only the federal and state governments to own radio and television.

Yet journalists are interrogated or jailed—fired if they work for government papers—for publishing articles said to oppose government policy, or embarrass authorities. There were several such occasions in 1987. On grounds of defending national security, officials impounded one issue of a magazine, limited religious broadcasting on radio and television and continued to require government authorization for interviewing civil servants.

The most chilling press-control action in 1987, however, was the closure by government decree of *Newswatch*, the popular weekly news magazine. The magazine and its three top editors were charged in April with violating the Official Secrets Act. The controversial edition was already on the street. It carried the long-awaited secret report of a presidential task force charting Nigeria's future. The ban, scheduled for six months, was ended a month early and the editors were never prosecuted. Professionals in many fields joined journalists during the time *Newswatch* was blacked out to call the ban illegal, undemocratic and a blot on the country's human rights record. While the incident symbolized the hazards of publishing here, it clearly also showed that Nigeria's lively press is resilient.

Of Nigeria's nearly 100 million citizens, about 10 percent avidly follow national and international developments as reported by the news media. In addition to information and entertainment, advocacy draws readers to the papers. Between 1936 and 1956 the national newspapers that were most successful were those that most actively advocated "freedom for all; life more abundant," a slogan for independence coined by one of the political parties. The papers that were prepared to "publish and be damned," such as the *West African Pilot*, were regarded as heroic institutions and their reporters, heroes in the struggle for independence. In their writing, they often referred to the American Declaration of Independence and the British Bill of Rights, among others, as their sources of inspiration. Responsible sensationalism and emphasis on the goal of independence were hallmarks of journalism in Nigeria. Nigeria attained independence on 1 October 1960.

The respect won by the media, then, enhanced their credibility in the years after independence. However, one government-owned and several privately-established newspapers that reflected mainly the political views of their proprietors have lost credibility through 1983. Writes our correspondent, "The public in my country can recognize a courageous, visionary, responsible, objective and progress-conscious newspaper when they see one, and will prefer to patronize it to others that they perceive as playing the role of 'his master's voice,' public or private."

Upon coming to office in August 1985, the present government, headed by General Ibrahim Badamasi Babangida, abrogated its predecessor's Decree No. 4 (1984) titled "Protection of Public Officers Against False Accusation Decree." This posed problems for journalists in securing and publishing information from official sources. General Babangida pardoned two journalists who had been jailed for one year under Decree No. 4.

Apart from self-censorship and the usual unwillingness of civil servants to talk to the press, there were no serious impediments to the free flow of information in Nigeria through the latter half of 1987 and 1988.

Norway
[||||||||||||||||||||||||||||||||||||||||||||||||||||||||||||||||||||||||||||___MoF

The news media play an important, traditional role in Norway. Journalists are conscious of this, and the public expects them to be. The credibility of the press, therefore, is very high in Norway. The country, moreover, has been spared the "dirty journalism" seen in other Western European countries. The public regards the news media as both the spokesman for and the tool of more powerful institutions or authorities (such as the government or large private firms). Yet Norwegian journalism is one of the world's freest despite its need for government subsidies since 1969. In recent years, the Norwegian media have also given greater prominence to what many see as unjust treatment of political-asylum applicants, and in some cases, the press succeeded in changing a "no" to a "yes" for political refugees. Politicians and government officials, aware of the impact of the media, often set their priorities by what the news outlets write at a given time.

Because of Norway's relatively small size, there is no great gap between journalists of the Oslo press and readers outside of the capital. Journalists of the Oslo dailies (considered national, while those outside Oslo are local) are far from isolated from general opinion in Norway. The content of the media, therefore, is an accurate reflection of what concerns people generally.

There are virtually no obstacles to the flow of news and information in Norway. The only restrictions pertain to covering defense issues and the fact that only the government has the right to transmit telecasts on a national level. The issue of coverage of defense subjects splits many Norwegians, and to a lesser degree, journalists of the dailies. Many agree that some topics are best kept secret. Others, however, feel that press freedom should be strengthened in this field. For example, Norwegian journalists get more information about the defense of Norway from American civilian and military officials than they get from Norwegian officials on the same topics.

There are no known examples of intimidation of journalists or threats by government officials. The climate of freedom of expression continues at the same high level as it has been since the end of the Nazi occupation in 1945—although there are still demands from center-right politicians for the abolition of the government monopoly on national television transmissions.

Panama
[IIIIIIII_____MoF

Independent journalists called 1987 their worst year—until 1988.

For twenty years Panama has lived under autocratic military rule. For the first twelve years, the country suffered a virtual information blackout in which no independent voice could be heard. Credibility of the media dropped markedly, as did the circulation of newspapers.

In 1979, as the Panama Canal Treaties were signed, a "democratic opening" was announced by the regime. Though little changed in the autocratic composition of the government, some 700 Panamanians took advantage of the "opening" to create *La Prensa*, the first independent, daily voice to be heard in the country in a long time. The paper was an immediate success and became Panama's newspaper of record. Soon after, *Quiubo* (an independent weekly tabloid), *Extra* (an opposition daily tabloid), *El Siglo* (a daily tabloid), and two radio stations—KW Continente and Radio Mundial—started up.

During this period, the regime used many methods of repression. Some examples: *La Prensa* suffered physical attack by government goon squads (Oct. '81); libel suits with arrest warrants for editors (Nov. '81); personal threats to its directors by the commander of the Panama Defense Forces (Dec. '81); personal attacks against the publisher (June '82 and later); storming and violent takeover of the paper's premises by Panama Defense Force troops and destruction of equipment (July '82); government-related mob attack (June '83); physical beating by government goons of a daily columnist and shooting at a woman reporter and a photographer (Apr. & May '84); ordered shut down by the Labor Ministry (Sept. '84); physical attacks on reporters (Dec. '84); death threats to editors (Oct. '85); jailing of an associate editor (Feb. '86); death threats, personal attacks and forced exile of an editor (May '86); rubber-stamp legislature resolution condemning an editor as "traitor of the nation" (June '86); associate editor exiled (July '86); harassment by Ministry of Government and Justice (Oct. '86); order of prior censorship (June '87); takeover and occupation by Panama Defense Force troops (June '87).

By September 1987, all independent newspapers—*La Prensa, Extra, El Siglo* and *Quiubo*—were closed and their plants occupied by Panama Defence Force troops. Independent radio stations—KW Continente and Radio Mundial—were closed by government order. All television stations, though sympathetic to the regime, are under orders of prior censorship. The editor and associate editor of *La Prensa* are exiled, and many other journalists are under threat. Panama is again under a total blackout of independent media. The people's only method of securing independent information is through the telefaxing of news reports received from abroad, and reproduced on copiers in Panama as single-sheet handouts.

A survey conducted in Panama by the Gallup International Research Institute (August 1987) elicited—in response to questions concerning people's fears and inhibitions—answers such as the following: "The telephones are bugged." "In this country, we can be thrown in jail for any reason at all." "The conversation we are having could be classified as subversive."

During the first half of 1988, the image abroad of General Manuel Antonio Noriega, Panama's de facto leader, suffered dramatically. He was charged in the United States with drug trafficking and many related activities, and implicated in at least one bloody assassination by live decapitation. The general and his associates blamed the press for his blackened reputation. Five foreign journalists were expelled or barred from Panama for their reporting of demonstrations.

In March 1988, police suddenly took vicious actions reminiscent of the harsh crackdowns eight months earlier. The police raided the Marriott Hotel in Panama City which headquartered reporters working for the foreign press. As a press conference of the Civic Crusade, the opposition alliance, was about to start, police beat several journalists, arrested some dozen reporters and representatives of the Crusade, and confiscated tapes and film of newspersons. More than fifty persons were arrested, and some taken to hospital for treatment after beatings.

Also in March, security agents ransacked Radio Mundial. The station was totally destroyed. The owner and six station employees were beaten before fleeing. The owner's father was beaten, kidnapped, and later found injured where he was tossed out.

And, at this writing, winter 1988—Panama's worst year for news blackout—was not yet over.

Perhaps a hopeful, long-range development, however, was the wide use of facsimile machines by Panamanians in the U.S. and Europe. The exiles used fax to send news reports and documents concerning Panama by direct telephone lines to anti-government friends in Panama. The faxed copies were reproduced and widely distributed in beleaguered Panama City. One four-page publication, *Alternativa*, is created every week or two on a desk-top publisher in Pennsylvania and sent by fax to Panama. Fax may soon become as important as pamphlets, audio cassettes and VCRs in breaking the absolute control by censors in many places.

Papua New Guinea
[IIIIIIIIIIIIIIIIIIIIIIIIIIIIIIIIIIIIIIIIIIIIIIIIIIIIIIIIIIIIIIIIIIIIIIIIIIIIIIIIIIIIII_____MoF

This largest Pacific island-nation with a democratic parliamentary system had tranquil press-government relations during most of 1987. The government charged an Australian television journalist with breaching Papua New Guinea security on its border with Irian Jaya. The prime minister later apologized. He also decided to delay the inception of television by two private networks, explaining, "Our cultures and languages are too precious for us to allow television to operate in PNG without strict policy guidelines."

That statement forecast what journalists regarded as a far more threatening action by the PNG communications minister late in 1987. He told the parliament that his government proposed to introduce a bill to control and regulate all mass media—radio, newspapers and television—in the country. Opponents of the bill, mainly the press and the opposition parties, claim the bill, if enacted, would damage democracy. The minister and the Catholic church say the bill would only curb foreign domination of business enterprises, particularly the news media. The bill would regulate all the news and information systems under a single quasi-judicial body to be called the Mass Media Tribunal.

The body would include representatives from the media headed by a political appointee. It would set uniform codes of fairness and standards of journalistic conduct for all PNG media organizations which would be licensed. Each license would be renewed on the basis of a public inquiry proving that the organization had fulfilled the conditions of licensing. Regarded as most threatening by the media was giving the tribunal power to direct a news organization to publish or broadcast matters that the tribunal or the Communications Ministry deemed to be of national importance. The tribunal could also direct a licensed organization not to publish matters prohibited by the censor, the tribunal or the minister.

The general manager of the PNG *Post Courier* said the scheme is "the sort of legislation you would expect to find under a dictatorship; not in a democratic country such as PNG." Despite government assurances that freedom of the press would not be interfered with, he said, "such a proposal provides a mechanism by which control of the press can be implemented." Oseah Philemon, one of PNG's most senior journalists, said, "A government that legislates to control the media is a government that does not want public scrutiny of its decisions. It is a government that does not intend to be accountable for its policies and action."

Paraguay
[IIIIIIIIIIIIII_____MoF

Censorship in Paraguay is pervasive and, to make journalism more nerve-wracking, controls are arbitrary. A government ruling in November 1987, for example, allowed a press conference to be organized at Radio Nanduti—"esta permitido, pero"—but, said the official decision, "no se puede entrar"—no one may come!

This writer was to have been the speaker at the conference. I had come to Asuncion to lend support to Humberto and Gloria Rubin, directors of embattled Radio Nanduti. The closing of the radio station in January was as crude and bizarre as life under the dictatorship of General Alfredo Stroessner. Radio Nanduti became Paraguay's major opposition voice three years earlier when Stroessner shut down *ABC Color*, a daily newspaper which was independent and critical of the regime.

Humberto Rubin, who created Radio Nanduti twenty-five years earlier, turned it into the country's most interesting purveyor of news and information. "Nanduti," in the language of the indigenous Guarani Indians, means "spider web." Using live news reports and comment from mobile units spread across Paraguay (an innovation here), coupled with listeners' phoned reactions, Rubin spun a web of fresh and uninhibited ideas. One-third of Paraguay's 3.7 million citizens were said to listen to some part of Rubin's two-hour morning news program. Such Nanduti programming challenged the thirty-four-year-old military regime, which forbids all criticism of the president, members of his family, government officials and civilian leaders.

Rubin became a personal as well as professional target of government and progovernment broadcasters and newspapers. His ample physical appearance, sports shirts and full beard drew vicious caricatures, and such references as "that ungrateful Jew." Anti-Semitic catcalls were mixed with other vulgar shouts, shrill blasts and the sounds of gunfire and rocks aimed at Radio Nanduti studios on 28 April 1986. A mob, which was later traced to a government agency, attacked the station. A week later, its transmitter was targeted. When Rubin persisted, jammers placed a shrill whistle over every news transmission on Nanduti's frequency. Early in January 1987, with Nanduti's advertisers shamefacedly admitting they had been pressured into dropping their advertising, Rubin silenced the station. He then held unbroadcast seminars and published archival material in the Nanduti building. The press conference in November 1987, to be addressed by this writer, would have urged the reopening of the facilities to daily broadcasting.

Radio Caritas, the courageous voice of the Roman Catholic church, repeatedly carried an interview with us which had been taped earlier in the day. A mobile unit of the radio broadcast live as the car drove toward the Nanduti building. The car was stopped at a paramilitary blockade set up outside Nanduti. Armed forces had shut off all access to the building, and guards ringed the structure. Some broken glass from the mob attack months earlier—an attack not stopped by police or the military—still remained just inside the entrance. No reporters were allowed to enter. When I emerged, I was briefly detained, and my passport examined. I walked to a hotel and held a press conference. I said that in Vietnam, the Soviet Union and South Africa I had never been barred from speaking as I had in Asuncion. Such action, I said, reflected the government's fear of the radio, of free speech. These remarks were published the next day in several Paraguayan newspapers—another indication of the arbitrariness of the system, and the choices editors make on how far they can test the officials.

That day, too, a delegation from the Inter American Press Association met separately with the information minister and President Stroessner. The IAPA urged the reopening of the newspaper *ABC Color* after four years. They came away feeling the paper might be allowed to publish again. But when they next went to the *ABC Color* building to hold an announced press conference, police prevented reporters from assembling. This ended another year of intermittent threats and violence directed at journalists and the news media.

As 1987 began, two radio journalists were in prison; five people were injured, in March, in clashes with police after a Roman Catholic mass protesting the continued closure of *ABC Color*; a Radio Nanduti and an El Pueblo journalist were arrested in April; the editor of *Hoy* was imprisoned in June; a photographer for *El Diario de Noticias* was beaten and his hand broken; a photographer for *El Pueblo* was arrested in July; the publisher of *El Pueblo* was imprisoned in August, and the weekly indefinitely suspended.

Yet 1987 was different from preceding years. The selection in August of a new governing body of the ruling Colorado party produced two contending factions for the first time. An intensive electoral campaign between the "traditionalists" and the "reformists" broadened freedom of discussion in the country. Colorado party leaders used their publications to attack opposing candidates in ways rarely seen in Paraguay. Charges and proofs of dishonesty were publicly aired. Self-censorship in the general news media markedly decreased. Opinions of the Stroessner opposition made it into print. But this less restrictive period ended 1 August the day of the Colorado party convention. Repression immediately escalated.

In April 1988, a journalist for *Hoy* and *La Tarde* was detained and severely beaten. During the pope's visit to Paraguay in May, the government imposed additional restrictions on the news media. All journalists had to obtain special accreditation, and no reporter could cover more than one or two of the pope's numerous activities. Moreover, radio and television could transmit only the government's coverage of the pope. Late in 1988, the government detained six staff members of Radio Caritas and forced the station to reduce its signal to just one kilowatt, so it could be heard only in the Asuncion area.

The press/government relationship in Paraguay suggests the utility as well as the futility of censorship under even a relatively benign dictator. The news media have no direct influence on the government. A dictator does what he believes he has to do to maintain power. The news media serve as an irritant, yet provide a clue to actual public opinion. The government, however, strongly (often violently) influences the content of the news media. Yet the more successfully it does this, the less credible become the news media in the eyes of the public. Indeed, once the public understands that news and comment are under government control, people usually conclude they have not been told the whole truth— no matter how much a courageous journalist may risk in revealing *some* truth. This disbelief is sustained by the iron hand of dictatorship clad in the velvet glove of journalistic self-censorship.

Stroessner was overthrown early in 1989 in a bloody coup engineered by General Andres Rodriquez. Radio Nanduti and *ABC Color* started up again, but the limits of freedom were not clear.

Peru
[IIIIIIIIIIIIIIIIIIIIIIIIIIIIIIIIIIIIIIIIIIIIIIIIIIIIIIIIIIIIIIIIIIIIIIIIIIIIIIIIIIIIIIIIIIIIIIIIIII__MoF

The military who ruled Peru for twelve years ending in 1980 nationalized the newspapers, controlled radio, repressed magazines and left a legacy of increasingly murderous terrorism. The media have been returned to their private owners, but confront diverse dangers in a climate of violence and economic crisis. In 1983, eight journalists were lynched by an enraged peasant community. Late in 1988 and early in 1989 three reporters were assassinated. Around the same time, radio stations and news agencies were occupied by guerrillas, and some newspapers were bombed.

Peruvians, however, may read or hear news and opinion from every conceivable viewpoint, including that of terrorist groups. The four major private television stations voluntarily agreed in July 1988 to reduce to one minute the coverage of news of terrorism and acts of violence in their daily newscasts. Called the "minute accord," it is supposed to reduce the public impact of the terrorist groups, and is a response to criticism of the press for dramatizing those who break the law and exploit violence and its morbid attraction.

The government owns one of the three national television networks, a national radio network and two newspapers. Opposition parties control their own newspapers, and also have access to government news media.

Our correspondent writes: News media in Peru have traditionally swung from restrictions imposed by dictatorial governments and the losing battles led by stubborn journalists, to the fiercely competitive and quite abusive free-for-all atmosphere of democratic regimes. Both crude sensationalism and a healthy rebellious spirit are long-standing traits. Revolutionary pamphlets played a heroic and unifying role in the war of independence with Spain, but during the early nineteenth century civilian and military groups used notoriously libelous publications as weapons of personal aggression and national factionalism. And today's style somewhat reflects the past. However, historian Raul Porras Barrenechea put things in a positive perspective: "Despite all outside criticism and inherent faults, one single valiant merit redeems our journalism: its obsessive love for freedom." Sixty years later, the public probably still views the news media with Barrenechea's eyes.

Credibility, in other words, may not be Peruvian journalism's strong point, and many of its sins are evident, but the willingness to radically criticize official policy when a government's honeymoon period is not even over, and to denounce corruption and human rights violations, and attack real or imaginary flaws, is greatly appreciated. Possible inaccuracy and injustice seem at times to be of secondary importance. In a society in which political power can often be wielded with an iron fist and a predatory glove, the man in the street first enjoys the feeling that the big "taitas" (chieftans) can get it in the neck, as bulls are pricked with "banderillas."

It must be added, however, that a change is taking place. Democracy, in its tenth year and still serving as inspiration, is pushing Peruvians and, in particular, Limenos to demand more objective information or, indeed, "banderillos" with a better aim.

This is especially understandable in Lima where fourteen dailies somehow survive along with sixty-two radio stations, eight TV channels, and at least seventy-five magazines. In this pluralistic orgy of conflicting views, more and more people want to know what is really going on—and the times are full of momentous, dramatic events. Lima's newstand tabloids are garnished with corpses, outrageous headlines, and a definite predilection for female buttocks, but in terms of readership the sober, rather staid, and now quite cautious *El Comercio* holds a first or second position rather handily.

Similarly, in the magazine field, investigative reporting is becoming an indispensable ingredient for an increasingly discerning middle class readership. To be critical and aggressive is now definitely not enough in this field.

Radio journalism has traditionally lacked an editorial policy in Lima. Radio can be quite strong in the provinces on local issues, though its impact has been secondary. But now "Radioprogramas," twenty-four-hour news and chat radio, is at the top of the charts. In television, five of the city's eight channels are engaged in a newscast war which enriched the journalistic content of their scheduling, and at times provided elaborate programs along the lines of "60 Minutes" in the United States. Finally, the word "credibility" has been popping up in media surveys conducted in the Lima-Callao metropolitan area, and the top credibility and readership ratings of newspapers and magazines have been coinciding.

But how deeply is this new virtue appreciated by the public, and how really influential are the news media in Peru? In the daily political infighting, they are powerful, and a journalistic exposé can be devastating. But news media are not as successful as they would like to believe in pushing people toward specific political goals. There are humbling experiences.

During most of the 1930s, '40s and '50s, the APRA kept its position as the major political party in the country despite opposition from practically all of the news media. In 1983, Marxist Alfonso Barrantes was elected mayor of Lima while most newspapers, magazines, radio and television backed other candidates. And violence is still spreading today despite generalized—if not universal—condemnation.

On the other side of the coin, APRA President Alan Garcia's landslide victory in 1985, and the extraordinary popularity he enjoyed until 1987, did not expand the readership of several Aprista tabloids and magazines created with massive financial support. Thus, the news media in Peru may be a strong factor in the mechanics of political power and among the more stable strata of society, but their influence on the wider, socially and ethnically diverse public is smaller. The credibility of the media is probably low.

Yet, Peru enjoys a high degree of freedom of expression. It shows not only in the diversity of its media, but in its extremes. Two Lima dailies, for example, have supported terrorist groups and justified their killings, and though some yellow periodicals are incredibly vicious— commonly attributing homosexuality, drug addiction and narcotic-dealing to cabinet ministers and presidential aides—it is practically impossible for the average citizen to win a libel suit. Suits by politicians are frowned upon, and sanctions are minimal.

There are, however, some impediments to the flow of information on governmental affairs. Sensitive material can be difficult or impossible to get, while the government has a tendency to disseminate false or self-gratifying information. Friendly journalists are favored with scoops while critics may be left in the dark. Sources within the regime are diverse, and leaking hard information is the national pastime. The main attempt to manipulate at least part of the media derives from the way advertising budgets are used by the public sector. Investment does not coincide with ratings, and advertising can buy editorial support for ministers and public officials. It is, however, more a policy of personal promotion than a centrally planned scheme. The president has been critical of it, and an investigation was launched when the previous prime minister resigned. Most of the major media, furthermore, do not condition editorial content on government advertising.

In 1987 most journalists opposed a proposed state monopoly in banking because of its potential threat to the independence of broadcasting and publishing. The law, now weakened by court actions, could have been viewed as a setback. Incidents occurring during the takeover of the banks, in which reporters were barred from the premises or were forcibly expelled, suggested a relative deterioration in the status of Peruvian journalism. Newsmen were also pushed around in May during a police strike.

Perhaps more important, two dailies were bombed in 1987, *Expreso* by the MRTA group and *El Diario* by what seems to be a new right-wing paramilitary outfit. And propaganda raids on media plants increased to thirteen, all carried out by left-wing terrorists. The most specific attempt to hamper the free flow of information occured in the emergency zone of Ayachucho, where Shining Path (a Maoist terrorist group) is particularly active. The military authority there forbids journalists to travel outside the main cities. The ban is applied inefficiently—several newsmen have sneaked through the roadblocks—and no sanctions have been contemplated for those who publish stories. But the difficulties and harrassments are there.

In August 1988, the president proposed laws to provide stiff punishment for those who use the mass media to incite or praise terrorism. "We all know," he said, "that terrorism takes advantage of the freedom of the press that exists in our country." The law was passed after considerable debate. Through March 1989, however, it has not been invoked to repress legitimate information or opinion.

As though to emphasize the urgency, three journalists (and the family of one) were murdered within ten weeks (November 1988 to January 1989). In nine years, thirteen journalists have been killed in Peru.

Philippines
[IIIIIIIIIIIIIIIIIIIIIIIIIIIIIIIIIIIIIIIIIIIIIIIIIIIIIIIIIIIIIIIIIIIIIIIIIIIIIIIIIIIIIIIII_____ MoF

In this volatile society, still troubled after fourteen years of martial law, and enduring a bloody leftist insurgency, every conceivable political view may be published or broadcast, including strongly oppositionist statements and communiques from the guerrillas in the hills.

The heady atmosphere following Corazon Aquino's election victory in 1986 has largely dissipated in the difficult reconstruction of the divided, troubled land. Yet the print and broadcast media appear with full freedom, tempered by traditional self-censorship. The government has confiscated the assets of some publications owned or controlled by those close to the deposed Ferdinand Marcos. Although Communist publications are illegal, views of the Party appear in print elsewhere.

In this climate of insurgency, however, violence has been turned on journalists. Eleven newsmen—six broadcasters, a TV cameraman, two still photographers and two newspaper journalists—were killed by gunmen of the political left and right in 1987. (The first killing of a journalist in 1988 occurred in March when an editor-publisher who also headed the local press club was murdered in Mindanao. Two more journalists were murdered in July and August.) In January 1987 rebel soldiers seized a television and radio station, and in August they raided two other television channels. Our correspondent reports that these killings have little effect on the quality of Philippine news coverage—though this is difficult to understand. One would expect greater journalistic caution in the aftermath of such events. Instead, the government responded in October by closing one radio station and warning others, and shutting a television channel which had carried programs that seemed to support the insurgents. Earlier in the year, the government announced the closure of an opposition paper, the *Philippine Daily Express*.

President Aquino was credited by the editor of the Manila *Bulletin*, the largest newspaper, of supporting press freedom but being wary of misquotation by journalists. She permitted fewer and fewer interviews and was reached mostly by "ambush" interviews reminiscent of White House shouting matches with President Reagan enroute to the helicopter pad. She cautioned journalists at midyear to be careful not to harm a person's reputation. In March 1988, charges were filed against a columnist, editors and the publisher of the *Philippine Star* for "maligning the president." The government's press secretary said that officials make no effort to control or influence the press but, remarked a *Manila Times* columnist: "In the Philippines, you don't have to control the media to influence it." Interpersonal relationships between publishers and government officials often help determine how certain stories are handled.

In December 1987 the Philippine Press Institute called the local press the freest in Asia, but possibly the most licentious and least credible. Competition for readers often leads to irresponsible reporting, the survey stated. In Manila alone, twenty-three daily newspapers struggle for readers in a market which can serve far fewer than that number. "There is a strong perception that Filipino journalists are courageous, but there also exists a pervading doubt about their honesty," said the institute, composed of newspaper publishers and editors. The group worried that a public perception of irresponsibility could generate support for government restrictions.

Complaints of government restrictions are sometimes exaggerated. The radio station that was shut down in October 1987 during fear of a possible coup, had actually broadcast encouragement to the plotters while vilifying the government. No other closures were attempted and many broadcasters opposing government policies remained on the air. Some journalists say they had been asked by President Aquino to reduce criticisms of government. Three highly critical columnists were silenced by one newspaper, but journalists generally say their freedom is basically unrestricted.

Poland
[||||||||||||||||||||||||||||_____MoF

As is usual in Communist, and as well in authoritarian non-Communist, countries, the constitution promises freedom of speech while citizens can be arrested and fined for writing, printing, distributing or even possessing material critical of the government. The official censor is the Office for the Control of Press, Publishing and Public Performances. With the winds of glasnost blowing from the east, it became more difficult for Polish censors to control the proliferating underground publication of books, journals and pamphlets. Unofficial cassettes conveying political and social commentaries receive such wide circulation inside Poland that the official radio at times finds it necessary to respond to the underground speakers. Official television programs depict life in Poland so at variance with reality that alternatives are welcomed in the form of videocassettes. Underground news tapes are widely circulated, as are videos from the West. The underground frequently interrupts official, prime time TV newscasts with reports from clandestine stations. Dissidents sometimes manage to broadcast anti-government text "crawls" beneath official pictures on TV screens. The underground often announces these messages beforehand, but authorities cannot trace the transmitters until they have broadcast.

Our correspondent writes: In Poland, where the state has a monopoly of ownership of newspapers, radio and television stations (with the exception of thirty-two Roman Catholic magazines, and one independent magazine), news coming from the state-owned media is strongly biased. Although the public is aware of this bias, the state media have a strong influence because it is not always easy to check the truth. There is, however, considerable help in this regard from underground publications and Polish-speaking foreign radio stations [Radio Free Europe, the BBC]. But it is not always possible to compare official and unofficial news; therefore, comprehension of international events is low. Official news is always presented from the "socialist camp" point of view. While the credibility of the official organs is low on international issues and ideological questions, credibility is much higher on issues concerning the economy and social problems.

State ownership of the media leads to practically total dependence on the political and strategic decisions of the ruling class. There are some editorial differences among newspapers, and much larger differences in content among magazines. The possibility for limited independence exists only in a few magazines that are politically strong enough to fight for some marginal freedom.

Several levels of control are applied by the state: some is informal, as influencing the editor by approaching him through the party apparatus; some is formal, by outright censorship. In the case of Catholic and private magazines, there is only formal control applied by the censors. Catholic papers, however, have church assistants who assure that religious issues are handled in accordance with the doctrines of the church. Catholic and private papers cannot always publish what they want, but they cannot be made to publish what they do not want. Precisely this fact makes their credibility enormously higher than the state-owned papers. In other words, these papers don't lie, but they cannot always tell the truth.

There is also internal censorship. Every journalist living in a country such as Poland, and trying to tell the truth, must make compromises. These compromises may be successful, but they are always dangerous. Only those writing for the underground press do not have to make compromises. Yet the underground press, while very important for the social life of Poland, cannot become totally professional.

1987 was not a bad year for Polish news media. Changes made were quite important. They should be called a liberalization, rather than democratization. The state permitted more publishing than since 1944 (with the exception of 1980-1981). Censorship was relatively liberal. A new monthly magazine owned and published by a group of private individuals who declared their anti-Marxist views was allowed to appear. Some "difficult" problems concerning the history of Polish-Soviet relations were finally allowed to be described—the Soviet invasion of Poland in 1939, the deportation of Poles that followed, and the Katyn Forest massacre of Polish army officers by the Red Army. Many other problems are still taboo.

In this climate, one feels that more important changes may follow. But we have yet to see structural changes in the government or party system. Only such changes can assure lasting effects. The more relaxed attituide of the censors profited first the Catholic and private magazines, but their policies puts pressure on the state-owned publications which must adapt to the new situation. Government spokesmen, too, are now providing more information, although it is still biased and fragmentary. Late in the year, it was announced that the censorship law would be modified, leading to the hope that structural changes and wider freedoms may result. These benefits for the Polish press were bolstered by the glasnost politics which continued in the Soviet Union.

* * *

IN FEBRUARY 1988, thirty prominent Polish independent intellectuals called on Communist authorities to respect the right of free association. The petition asked the authorities to allow the establishment of independent magazines "for free discussion of the advantages and disadvantages of the country's social and economic system." The petition, submitted to the Polish Parliament 1 February said "it is impossible to start overcoming [Poland's] crisis without real, and not only declarative, respect for the citizen's right to free association." Among the signers were eleven full members of the prestigious Polish Academy of Science, as well as Warsaw University Rector Grzegorz Bialkowski, former Solidarity Economic Adviser Ryszard Bugaj, and the Economic Society's Chairman Aleksander Paszynski. The official Polish news agency PAP responded, "The signers do not take notice of...very important changes which have occurred and continue to occur in Poland's socio-political life, including a relaxation of censorship to allow public debates on practically all internal problems."

Late in October 1988, after Solidarity leader Lech Walesa's meeting with top Communist officials, an independent monthly *Konfrontacje* printed an interview with Walesa which censors had previously blocked in another monthly. The editor of *Konfrontacje* said Gen.Wojciech Jaruzelski, the Polish leader, had authorized the change in policy. A liberal political lobby group is also forming around the paper. The Walesa/Jaruzelski rapprochement in April 1989 produced a further relaxing of information controls. [For further coverage of Poland see Chapter VI.]

Portugal
[IIIIIIIIIIIIIIIIIIIIIIIIIIIIIIIIIIIIIIIIIIIIIIIIIIIIIIIIIIIIIIIIIIIIIIIIIIIIIIIIIIIIIIIIIIIIIIIIIIIIIIIII|___MoF

Divestiture by the Portuguese government of four state-owned daily newspapers was begun in 1987—a major gain for press freedom. The government acquired the papers after the 1974 revolution when the country's banks were nationalized. The banks had owned the newspapers. Editors and journalists generally managed, however, to maintain their independence. The Lisbon *Diario de Noticias*, though a state property, is universally regarded as Portugal's best newspaper of record. Through the banks, the state indirectly subsidizes the major newspapers, though it does not exert editorial control.

It was also expected that during 1988 privately-owned television would be permitted for the first time. Unlike other European countries where such a decision is a political rather than a commercial one, in Portugal most seem to agree that one or two more television networks are needed. Portuguese Television (RTP) is a state monopoly broadcasting on two channels. To permit private television, the Portuguese constitution must be revised. The same legislation is also expected to privatize three or four radio stations with a national audience which have been state property. The new law may also legalize some fifty small FM stations spread throughout the country which operate without full authorization.

After regaining press freedom in 1974, following a half-century of official censorship, the Portuguese media have nevertheless lost some of their public credibility. This apparent inconsistency results from the prior uniformity of information and opinion imposed by censorship, followed by widely differing versions of events and contradictory commentaries. The negative public reaction was due to the excessive commitment of the newspapers to particular political party policies. Though the newspapers in recent years have become more objective, they have not regained much reader credibility.

This is one reason for the decline in the circulation of Portuguese newspapers since the end of the 1970s. The most important reason for this reduction was the sudden rise in the cost of newspapers. The price was formerly sustained by the strong economic groups to which the newspapers belonged and whose interests the papers defended without concern for their own economic viability. Thus, they sold newspapers at prices unrealistically low.

With the abolition of censorship in 1974 there came a proliferation of newspapers. Although some quickly disappeared, there are still more papers than the Portuguese market can sustain. The financial fragility of the newspapers, more than any factor of a political nature, is responsible for the lowering quality of information in Portugal. The government's consideration of economic assistance, which has already begun, provides some hope.

St. Kitts-Nevis
[IIIIIIIIIIIIIIIIIIIIIIIIIIIIIIIIIIIIIIIIIIIIIIIIIIIIIIIIIIIIIIIIIIIIIIIIIIIIIIIIIIIIIII_____MoF

Freedom of speech and press are provided under the 1983 constitution, but there is no daily newspaper. The organ of the labor movement appears three times a week and each of the major political parties publishes a weekly newspaper.

The government owns and operates ZIZ, the only television channel. But views of all political parties are carried on television, and on the commercial radio station.

St. Lucia
[IIIIIIIIIIIIIIIIIIIIIIIIIIIIIIIIIIIIIIIIIIIIIIIIIIIIIIIIIIIIIIIIIIIIIIIIIIIIII_____MoF

Press freedom is assured under the 1979 constitution. St. Lucia's two main newspapers, and several smaller labor and religious organs, represent widely differing opinions.

Radio St. Lucia is owned and operated by the government. It provides an English-language service. There is no censorship, but opposition parties have complained they have difficulty in gaining access to the government radio. Another radio system, owned by a French company, broadcasts in French, English and Creole.

St. Vincent
[IIIIIIIIIIIIIIIIIIIIIIIIIIIIIIIIIIIIIIIIIIIIIIIIIIIIIIIIIIIIIIIIIIIIIIIIIIIIIIIIIIIII_____MoF

There is only one privately owned newspaper, the *Vincentian*. Because of reduced advertising since August 1987, the paper has appeared weekly rather than twice weekly. During the year, writers in the government information service were instructed by the minister to publish news as well as information. The staff resisted and went on strike. While the order remained, the staff circumvented it but the information service of the government lost some credibility and St. Vincent's news flow deteriorated, though not significantly, in 1987.

There are political party organs which often take issue with reports in the *Vincentian*. Its editor maintains, however, that she has complete editorial freedom and tries to make space available to all points of view, while also expressing her own. There is no attempt at censorship by the government, though on occasion there is criticism of news coverage. A high percentage of the paper's advertising revenue comes from government notices and announcements. A previous government stopped advertising in the *Vincentian* because of criticism by the editor. This restriction ended when the government changed.

Fear of libel provides some hindrance to complete press freedom. Criticism of judges or policemen may be regarded as libelous. That sometimes deters criticism of unsatisfactory behavior and procedures. The sensitivity of politicians to criticism sometimes generates harassment of the columnist or the editor. Character assassination, rather than threat of physical violence, sometimes follows. Telephone calls to the home as well as office are the chief means of harassment.

The print media enjoy editorial freedom from both private ownership and the government. Public opinion, however, does little to sustain press freedom. For example, there was no protest several years ago when the national newspaper was deprived of advertisements by the government.

Senegal
[||||||||||||||||||||||||||||||||||||||||||||||||||||||||||||||_____MoF

The print press is relatively free in Senegal. The government does not practice censorship. The only limitation on the development of newspapers results from the lack of financial means of those who want to read and buy them. Limited circulation tends to limit news coverage.

Since there are seventeen political parties in Senegal, the print press reflects this ideological diversity. Almost 80 percent of the newspapers and magazines, however, are published by the political parties.

Though publications are neither censored nor banned, publishers must register with the central court, where approval is routine. Laws also set standards or codes of practice for journalists and publishers. The most influential newspaper in Senegal is controlled by, and favors, the majority Socialist party.

The audiovisual press is less free because radio and television are controlled by the state. The broadcast media criticize the government but not with the same vigor as the print press.

More and more, the print and broadcast media give priority to problems of economic development—the principal concern of the government.

Singapore
[llllllllllllllllllllllllllllllllllllllllllllllll_____MoF

In this small city-state, with a vigorous and internationally-oriented population, it is difficult to separate domestic news and opinion from international reportage. Many foreign publications are read here, and some of the magazines of international prominence either publish or have staffs here. For more than two decades, the government of Prime Minister Lee Kuan Yew has assiduously courted foreign correspondents. They have trumpeted worldwide the "economic miracle" of Singapore's steady development since World War II.

Restrictions on Singapore's press and other forms of expression have produced not only a controlled economy, but news and information media meshed with economic development. The government declared frankly in 1987 that it sets its own standards, and has no obligation to sustain a free and unrestricted press.

The channels of domestic and foreign news coverage converged critically, in the eyes of the prime minister, late in 1986, and continued through 1987 and into 1988. The issue: When foreign publications report domestic matters which displease the government of Singapore, and then do not publish an official response that satisfies the government, can the foreign publisher be accused of "engaging in domestic politics" under a law passed in 1986 for this purpose?

The first victim late in 1986 was *Time* magazine. It refused to publish in full a lengthy official letter replying to an article about Singapore. *Time* was gazetted, and severe curbs placed on its distribution until June 1987. Similar action was taken in February 1987 against the *Asian Wall Street Journal*. In October, *Asiaweek's* circulation was drastically reduced, as was that of the *Far Eastern Economic Review* in December. The *Journal* appealed to a Singapore court but the judge ruled in May 1988 that the government had the authority to restrict circulation on the allegation that not publishing an adequate letter from the government constituted "engaging in domestic politics" of Singapore.

A few days later, the prime minister defended his position in Washington before the annual meeting of the American Society of Newspaper Editors. He said that Singapore uniquely has four major presses in as many different languages: English, Chinese, Malay and Tamil. "Each," he said, "has different key values and world views." One value which does not fit Singapore, he added, "is the theory of the press as the fourth estate." Singapore, he said, cannot model itself on America. "It does not have the cultural, historical or economic base for an American approach to life and politics." When American publications, even if based in Hong Kong, begin to report daily on Singapore, the prime minister maintained, "they are no longer the foreign press [and] become domestic Singapore press." While "we allow American journalists in Singapore in order to report Singapore to their fellow countrymen," he went on, "we cannot allow them to assume a role in Singapore that the American media play in America, that of invigilator, adversary, inquisitor of the administration." To allow that would change the nature of Singapore society "radically." He doubted whether "our social glue is strong enough to withstand such treatment." As the controversy continued through 1988, several major foreign news organizations removed their bureaus from Singapore and others considered publishing elsewhere.

Meanwhile, in domestic journalism, the government continues to ban statements which it believes may create tensions among the several races. Indeed, discussion of race, religion and language is virtually forbidden. The private companies that publish newspapers are closely linked to the political leadership. Editors, therefore, have intimate knowledge of official attitudes and policies, and accept the guidelines under which they must operate. The Singapore Broadcasting Corporation, owned by the government, has a monopoly on domestic radio and television broadcasting. Editors there also follow guidelines set down for the newspapers. While Malaysian radio and television can be received uncensored in Singapore, Malaysian newspapers may not be circulated here. [For further coverage of Singapore see Chapter VI.]

Solomon Islands
[IIIIIIIIIIIIIIIIIIIIIIIIIIIIIIIIIIIIIIIIIIIIIIIIIIIIIIIIIIIIIIIIIIIIIIIIIIIIIII_____MoF

Though there are no daily newspapers, two independent weeklies and a weekly and monthly newsletter issued by the government (together with one provincial government weekly) provide a mixture of news and opinion that is free and open. Some political influences appear in the publications, but generally the judgments of editors and journalists prevail.

Opposition politicians have ample opportunity to have their views heard on the state-owned broadcasting system.

South Africa
[IIIIIIIIIIIIIIIIIIIIIIIIIIIIIIIIIIIIIIIIIIII_____MoF

Classic struggles for power—state power versus information power—

dominated the geopolitical landscape in South Africa during 1987 and 1988. Forced underground was the seething discontent of black South Africans based on their perception of mounting oppression by the state. Repressive laws and police actions increasingly made "propaganda" the target, rather than further separation of races. The realities of apartheid, the decades-old cause of bitter unrest and bloody violence, became less visible during the heightened struggle over information power—how, if at all, to report the basic issues of virulent racism?

The government systematically controlled the flow and content of information within the country and to the world at large. Earlier protests had turned into violent police crackdowns, all recorded in emotional television and print-press coverage. "The press has become the first casualty," one U.S. network producer stated. Journalists have been beaten by police, and protest films tampered with.

To darken the screens, early in 1986, the government forbade newsreporting, particularly television coverage, of violent scenes anywhere in the country. Reporters could not even be near places where confrontation might occur. It became a criminal offense to publish information on political unrest, detention cases, the treatment of detainees, and many forms of political activity without prior government clearance. Correspondents of three major American newspapers were sent home or not replaced. The government announced that 238 foreign journalists were refused new or renewed visas from July 1986 through May 1987.

Drawing the hard-news curtain immediately stripped the news media of their ability to cover "today's" story, for which they needed hard facts and first-hand sources. Most such sources had dried up, out of fear of reprisal from the government. It had become illegal to report any violence without government authorization. It was assumed, for example, that protests in the homelands were put down with greater severity in the absence of press coverage. But this could not be proved since all information had to be obtained from the government information office.

Coverage of South Africa appeared far less frequently on the front pages of major American newspapers, and on television. While some television coverage continued, the South Africans succeeded in reducing the temperature of the American debate because the nightly picture of conflict in the homelands had disappeared. The government apparently preferred to be attacked for censorship, rather than for repressive police activity against protestors. Coverage of South Africa on U.S. networks declined by two-thirds between December 1986, when tougher regulations were imposed, and fall 1987. The silence continued in 1988.

The government's power plays through 1987-88 took several forms. Severe new regulations controlling all news media were issued in December 1986 and strengthened several times during 1987 and 1988. With these regulations, and repeated threats to monitor the daily and weekly press, the government sought to drive a wedge between the small, weekly and mainly liberal opposition papers—the government dubbed them "the alternative press"—and the major dailies published in Afrikaans and English. By attacking the "alternative" papers, the government would weaken or eliminate this source of unremitting opposition. These papers generally reflected the attitudes and interests of the black homelands, and the liberal whites. The governmment maintained that these papers' orientation was radical (in favor of revolution, in support of the outlawed ANC, pro-Soviet, "Communist") At the same time, the government could assume that the mainstream press would further restrict itself, out of fear of suffering the fate of the threatened "alternative" papers.

At a conference in London in January 1987, I observed the split between representatives of the "alternative" press and some of South Africa's mainstream journalists. They were divided over whether a joint appeal should be made to the South African government, an appeal joined by an international assembly in London. The conference, called by the World Press Freedom Committee and others to challenge the censors, authorized its chairman to write to twenty-three South African newspapers. He reported the unanimous approval of the Declaration of London, and said that it strongly opposed all forms of censorship as well as self-censorship. "The strategy of divide and conquer is not new," said Harold W. Andersen, WPFC chairman, "and in some places it has worked." He urged the South African news media "to present a united front against censorship practices, including self-censorship."

Through Home Affairs Minister Stoffel Botha, the government threatened to establish statutory control of the press through a disciplinary body. It would use state power to punish editors and journalists with suspension from practicing, and could close down newspapers. The controlling body would be patterned on the newspaper licensing act which passed Parliament in 1982 but was never promulgated. The act was postponed indefinitely when the industry proposed the voluntary Media Council. Botha now complained that the council was resisting guiding those papers with "tendencies of negative reporting."

Coupled with warnings from State President P. W. Botha, in August, that left-wing newspapers and news agencies would be "investigated and dealt with," a series of new official regulations detailed the method to be used in censoring publications, and gave the home affairs minister the power to close a publication for up to three months. The ministry then appointed secret monitors to "judge" the press. The results soon became clear.

During my visit to South Africa in October 1987 I met privately with editors of the black press, foreign correspondents, and an editor of the "alternative" press. From all, I learned that sources of information had drastically dried up. Each editor, moreover, had on his desk many complaints from the Home Ministry. The editor of the *New Nation*, a black weekly funded by the Catholic church, received a formal notification from the minister regarding the newspaper's "subversive propaganda" under the state of emergency. "I am of the opinion," he wrote, "that there is a systematic and repeated publishing of matter...calculated to cause a threat to the safety of the public, the maintenance of public order and a delay in the termination of the state of emergency." The minister listed twenty-seven reports, photographs, poems and advertisements which, he said, violated one or another of the regulations. He issued "a warning by notice in the *Government Gazette*." This gave the newspaper two weeks' time to respond. Condemnation by the inquisitors is arbitrary, and may be based as much on the "tone" of the article as on the facts.

The *New Nation* took the minister to court but its challenge was dismissed by a Supreme Court panel. The editor of the *New Nation*, Zwelakhe Sisulu, had been in detention without charge for more than a year when the paper was closed down in March for a three-month period. He was released after two years, still not charged, placed under virtual house arrest, and forbidden to work on the newspaper. Days after the closure, a rival newspaper, the *Weekly Mail*, itself threatened with a similar ban, published stories which could not appear in the *New Nation* that week. Said the headline, "What *New Nation* would have said..." The *New Nation* returned at the end of June. By then another paper, *South*, had been shut down in May and reopened in June. Several international human rights organizations placed advertisements in *New Nation* welcoming its return.

The *Weekly Mail* and five other "alternative" papers were also threatened with suspension. The *Weekly Mail* said it had "a genuine problem" in dealing with the government's warning because the newspaper "cannot even understand it." The newspaper, an outspoken critic of the government's apartheid policies, said Stoffel Botha objected to an October article about ANC leader Oliver Tambo. The ANC supports a bombing and sabotage campaign to topple the white-led governnment. "Tambo has been for some years an important and widely perceived fact of the realities of South African political life," the newspaper said. "The article attempts to deal with that reality. It is not adulatory or uncritical."

The *Weekly Mail* noted that the minister has not issued warnings to two progovernment Afrikaans-language newspapers that carried descriptions of an ANC camp in Tanzania that were written in benign and positive terms. "The object...[is] not to point fingers at another newspaper," the *Weekly Mail* said. "The complaint is a more serious one: *Weekly Mail* just does not know what rational principle to apply in deciding what will make a particular report objectionable and what not."

The staid *Business Day*, South Africa's main financial newspaper, spoke out more bluntly in an editorial, declaring that Botha and his advisers "have neither the general education nor the wit to distinguish between revolution and radical chic."

A different analysis came from *Southern Africa Report*, published by Raymond Louw, former editor of the defunct *Rand Daily Mail*. Said the weekly newsletter, "Part of the government's problem is that its own highly compliant press presents information palatably, ignoring or glossing over unpleasant facts, unless it is forced to report them and then misrepresenting them in the government's favor. The government believes the opposition press uses the same propagandistic techniques but from an opposition viewpoint. It has no understanding of the meaning of a free and independent press." Louw concluded, "The government would prefer not to close newspapers. Instead, it wants to cow them into submission."

Through the latter half of 1987, government warnings attempted to cow all the major "alternative" publications. These included the large, daily black newspaper, the *Sowetan*; three weeklies, the *New Nation*, the *Weekly Mail* and *South*; and a left-wing monthly journal, *Work in Progress*. By April 1988, two community newspapers in Cape Town and Oudtshoorn, and a quarterly publication also received gazetted warnings. In May, mainstream editors as well as "alternative" journalists recognized that the survival of all newspapers was at stake and that this was a time of crisis. A petition sent to Stoffel Botha was signed by twenty-six editors, including most of the largest white and black newspapers. Editors of papers supporting the government did not sign.

The petition expressed the editors' concern at the "mounting political pressure on the press" and asked "in the public interest, that the unnecessary restrictions under the emergency regulations be eased. We believe the ministerial banning of *New Nation* and the threat to silence *Weekly Mail* and other newspapers and journals are against the national interest." The editors "opposed such authoritarian action" because:

"1. The public is being deprived of information and viewpoints which all South Africans need if they are to make sensible decisions on vital decisions.

"2. We have no knowledge of any recognized publication that is fomenting violent revolution. If the minister has any evidence of so serious a crime, we believe he should submit it to the courts for judgment. Such matters affect all of society, and cannot be left to the whim of officials, politicians or parties.

"3. We believe that the minister's concern may be less about press incitement to revolution than about coercing all the media into adopting a uniform attitude—a subservience not in the interest of South Africa."

The editors asked that the "rule of law be restored."

The series of threats and shutdowns served to institutionalize self-censorship, while finally drawing the white and black press into some coordinated, though limited response. The advice which the London Conference on Censorship gave to South African editors in January 1987 —urging them not to advance the government's policy of divide-and-conquer—may be viewed as a preamble to the petition to President Botha by twenty-six white and black South African editors in May 1988.

On 10 June the government announced it would impose on 28 July the emergency decree embodying stringent, new restrictions on the press. The decree seemed at first to require only small news agencies and stringers for local and overseas papers to register with the government. But the home affairs minister shocked the entire South African press as the deadline neared by stating that *all* newspapers and staffs would have to register, or face a fine and up to ten years in jail. The regulation implies the power to deny registration, and so determine who may work as a journalist. Those journalists who are registered may seem to readers to have been compromised. The new rules require that copies of all stories sent to newspapers must be sent to the Home Affairs Ministry within a day of dispatching to the newspaper. On the day the new restrictions were to take effect, the government suspended the regulations. The European Community (the twelve member nations) had expressed concern that the new rulings would require them to revise their trade and other publications.

In November 1988, the information minister suspended the *Weekly Mail* for four weeks, for "causing a threat to the safety of the public." The previous edition had condemned the recent municipal elections as "the gunpoint mandate." For some papers, a shutdown could mean bankruptcy through loss of advertising. Even confiscating a few issues of one of the mainstream newspapers could mean the difference between profit and loss on a full year's operation, and perhaps the shutting down of the paper. Most of the editors of mainstream papers in 1988 had received many complaints and demands for their sources of information. In view of the fact that seventeen organizations were added this year to the proscribed list, a newspaper could be asked to divulge the name of anyone from these organizations who provided information to the paper. A journalist refusing to divulge the source or to give evidence against that person, would be subject to imprisonment. And, as one editor put it, "In practice, it means a further, alarming, smothering of dissident voices."

There are other serious prices being paid by the South African society —black and white. For the newspapers, there is the cost in terms of money and time of dealing with these endless complaints. They usually demand the services of expensive lawyers, while the state's costs are covered by taxes paid by the public, black and white, including the editors.

Detention without trial and varied forms of harassment face journalists as a constant threat. Detainees also face restriction orders once they are released. At year-end 1988, five journalists were in detention under various security laws. Brian Sokutu had been held for two years, mostly in solitary confinement. Many journalists have been detained for brief periods, interrogated and threatened with far worse treatment. Half the country's editors have been convicted of some breach of censorship law. Telephone tapping and interference with mail is routine. Harassment varies only in degree of severity between white and black journalists. Assaults and detentions of journalists in the black homelands discourage papers from sending staff there. In June 1988, to close one further gap in the sourcing of information, the government's Bureau for Information asked all government officials to report all requests by foreign journalists for interviews, and to rebuff those who are not accredited by the bureau. The letter says this is done "so that the government can be informed of the attitude of the particular journalist and the media he represents, and the activities of foreign journalists can be monitored more efficiently."

As though to capitalize on the nearly universal opposition to the South African government's press policies, the African National Congress, long banned by Pretoria as revolutionary and terrorist, drafted constitutional guidelines for a multiparty democracy and a mixed economy. The ANC included press freedom among its pledges. The Communist party is among the constituents of the ANC, and its white leader formulated the bill of rights which referred to press freedom.[5]

The editors of three "alternative" newspapers appealed in April 1988 to editors of newspapers around the world. They asked for help "in opposing attempts to close us down permanently." Said the appeal, "We have no doubt that if the government deals successfully with us, they will then turn to other, more mainstream expressions of opposition. They have set out on a path towards silencing any opposition press, ensuring that they will be able to act without public scrutiny—and leaving the vast majority of South Africans without media that reflects their views...The minister has developed a strategy that draws out the process of closure for such a long time that the outside world tends to lose interest, enabling him to proceed without the international press paying much notice." The editors concluded their appeal by urging the foreign editors to monitor "further action against us" and "give full coverage of it in your newspaper; protest directly to the South African government; and encourage your government and your colleagues to join such a protest."

In September 1988 Minister Stoffel Botha announced a date by which some members of the press—it was not clear who—would be subject to licensing. Almost the entire press corps, joined by many overseas associations of journalists and press-freedom advocates, strenuously condemned the order. It was dropped. This was the first time journalists had won a struggle against a press crackdown. But they had no illusion the day of a freer press was coming just yet. They were right. Two months later President Botha threatened to force newspapers to disclose news sources.

In November 1988 some dissident Afrikaaners launched the first anti-apartheid newspaper written in Afrikaans, the language of the ruling party. They hoped, however, to open the eyes of fellow Afrikaaners. The first issue headlined: "Mandela: The New Era."

Early in 1989 the government warned four small anti-apartheid publications they may be suspended for "stirring up hatred," and about the same time the house of the coeditor of the *Weekly Mail* was firebombed in his absence.

Thus, the counterforce to censorship set in motion in January 1987 by foreign journalists, failed to halt—though it did illuminate—the overpowering advance of South African press controls during 1987-88. And the process came full circle with the victims of censorship appealing to the international press for their continued attention and support.

Soviet Union

In this era of glasnost, it is imperative to understand its origin and purposes. It is an instrument of Soviet power. Glasnost is a managment tool for the "reeducation of the masses"—at home and abroad. No less an authority than V.I. Lenin (not Mikhail Gorbachev) first used the term in its present meaning. Wrote Lenin in March 1918:

> We must transform—and shall transform—the press...from being a simple apparatus for the reporting of political views, from being an organ of the struggle against bourgeois mendacity, into an instrument of economic reeducation of the masses, into an instrument for acquainting the masses with the need to work in a new way. The introduction of glasnost in this sphere will of itself be an enormous reform and will facilitate the enlistment of the broad masses in self-dependent participation in the resolution of the problems that concern primarily the masses.

Lenin called for glasnost after signing decrees that placed the press under the complete control of the state. These decrees included the banning of all non-Bolshevik publications and making advertising a state monopoly. The freedom of speech and press for which Russian revolutionaries had long fought, was denied them just a few months after the Bolshevik victory. Andrei Vishinsky, the attorney-general under Stalin, put it clearly, "In our state, naturally, there is [not] and can be no place for freedom of speech, press and so on for the foes of socialism.[6]

Early in 1987, an unprecedented press law was under discussion in the Soviet Union. The new Law on the Press and Information would be the first since Lenin's decrees to institutionalize glasnost, not only as a propaganda tool but as a means of checking the highly centralized bureaucracy. Although all mass media in the USSR are controlled by the Communist party and the government, there is no law governing the uses of these fast-expanding communications systems. For the first time, in 1987, prior censorship of all publications, broadcasts and films was no longer required. It became the responsibility of individual editors—all civil servants, to be sure—to make certain that their products are within party policies and directives. "The censors are in our heads and hearts," Novisti's Igor A. Schwartz told the *Statesman* of India late in 1988. While the Soviet press represents "new thinking" in domestic issues, it tends to rely on "old thinking" for foreign policy and analyses of issues not yet resolved by the Kremlin.

In the absence of a press law and specific, day-to-day guidance, reports and commentaries in the press, magazines and television inevitably reflected a variety of approaches and objectives. The main purpose of glasnost, in the daily reiterated words of Mikhail Gorbachev, is perestroika: the restructuring of Soviet society to make it more productive in all aspects of a modern state. Moving perestroika from a campaign slogan to massive social change entails the fundamental reeducation of 280 million people. That is the task of glasnost.

Enlisted in this campaign are some 8,500 newspapers with a circulation of nearly 200 million published in fifty-five Soviet languages. It is now possible to send a message from Moscow's Intourist Hotel to the U.S. in twenty-four hours, using telex and telefax. Television, however, is becoming the most important arm of the regime's propaganda apparatus (known for years as Agitprop). Changes in 1987-88 are vastly increasing the influence of television on Soviet society. Central television now airs 148 hours of programming a day in the eleven time zones of the USSR. The key factor in this extensive development is satellite broadcasting.

A growing network of telecommunications satellites is also making Moscow television a factor on TV screens in many countries abroad. Satellites keep Moscow's allies and clients within television reach for clear ideological purposes. Soviet television is now seen in Warsaw more hours each day than Polish TV. The USSR's "first program" is seen throughout Czechoslovakia, and in Hungary and the German Democratic Republic as well (though some 85 percent of all GDR households watch West German TV). The Soviet-bloc satellite system, Intersputnik, is similar to the U.S.-created Intelsat—with the significant exception that member countries of Intersputnik must communicate with each other only through Moscow.

By intelligent use of satellite facilities, Soviet television "bridges" appear frequently on American commercial networks. In May 1988, a radio version of such a bridge was broadcast over 100 stations in sixty of the top sixty-five U.S. markets—the first live U.S.-Soviet call-in show to air nationwide in both countries. Television networks in both countries gave saturation coverage to the 1988 Reagan-Gorbachev summit in Moscow. After several years of viewing Gorbachev and several Soviet telecasters on U.S. screens, one American press analyst wrote, "Before Gorbachev, the United States had only to worry about Soviet military might. Now it has to take into account its video power."

Even the conflicting uses of glasnost within the Soviet Union created a positive impression abroad. The proliferation of criticism of middle managers of Soviet farms and plants, the publication of hitherto banned books, attacks on Stalin's murderous regime, including his friendship pact with Nazi Germany, questioning—finally—of the Soviet invasion of Afghanistan, and, perhaps most significant, the personal and institutional implications of restructuring the nation's political, industrial and social systems—all of this gives the impression of a society democratizing itself. Yet the eventual outcome of these undeniable changes is by no means clear, except for Gorbachev's frank declaration that democratization in any Western sense cannot be permitted. He also has chided the press for being irresponsible at times. The circulation of pioneering publications such as *Ogonyok* and *Moscow News* has been restricted.

In Moscow in July 1987 I asked the editor of *Moscow News'* English edition, Vladimir Pichugin, how different he felt coming to work that day than a year earlier. This day, he said, he looks out the window and sees crowds in the streets reading his paper. Up the street, in front of another newspaper office [*Pravda*], nobody! Clearly, readers like the new, critical tone of *Moscow News* and are tired of the boring polemics in the conservative press. We asked Pichugin how far this self-responsibility or individual editorial judgment can go. "Nobody knows," he said, adding, "I read *Pravda* to get the party line and *Pravda* reads *Moscow News* to see how far *they* can go."

In 1988, *Pravda* as well published, particularly in its letters column, reassessments of the Soviet past, and reports of current struggles between those supporting perestroika and the conservative dissenters. All of this appeared in the official press. But there are also the small, embattled unofficial publications. My son and I visited Sergei Grigoryants, editor of the magazine called *Glasnost*, the week he launched it in mid-1987. He had been released from prison just four months earlier, a victim since his youth of laws prohibiting unofficial writing which allegedly maligns the Soviet Union. In earlier years, Grigoryants published samizdat, the self-published, secretly distributed and often penalized writings of dissenters. Now, with *Glasnost*, Grigoryants is testing the policy of glasnost. He sought, but was denied, official permission to publish his magazine. It includes articles, documents and personal critiques of aspects of Soviet life not covered in any of the official publications. Grigoryants's authors include former political prisoners, Andrei Sakharov, many religionists and labor people. In the beginning, *Glasnost* was handtyped on onionskin paper with a few carbon copies providing the sole form of distribution. (I published in an American newspaper the first English translation of the introduction to Grigoryants's first edition of *Glasnost*.)

"The future of the magazine *Glasnost* may indicate the future of the policy of glasnost," Grigoryants told us. Would he continue without permission, and what would be the penalty? "They're very inventive. It is difficult to make assumptions. It's all so new. But this magazine will be taken as an example," Grigoryants added. By mid-August, a blistering attack on Grigoryants appeared in *Vechernyaya Moskva*. The author told Western reporters, "If glasnost continues, if it goes deeper, what is the need for an unofficial press?" When glasnost has triumphed, he said, the Party will publish everything worth publishing.

But that was not the end of it. Grigoryants and his colleagues were increasingly harassed, detained by the police, and on 9 May 1988, Grigoryants was imprisoned for one week and his entire archives and printing equipment confiscated by the authorities. Said Grigoryants upon his release, "The actions show that Gorbachev's drive for democratization of Soviet society and glasnost, for openness, have not meant more tolerance of dissent." He added that lawlessness and violations of civil rights continue in the Soviet Union. In November, Grigoryants was again imprisoned, this time with his deputy editor, for one month. They were accused of taking photographs in troubled Armenia. He was in prison when the earthquake struck. Grigoryants reported by telephone from prison that Soviet troops were acting as occupiers, "imperialists," attempting to "play off the Armenians against the Azerbaijanis." Ironically, U.S. intelligence satellites had better, faster pictures of the devastation in Armenia than Soviet satellites, which are focused on American sites. The Soviet press strongly criticized local officials, probably welcomed by Kremlin leaders who were already dissatisfied with the nationalist dissension in Armenia all year.

A different view of the policy of glasnost was provided by two Soviet journalists addressing the International Press Institute meeting in Istanbul in May 1988. Alexander Pumpyansky of *Novoe Vremia* (New Times) said that while glasnost has produced some improvement in domestic coverage of Soviet news, almost no change is apparent in the foreign field in which he works. His own stories are sometimes withheld because they are too critical of the United States. "This is not a good time to be critical," he is told. "You are good for bad times, but these are better times," he said he hears. Glasnost, he said, is a developing process, "the prize of the present power struggle," which he called the movement toward "democratization." The climate of the country is being changed, he added, and the "myths" in Soviet affairs—the "religious"-like beliefs of decades—are being killed. Glasnost, he added, is realism, not dogma, not falsities. He said, "We are not the shining house on the hill, not the only winner in history," and "we should take lessons from the West." The Soviets, he said, are trying to kill the "enemy image" and the "trenches psychology." He continued, "Our system of elections is not perfect." It is, he said, "selection, not election."

His Soviet colleague, Vladimir Milyutenko of the government's Novosti press agency, provided recent polling data on Soviet public attitudes toward glasnost. Eighty percent of those responding said glasnost is a reality in everyday life, one-third saying it is irreversible, two-thirds saying it must be made permanent. But then, one must assume, two-thirds believe it *is* reversible. Perhaps that, too, is glasnostian realism.

An Italian editor commented that only pragmatic and not principled reasons had been given for opening the Soviet system to glasnost. No one is saying, with Voltaire, "I disagree with what you say, but I will defend to the death your right to say it."

From another perspective, a Soviet columnist interviewed in the *Statesman* of India noted that there is a great difference between the "central Moscow press and the provincial press, even including Leningrad, where the mass media have been left behind by perestroika." The reason? "Those who try to keep control over the mass media find it easier in the provinces. So we journalists must restructure the mass media itself. People at the official level try to influence and restrict, and when we struggle we find inertia." Another Soviet journalist told the Indian reporter: "I am maximally skeptical. There is a difference in the concept of perestroika and its implementation about what we should or should not publish in our everyday work." He said, "Our readers believed less and less in what was published." Everyone has come out of "our past," he said, "but we have not broken with our past."

Quite different, however, was the optimistic blueprint for the future put before the Nineteenth Communist Party Congress in June 1988. It stated that perestroika had already produced a "fundamentally new ideological and political situation in [Soviet] society...a real pluralism of opinions, open comparisons of ideas and interests" and generated increasing popular support for "revolutionary changes in the country." The blueprint invited the delegates to reappraise deeply and fearlessly how to combine the one-party system "with processes of democratization," without political confrontation and disunity of social forces.

And, one may hope, how to remove not only the searing stains of the Stalin era, but those that still persist: Irina Ratushinskaya, the reknowned poet, released late in 1986 from a Soviet prison camp, said a year later that glasnost "means that some official writers are allowed to speak a little bit more. It is important, but what about the other people—our people?" She said "there is still a long list of forbidden literature in the Soviet Union. It forms a big, thick book which the KGB uses when they conduct searches." She added, the book "has not become much lighter because a few names have been scratched out."

Nor was the use of the news blackout ended. The news-making first visit of Andrei Sakharov to the United States in November 1988 went virtually unreported in the USSR.[7]

Spain
[IIIIIIIIIIIIIIIIIIIIIIIIIIIIIIIIIIIIIIIIIIIIIIIIIIIIIIIIIIIIIIIIIIIIIIIIIIIIIIIIIIIII_____MoF

Judicial pressures on Spanish journalists continued during 1987. There were also new initiatives on the legislative front. These, together with growing cases of intimidation of journalists by officials and public figures, aroused concern among journalists.

One widely reported case involved the news magazine *Tiempo* which appealed to the European Court of Human Rights in Strasbourg, after the Spanish Supreme Court ordered the paper to publish a reply by the chief executive of a state-run organization of food markets. This occurred even though it had been shown that the reply was based on false data, and the published information was accurate. This new judicial trend of giving precedence to the right of reply over the dissemination of "truthful information," as stated in Spain's constitution, has created deep concern among professionals. It is another step in the direction of many court rulings on the "right to honor and privacy." Such rulings have levelled stiff punitive payments for publishing reports on well known public figures.

In recent years, newspapers have had to pay more in damages to the "offended" family of an airline pilot than the money received as compensation by the families of passengers on the plane, flown by that pilot, which crashed near Bilbao. The newspapers *El Pais* and *Diario 16* had published details of the pilot's past psychological and alcoholic problems. While not refuting the information, the court ruled that the pilot's honor had been injured. Another court ruled that show business personalities were entitled to payment for news reports about them. Such developments mark a trend toward drastically reducing the editorial independence of newspapers.

In Parliament, two controversial bills were introduced by the Socialist government. One would legalize private television, providing three nationwide networks. But it would penalize existing media companies, which could be reduced to owning no more than 15 percent of the stock of a new network (though other companies may own up to 25 percent). The bill also places harsh conditions on the programming content of future networks. (A law permitting private ownership of TV channels passed in April 1988, after bitter opposition. The act then faced court tests.)

A second bill, on telecommunications, increases significantly state control over the existing, and heretofore rather unfettered, network of independent radio stations. These private stations would now be compelled to broadcast "important" government messages.

The third bill, being prepared at year's end, was intended to regulate the right of journalists to professional secrecy, which right is included in the constitution. Draft versions caused immediate alarm as they virtually suppressed this right, compelling journalists to reveal their sources to judges. Professional organizations still hope to avert such legislation.

The politician who most notably coerced journalists was Inaki Aldekoa, a leader of Herri Batasuna (the political front of the ETA Basque terrorist group), who accused Carlos Yarnoz, a reporter for *El Pais*, of being a confidant of the military intelligence service.

Although press freedom has been restored in Spain for a decade now, one experiences here the same kind of pressure, through courts or through legislation, which is developing elsewhere in the West. This is a new, subtler form of censorship or indirect censorship.

Sri Lanka
[|||||||||||||||||||||||||||||||||||||||||||||||||||||||||||||_____MoF

The normal role of the news media in Sri Lanka as in all democracies is to inform and mold public opinion. But these are not normal times.

Until 1971 the print media were relatively free to function under normal law except in time of emergency. Since 1971, except for a few brief periods, Sri Lanka has been in a state of emergency. The first emergency declaration resulted from the 1971 insurgency followed by problems related to drastic controls over the economy. Ethnic uprisings ravaged the country and brought on another state of emergency from 1983 to August 1987. The latest declaration reimposed very strict censorship on all the media. It was unlikely that these press controls would be lifted soon owing to resentment by the opposition to the terms of the newly-signed accord with India.

The government controls Sri Lanka's largest newspaper chain and also owns the radio and television services. There are, however, many independent newspapers and journals offering diverse views on domestic and foreign issues. Opposition political parties publish periodicals which are not restricted by the government. Some journalists claim, however, that the government indirectly intimidates the press. Since the government some years earlier had nationalized a major newspaper group, any threat, however veiled, to take over independent publications alleged to operate against the public interest, is taken seriously. There is a notable credibility gap between the public and the government-owned press and the electronic media. At times when all news is censored, the public suspects that much information is being suppressed. The greatest impediment to the flow of news and information, therefore, is from the politicians and bureaucracy, though there may be a few instances of suppression or angling of news from within the print media due to political or economic considerations.

There was, then, a major deterioration in the news flow within the country during 1987. (Foreign correspondents, of course, were free to file their reports without hindrance as the censorship regulations do not apply to them.) The effects of the restrictions on the news media are general disaffection among media men and the general public, the rise of rumor, and the resentment of youth who feel that their democratic rights, and particularly the right to be heard, are trammelled.

Sudan
[||||||||||||||||||||||||||||||||||||||||||||||||||||||_____MoF

This country of one million square miles is the largest in Africa. It borders on eight states and the Red Sea, and bridges large Arab and Moslem populations, as well as the desert region of northern Africa and subtropical, Black Africa to the south. The role of the mass media in this strategic, developing country falls into two distinct categories since the overthrow of the Nimeiri dictatorship in 1985. The Transitional Military Council (TMC) ruled the country after the uprising until an elected government took power in May 1986. For most of the remaining period, Prime Minister Sadig el Mahdi of the Umma party has served in coalition governments with the Democratic Unionist party.

The TMC, in the spirit of the popular uprising, allowed complete freedom of the press to flourish. A host of privately-owned daily Arab-language newspapers appeared on the newsstands. This was in contrast to the two nationalized daily papers the Nimeiri regime had allowed to be published. Also during this period, one or two English-language weeklies were published by the regional governments of the south. With the coming to power of Prime Minister el Mahdi in 1986, freedom of the press had ups and downs but newspapers were free of outright control.

First, the government tried to maintain the state ownership of the two papers taken over by the Nimeiri regime, and to use them as government mouthpieces. To accomplish this, the government established a committee which it hoped would recommend the continuance of state ownership. However, the group voted unanimously to denationalize the papers and return them to their original owners. The government retaliated by restoring the titles to their original owners, while retaining the nationalized printing presses of the two papers. This was clearly an attempt to stem the free flow of information by preventing the two prestigious papers from printing. One daily has since found an alternative printing press, but the other has yet to reappear since it was denationalized.

Since then, the government has tried to introduce wide-ranging controls over the press by drafting and attempting to enact restrictive press laws which are widely interpreted as the first step toward state censorship of privately owned newspapers. Opposition to the proposed laws has been intense, especially since the government drafted the measures without consulting people in the news media. The government has amended its draft at least three times and has agreed to submit the measure to media organizations and unions for discussion. These moves suggest an attempt to control and restrict information especially that reporting on the civil war in the south. The proposed laws would mean that the only information to be published on the war in the south would be that emanating from the Ministry of Defense. For a newspaper to use another source would be to court closure and prosecution.

A more discreet and less visible way in which the government has attempted to control the press has been through the artifically created newsprint shortage. Although government authorities continue to issue licenses for the import of newsprint, the financial authorities have virtually stopped issuing approval to secure the foreign currency needed to use the license. As a result of this backdoor control, at least three newspapers have had to shut down due to lack of newsprint.

Despite the government's efforts, many Arabic-language dailies and one English-language daily are still publishing, together with a host of sports papers that are published twice weekly. There are several Arabic- and English-language weekly newspapers.

The several governments of Sadig el Mahdi have been very concerned about the publication of opposition views in the national press. The prime minister has on several occasions publicly condemned them as anti-democratic. One paper, *Al Siyassa*, was closed down on the order of the interior minister. The courts, however, ordered the paper reopened pending a thorough investigation of the charges.

Radio and television are largely government controlled and maintained.

During the transitional period, again in the spirit of the popular uprising, the airwaves were made available to any person who had a valid opinion to express. The right of reply was granted to those who disagreed with any opinion so expressed. This practice ended when the elected government came to power. Since May 1986, the government has been accused of abusing radio and television by using them for partisan political purposes in favor of the UMMA and DUP. This has given the government parties unfair access to public opinion which other political parties do not enjoy.

More ominous than the management of the press has been the management of news. The government has become increasingly sensitive to charges of its failures in governing the country, so that radio and television news has provided only a standard list of government "successes." Members of the cabinet have become increasingly reluctant to talk to the press, except to recite a prepared brief.

The government turns to the press as an easy scapegoat for its failures. Early in September 1988 the minister of information and culture announced controls on the foreign press. The government had deplored international press coverage of Sudan's relief effort following floods and famine in the south. The *Sudan Times*, the English daily, objected to the threats. The minister quickly retracted the plan to muzzle the foreign press. The *Times* had run a six-column, front-page headline, "Flood Relief Shambles Goes On As Government's Capacity To Lead Wanes," which was hardly the sign of a thoroughly repressive governmental press policy. Indeed, the *Times* demonstrated repeatedly in 1988 its independence and considerable freedom. Workers in the State-run Sudan News Agency, moreover, called a strike in January 1989 to demand removal of government control over the content of news stories. [For further coverage of Sudan see Chapter V.]

Sweden
[|||||||||||||||||||||||||||||||||||||||||||||||||||||||||||||||||||||||||||||||||||||||___MoF

Principles of noninterference by government in the press dates to the 1700s here, and was most recently embodied in Sweden's Mass Media Act of 1977. This applies to all information media.

The government, since 1966, has provided state subsidies to political parties which need assistance in publishing their papers. Press subsidies now total hundreds of millions annually, and it is likely that the subsidy system could not be ended without jeopardizing many of the country's newspapers.

Some 40 percent of the newspapers are owned by parties and trade unions, about half are privately owned, and 10 percent are controlled by foundations. On the private papers, often owned by families, the editor is not always a member of the proprietary family.

Only the publication of sensitive national security information, and excessive violence on television are subject to government censorship.

Switzerland
[IIIIIIIIIIIIIIIIIIIIIIIIIIIIIIIIIIIIIIIIIIIIIIIIIIIIIIIIIIIIIIIIIIIIIIIIIIIIIIIIIIII__MoF

The country's high-quality press and tradition of full journalistic freedom is suggested by the 200th anniversary in 1980 of the *Neue Zuricher Zeitung*, the world-class daily distinguished for its reporting and analysis.

There are only seven newspapers with a circulation of more than 100,000 copies, but they, too, generally provide objective news coverage free of governmental interference. The papers are privately owned, and editors customarily use discretion in handling sensitive security matters. There is some selective leaking of government information to favored journalists, and an effort to place favorable stories in the press.

The broadcasting services are operated by the postal administration. Television news is criticized by the public but less than in previous years. All in all, the credibility of journalists is moderately high.

Taiwan
[IIIIIIIIIIIIIIIIIIIIIIIIIIIIIIIIIIIIIIIIIIIIIIIIIIIIII_____MoF

Sixty dissidents, beating drums and gongs, marched through downtown Taipei, 14 July 1987, to celebrate the end that day of thirty-eight years of martial law. Elsewhere in Taiwan the historic occasion drew little public attention. While this relaxation was welcome, it would not mean the end of Kuomintang (KMT) controls, which most Taiwanese sought. This continuity was symbolized in the simultaneous ending of censorship by the military, and the turning over of that power to the Government Information Office. Yet in the early months of 1988—during which President Chiang Ching-kuo died and was succeeded by the first leader born in Taiwan—central controls over the news media were significantly relaxed.

Sharply diminished was the heavy censorship, confiscation and suspension of newspapers and magazines with controversial articles. A major press ban was lifted 1 January 1988 when the government began accepting applications for new newspapers for the first time in thirty-six years. The government also lifted the page limit of newspapers from twelve to twenty-four. The press licenses had been limited to thirty-one, of which seven were run by the army, five by the KMT, and two by the government. Of the seventeen private papers, several are owned by KMT leaders or others with close KMT ties.

Under the new publications laws, it is still illegal for newspapers to "advocate communism," or support independent statehood for Taiwan. Reporting on the personal activities of the president and his family is discouraged, and official guidance is often provided by the Government Information Office and the KMT.

The most censored publications continue to be the opposition magazines. These were created mainly by the democratic opposition who sought to reach a broader public through monthly, and later weekly, periodicals. The rapid growth of these magazines in 1983-84 was followed by an increasingly severe clampdown in 1985-86. By mid-1986, virtually all of the dozen magazines were banned or confiscated. Only in the beginning of 1987, when it became apparent that political change was due to take place, did new magazines appear, and bannings and confiscations diminish. It was not until martial law was lifted in mid-July, however, that a significant drop in censorship occurred.

The authorities, however, were never able to control the flow of information in the magazines as tightly as with daily newspapers, radio or television. In the late 1970s and early 1980s a lively opposition press came into being, despite frequent arrests, confiscations and bannings. These magazines—"tangwai," literally "outside the party"—increasingly raised issues which until this year had been taboo, and set new norms more in line with those held in democracies in the West. They focused on democratization, reform of the anachronistic political system, adherence to human rights, and on social and environmental issues. Now, under the new relaxation of controls in 1988, they could engage even in moderate discussions of statehood and the mainland. Magazine writers, however, still find it difficult to secure interviews from officials, though banning and strong censorship of magazines was significantly reduced.

The authorities also loosened restrictions on the dissemination of publications from the mainland. And when two Taiwanese journalists visited mainland China in September 1987, they returned home to threats of imprisonment for an act illegal under Taiwanese law; in March 1988, the Taipei District Court acquitted the journalists of charges.

Large sections of the public, however, perceive the two major newspapers, the *United Daily News* and *China Times*, as little more than power tools of the ruling party and government. The owners of these papers are both members of the Central Standing Committee of the Kuomintang. With new freedom to own newspapers, several prominent opposition leaders indicated they plan to start a daily newspaper.

Television coverage of matters deemed politically sensitive is even more controlled than coverage by the newspapers and magazines. All three television channels and all but one (the English-language station) of the thirty-three radio stations are owned and operated by organizations or persons associated with the KMT, the government or the military. No new licenses for television or radio broadcasting will be issued—"for technical reasons." This is generally interpreted as an attempt to keep the opposition far removed from the electronic media. At present, they pay scant attention to the opposition, and when they do they portray it negatively.

All in all, in the latter half of 1987 and 1988, restrictions on the newspapers and magazines of Taiwan were notably reduced.

Tanzania
[IIIIIIIIIII_____MoF

Journalistic standards and ethics in Tanzania deteriorated in 1987. The mass media, controlled by the government and the CCN (Revolutionary Party of Tanzania), have acquired the role of promoting "socialism" in Africa.

In September, the Tanzania Journalists Association, composed of members of the government and Party-controlled media, convened a meeting of journalists from the southern Africa countries and the liberation movements of South Africa and South West Africa—with the participation of the International Organization of Journalists, headquartered in Prague, and delegates from Communist countries of Eastern Europe. The concluding communique formed the Southern African Journalists Association to employ government and party-controlled media to promote "socialist" ideology throughout the region.

The mainland government of Tanzania owns the only English-language daily newspaper, the national news agency, and the mainland radio station. On Zanzibar, the government runs a radio and television station. The Swahili-language newspaper is owned by the party. The Newspaper Ordinances of 1968 empower the president to ban any newspaper he considers flouting the "national interest." The government may search and seize any paper without a warrant, and withdraw its license to publish at any time. This makes almost impossible the launching of an independent newspaper or magazine. Restrictions on newsprint are another form of limiting publications. The only two effective printing presses in the country, one owned by the government and the other by the party, can refuse work on the pretext of a shortage of newsprint or pressure of other work.

Two religious newspapers, one published fortnightly by the Roman Catholic church and the other put out monthly by the Lutheran church, enjoy some freedom because their journalists and publishers apply strict self-restraint. The Catholic paper sets its own type and supplies newsprint to the printer to ensure continuous publication.

The Tanzania News Agency (Shihata) was created in 1976 by the government and placed under the Ministry of Information. Only Shihata or its agents can disseminate any news to or from Tanzania. All incoming foreign news agency copy must pass through Shihata for perusal before the reports can be sent to publishers or broadcasters. The same act that formed Shihata allows it to license journalists, both local and foreign. This clause was not implemented until 1986 when journalists were notified that they must possess licenses in order to practice their profession. Since the license fee is high, many journalists face the risk of prosecution for not having obtained the license.

The government and party-controlled media—newspapers, radio and television—are geared to promoting the cause of Tanzania's socialist ideology popularly known as "Ujamaa." All journalists are required to be well versed in the party's political ideology and most have attended a party school for indoctrination. Straying from ideological lines may mean dismissal. Journalists must also attend the Tanzania School of Journalists where they spend two years in combining professional skills with ideological indoctrination. At the Journalists Association and Press Club they continue to solidify their support and promotion of Ujamaa. As editors in the news media, they must daily satisfy the objectives set forth by the Ministry of Information. Criticism of government policies may result in dismissal or even detention without trial.

Local as well as foreign journalists find it difficult to secure accurate information directly from government or party sources, or the police or military. Such information usually comes through the ministry or from Shihata, where information is placed in governmental perspective. Some highly placed government and party officials occasionally denounce journalists for distorting information when the truth about sensitive issues is published. Once denounced, journalists may be severely reprimanded, and editors of the publication warned or replaced. Foreign journalists on arrival must report to the Ministry of Information, and disclose the purpose of the visit. Even so, a foreign journalist may find himself harassed by local officials, the police or even the military. Movement about the country, as well as access to news, is extremely limited. The Ministry of Information reports that more than fifty foreign journalists were barred from entering Tanzania in 1987 because they were regarded as "security risks."

Because of the stringent domestic controls, Tanzanian citizens tune in foreign broadcasts and purchase foreign publications though they are expensive. The government occasionally bans foreign publications on the pretext that no foreign exchange was available to import them. Some foreign publications are impounded when it is discovered that they carry sensitive news. Such information is usually provided by Tanzanian journalists who live abroad in self-exile, in order to write the truth about their country and avoid persecution.

Togo
[IIIIIIII_____MoF

All the news media are government-owned, -operated and -controlled. No independent newspapers or magazines are allowed to exist, nor can any independent views be expressed. The purpose of the media is to project a positive image of the government and its policies. In particular, the media galvanize public support for the head of state, President Gnassingbe Eyadema. The media are mainly responsible for projecting the official cult of the personality of Togo's president and his twenty-two-year role.

The media consist of a daily newspaper *La Nouvelle Marche*; a monthly magazine *Togo Dialogue*; one television station; two radio stations, one in Lome and one in Kara; and a news service Editogo.

Domestic news coverage lavishes praise on the president and reports the gratitude expressed at home and abroad for his policies to promote peace, stability and development. The news also emphasizes the constructive role the president plays in regional and world affairs, and the esteem in which he is held by foreign leaders. No critical evaluation of the president or his policies is allowed. The largely ceremonial activities of other government officials are also reported without critical comment. Foreign donations and development aid are emphasized in the media to show that the hundreds of millions of dollars Togo receives each year from the international community is a mark of confidence in the regime.

Only in the cultural realm can individual opinions be expressed by journalists, but these must be generally positive. Negative comments about theatrical productions or other cultural events are not aired. The government does not, however, hide all "bad news." It permits cases of corruption in its ranks to be reported. In 1987, the case of embezzlement at a government-owned agricultural bank was fully described with those involved named, though none held high government office.

The Togolese government makes a great effort to influence foreign journalists' coverage of events in Togo as well, with the result that foreign commentary on Togo is generally positive. The government has hired public relations firms in the United States and France which seek out reporters to write "positive" stories about the country. The government is also reported to have paid directly several foreign journalists, notably French, to write favorable reports. It uses these positive accounts to demonstrate to its own public the strong international support Togo's government enjoys. At the same time, all foreign press are welcome in Togo and are allowed to travel freely. But because Togo is not a country journalists visit in abundance, and because Togolese officials and citizens are afraid to speak critically of their government or its policies, few negative reports are to be found in the international press.

Media coverage does not reflect the broad concerns of the public. The major economic, social and political problems facing the nation are simply not discussed. Opinion columns and call-ins do not express dissenting or independent views. Coverage is given only to views that support the government. Much media attention is given to government organized mobilizations of the public that show support for government policies. On occasion an editorial does appear in explanation of a particular decision. But in general, little justification is provided for policies adopted by the government.

Coverage of foreign news is much freer than of domestic news because Togolese media publish accounts of foreign events from Agence France-Presse and Reuters. And nothing is done to prevent Togolese from listening to foreign broadcasts, such as the Voice of America and BBC. Foreign magazines and newspapers are also allowed to circulate freely. However, Togolese journalists are not permitted to comment critically on current events. When Togolese media report international news, they must do so free of opinion or criticism. The government's policy of friendship with all countries and support of peaceful solutions to all conflicts, moreover, effectively precludes the media from taking sides on any issue. Criticism of foreign governments or of internal developments in other countries is allowed only in regard to apartheid in South Africa and terrorism emanating from Ghana. *Togo Dialogue* on occasion does print analyses of major international issues and of political developments in foreign countries with differing viewpoints.

Togolese mainly learn what is going on abroad by traveling, listening to foreign broadcasts, reading foreign magazines, and working for international companies. Togolese look to their own media to find out what opinions they are supposed to hold and what issues it is better to say nothing about. Togolese news has no ideological slant although it is clear the country is part of the "French community" and pro-Western. No criticism is allowed of the heavy French influence in the country. The main problem is not the credibility of what is reported but the information omitted.

Because Togolese working in media are government employees, they toe the official line without question. Moreover, as citizens of Togo, they are aware that even indirect criticism of the government, its political or economic system, or its policies could result in loss of work, arrest or imprisonment; they engage in effective self-censorship. Authors of books or articles deemed critical of the government have been jailed, sentenced, beaten, and their works confiscated or banned. All writing for publication is subject to censorship.

In 1987, the issue of human rights came to the fore in the media following the creation by the government of a national human rights commission. It is empowered to receive individual complaints from Togolese citizens and foreign residents. For the first time, statements and speeches by Togolese officials and foreigners in support of human rights were publicized. This new trend of focusing attention on human rights could have effect on the role of journalists and the media in future, although no direct references to freedom of expression or free press have as yet been made. The Human Rights Commission in its first year did not receive or deal with cases pertaining to free expression—an area that needs the commission's attention.

Tonga
[||||||||||||||||||||||||||||||||||||||||||||||||||||||||||||||||||| MoF

This 270-square-mile island, with a hereditary constitutional monarchy, has no political parties. The government publishes a weekly newspaper in both Tongan and English, and several independent papers have been launched. The government operates the radio station which carries commercial programming in four languages (Fijian and Samoan, as well as Tongan and English). There were no reports of violations of press freedom.

Trinidad and Tobago
[||||||||||||||||||||||||||||||||||||||||||||||||||||||||||||||||||||||||| MoF

The newspapers and magazines are privately owned and operated, and traditionally independent of the government. In a gesture of goodwill toward the news media, the president, who began his five-year term in February 1987, invited journalists to sit in on his first cabinet meeting. The government also removed an onerous tax imposed on newsprint by the previous government.

The news media sometimes play a critical role, but without retribution or offense shown. Journalists are occupationally wary but generally approve the new government's policies of increasing the flow of information.

The government operates and owns the only television channel in Trinidad and Tobago. The station's news coverage is generally unbiased, though disputes arise over the allocation of time to political parties. The station assigns programming in proportion to the share of votes each party received in the previous election. Of the two radio stations, one is privately owned and the other government-operated.

Turkey
[||||||||||||||||||||||||||||||||||||||||||||||||||||||||||||||||| MoF

The Turkish press remains vigorous, often opposing government policies, in the face of lessening but still substantial taboos and controls.

The country's eight main daily newspapers carry varying degrees of sensational reports and photographs—a robustness in stark contrast to the papers published under military rule which ended in 1983. Nearly all are anti-government, but recognize the three principal taboos which prevail under civilian as well as military governments. These taboos preclude references which may be considered favorable to "separatism" (a euphemism for the struggle of the Kurds for self-determination), Islamic fundamentalism or communism. But the papers have begun to test the limits of these taboos.

In December 1987, authorities confiscated copies of the weekly newsmagazine *2,000'e Dogru* for testing the right to report on the Kurds. The article described government documents prepared secretly seventeen years earlier, revealing which Kurdish tribes were loyal to the Turkish government. The editor was charged for inciting "hatred and animosity" by "emphasizing regional differences." The same editor had been convicted and sentenced to prison for two earlier articles. By May 1988, her paper had been confiscated four times and received eighteen prosecutions—the latest delivered on the day the International Press Institute met at its annual assembly in Istanbul. The editor then faced more than a century in prison.

Officers of the IPI took that occasion to confront Prime Minister Turgut Ozal in person. They acknowledged that he had removed some restrictions on the news media since he came to office in December 1983. But they pointed out that Turkey is still a "limited" democracy with journalists in prison for expressing their opinions. Economic pressures on the press have increased. Newsprint prices were raised by more than 200 percent in the previous sixteen months. State banks have placed an embargo on advertising in some newspapers regarded as critical, and excessive fines have been imposed by the judiciary on publications. They also acknowledged that an onerous draft law on the press had been introduced and dropped after strong protests by publishers and journalists organizations.

There was, however, some indication of how far democratization had come when the editor of a major Turkish daily stood before the prime minister, in the company of hundreds of foreign journalists, to voice a challenge and invoke the need for criticism as a "driving force in the advancement of civilization."

Criticism has not fared well in Turkey's checkered history. A walk through the marble streets of partly restored Ephesus, laid out 3,000 years ago, suggests Turkey's variegated past, and her uncertain present. Ephesus was conquered by Amazons, Greeks, Persians, Romans, Lydians under the rule of Croesus, Romans a second time, and under the Ottoman Empire left to decay. These are the streets Alexander the Great trod, which saw Anthony and Cleopatra, and heard St. Paul preach.

Today, the common visitor can walk uninhibitedly (for such an important restoration) among the magnificent ruins. The sense of millenia, and of change following change, continues to permeate Turkish politics. Since the first free elections in 1950, the army has intervened whenever civilian authorities ran the country into debt, or became increasingly corrupt and authoritarian. Journalists suffered or were relieved in the ebb and flow of authoritarianism. Now, with Turkey seeking entry into the European Common Market, journalists may fare better. The EEC expects its members to sustain democratic processes, and has machinery to examine and perhaps expel recalcitrants. Indeed, there is probably more criticism of Turkey by Europeans because the country does seek to join the club.

Uganda

More than fifteen newspapers publish widely divergent views covering the Uganda political spectrum. Under President Yoweri Museveni, who assumed office in January 1986, there was more freedom for journalists than at any time in the preceding decade. The honeymoon lasted until some journalists began pointing out the shortcomings of the government in 1987.

The news media continued to report violations of human rights, a civil war and corruption by government and army officials. But both the president and the minister of information, while publicly supporting press freedom, increasingly reprimanded journalists for reporting irresponsibly and inaccurately. At times, the government issued warnings to the press in general, publicly rebuked individual journalists, and on several occasions threatened to close any paper that defied a government directive. On one occasion officials refused to register a newspaper, though all the formalities had been completed. Some journalists were hauled before military men who made veiled threats on their lives. Some journalists went underground for fear of a violent assault.

Not all journalists suffered this fate. Some papers reflect certain political trends, and are generally mouthpieces of political parties. Some papers, therefore, support the government coalition, while others do not. Those supporting the government serve as a parliamentary platform where government policies are defended and amendments proposed.

Some newspapers report events as they come. This is the most popular segment of the media, and serves to keep the public aware of what is happening in their country. These papers have uncovered scandals involving highly placed people, and have directly influenced events during 1987. Another segment of the press engages in disinformation, especially where the facts are not favorable to the government. This press has managed through disinformation to influence public opinion on certain issues.

The news media in Uganda are relatively young, and thus face numerous impediments. Special interests within the media influence the selection and projection of news and information. Reporters take into account the interests of the owners and editors of the newspapers. But the most important impediments come from outside the media, more or less influenced by government officials. On some occasions, officials have ordered journalists not to comment on particular subjects, or directed them how a particular issue should be approached. Journalists, consequently, are wary of government handouts.

News sources generally are hostile to journalists. Not only officials, but private individuals are known to issue threats or offer bribes to editors or reporters. It is difficult for a reporter to rebuff such an offer, particularly if his assignment comes through an editor who has been compromised.

In 1988, President Museveni struck at Ugandan newspapers for "undiplomatic coverage" of Ugandan relations with Kenya. He chided them for "supporting dissidents in other countries." A Ugandan paper had erroneously reported the massing of troops on the Uganda side of its border with Kenya. This led to border skirmishes.

These diverse pressures inhibit journalists even in this new environment of relative permissiveness. "Freedom of the press is not absolute," said the minister of information.

United Kingdom
[||||||||||||||||||||||||||||||||||||||||||||||||||||||||||||||||||||||||||||||||_____MoF

The years 1987-88 were a time of conflict between the government of the United Kingdom, and the print and broadcast organizations. The independence of the news media was repeatedly challenged by acts of the government in the name of national security. Such acts, in a parliamentary democracy without a written constitution, strain the social contract, no matter how old its tradition. The UK remains one of the freest nations, and its news media one of the most independent and sophisticated, but two years of government challenges have generated serious tensions in the press-government relationship.

Indeed, repeated governmental pressures on the news media have led to the formation of Charter 88, a group of prominent journalists, academics, judges and novelists. They said that "300 years of unwritten rule from above are enough." They called for a written bill of rights with guarantees such as the freedom enshrined in the U.S. Constitution and amendments. Such rights can be better secured and guarded, they wrote, "once they belong to us by inalienable right."

The Official Secrets Act, while not the sole cause of controversy, symbolizes the problem. The act sets criminal penalties for the unauthorized disclosure and publication of official information, no matter how sensitive or trivial that information may be. The law is designed to protect vital secrets, but journalists claim it is used to discourage journalistic inquiry and silence government whistleblowers. An attempt to revise the seventy-six-year-old law was defeated in 1987, and revived in 1988.

The government in May 1989 limited criminal penalties only to six categories of national security, rather than to information in general. There would also be tests of damage the government would have to prove. It would no longer be an offense just to receive sensitive information. Yet, journalists could still be prosecuted for publishing security material even if it had been first published abroad. This has occurred in the wake of two strenuous efforts by the British government to censor books deemed injurious to national security.

The first case continued through much of 1987 and 1988. It involved the government's demand that no reference be made in British publications to *Spycatcher*, a bestselling memoir by a former member of the British domestic intelligence service. The book had already been published abroad and was widely read in the UK, but the government sought a court order banning references to the book. The government eventually lost the case, as it did in its effort to ban a second spy book, *Inside Intelligence*. The government had banned publication of the book in the UK but the author had it printed abroad privately. The government then banned in the UK magazines with excerpts from the book published in the United States. The publisher flew to public protest meetings in London, just as others protest censorship in far less free countries. In both censorship cases, press and political critics claimed that the government's "embarrassing vendetta" had made a mockery of press freedom in Britain.

There were other instances: The police in January 1987 raided the offices of the BBC in Scotland over its plan to broadcast a program on the Zircon spy satellite. After examining the material to be televised the government decided the broadcast would not breach national security. The program was scheduled for fall 1988.

The government banned Radio 4's proposed program on the security services. It also banned Channel 4's broadcast concerning a pub bombing in Birmingham. The BBC was repeatedly harassed to alter its programming of controversial subjects such as the Irish Republican Army. In April 1988, detectives from the Royal Ulster Constabulary went to the Belfast offices of the BBC and ITN, and seized untransmitted film of the IRA mob killing two British soldiers. The BBC and ITN had refused to hand over the film, so as not to appear to be part of the police action and perhaps lose access to further such reporting. Prime Minister Thatcher condemned the refusal, saying that "everyone, the media included, has a bounden duty to do everything they can to see that those who perpetrated the terrible crimes, which we saw on television and which disgusted the whole world, are brought to justice."

The line is hair-thin between a citizen's "bounden duty" to assist the police in such an inquiry, and the news media's responsibility to preserve their ability to gain access to information, on the basis of which all other citizens may do *their* "bounden duty."

In October 1988 the government banned British radio and television from broadcasting interviews with members of outlawed paramilitary groups in Northern Ireland. The ban applied to one elected member of the British Parliament who is also a leader of the Irish Republican Army. The outlawed leaders could be quoted, but not allowed to speak directly over the media to "propagate terrorism," said the Home Office. The BBC said this ban on elected officials "sets a damaging precedent." The *Independent* called it "the biggest permanent gag on broadcasters in peacetime."

None of these incidents, in themselves, produced a loss of information for the public. Indeed, for the most part, the government suffered tactical defeats. But the tension between the press and the government, and the climate of suspicion generated in the public, is potentially harmful in a democratic society—particularly when there is legislation which tends to restrict the independence of the news media; the Contempt of Court Act, the Criminal Justice Act, and the Law of Confidentiality. And, above all, the Official Secrets Act.

Vanuatu
[IIIIIIIIIIIIIIIIIIIIIIIIIIIIIIIIIIIIIIIIIIIIIIIIIIIIIIII_____MoF

Press freedom is protected in the constitution. The government, however, publishes the only national newspaper, a weekly, and operates the only radio station. Both the paper and the radio are strongly influenced by political and ideological concerns.

Western Samoa
[IIIIIIIIIIIIIIIIIIIIIIIIIIIIIIIIIIIIIIIIIIIIIIIIIIIIIIIIIIIIIII_____MoF

Government business is conducted through a combination of the British parliamentary system and Samoa's traditional social structure. Some political influences and occasional acts of censorship are directed to the independent newspapers. The government controls the radio system and permits commercial broadcasts.

Zaire
MoF

The news media of Zaire, with the exception of an independent biweekly, are owned by the state. This includes two national daily newspapers, the radio and television systems, and the Zairean News Agency (AZAP). AZAP is the channel that provides all domestic and international news to all the media.

The government's main concern now is national development. Before the 1980s the news media concentrated mainly on ideological propaganda, and prescribed the attitude every "good citizen" was to take in national affairs. This excessive concern for ideology, then, was the major hindrance to the free flow of information. Anything that did not comply with the ideological views of the time was invariably labelled "subversive." Today, government officials and the people generally no longer see subversion all about them. Perhaps the prevailing economic crisis has made them more objective.

The main impediment to the news flow in 1987-1988 is linked to the socio-linguistic profile of the country. For many years, most important news items have been communicated in French and four local languages in order to reach the broadest mass of the population. Toward this end, the governmment created regional broadcasts on radio and television, and regional newspapers. But all languages in the country could not be used on the broadcast media, and the high rate of illiteracy made newspaper reading impossible for many. The high cost of radio and television sets further reduced the potential broadcast audience. Thus news and information media do not reach large segments of the population.

Yet there have been noticable improvements in the performance of the media. Some years ago it was unthinkable to hear or read any serious criticism (whether constructive or destructive), or any controversial treatment, of the government and its actions. Now this is possible. Criticism seems to deal, however, only with some limited issues, not concerning the basic operation of the government or the management of the economy. Yet the government seems to welcome constructive criticism. The fact that most of the media are owned by the state may explain journalists' reservations in dealing with sensitive issues.

1987 was nevertheless a year of improvement. The government assumed a positive attitude toward enlarging the flow of information, broadening the areas of freedom within which journalists may operate, and maintaining reasonable restrictions, without which freedom would be meaningless. Yet in 1988 local officials shut down one paper and arrested for three months the director of a major newspaper who criticized the officials for drug trafficking.

Zimbabwe
MoF

This country is still only nine years away from the pre-independence UDI regime during which censorship was routine. Since independence in 1980, there has been no formal government censorship, and a significant widening of the area of press freedom. However, a state of emergency still obtains and this imposes severe restrictions on the reporting of any matters touching on "defense, public safety, public order, state economic interests, public morality and public health." In these areas, journalists are largely dependent on government press statements.

The media, particularly the print media, with their unique blend of ownership by private capital and independent public trust (Zimbabwe Mass Media Trust), largely determine the parameters and subjects of public discussion. In most areas, the news media enjoy a high degree of credibility but this is somewhat undermined by the security restrictions. As a result of these restrictions, reports of domestic events are sometimes heard by the public on international radio broadcasts (e.g. BBC World Service) before they appear in Zimbabwean news media, or are ignored by them.

Since independence, the elected government has sought to enlist the support of the media in its drive for national development. On the whole, media professionals have responded enthusiastically to this, and the most obvious effect has been to give far greater prominence to rural affairs than in the past. So far, however, because of economic limitations, only a minority of the public has access to the news media. Government policy is to widen and increase media distribution and penetration. Most of the privately owned press are small circulation papers based in small country towns.

There is only one privately owned newspaper of national importance, the *Financial Gazette*, a weekly business newspaper. It is often critical of government and, within the sphere of economic reporting, bolder than the other national newspapers in which the Zimbabwe Mass Media Trust has a controlling interest. It provides the latter with the only internal competition they face.

There is room to improve government's attitude and policy on the dissemination of information. The relations between government officials and journalists are frequently strained by unnecessary suspicions. Thus it is often very difficult for reporters to cover government activities fully and accurately. In 1987 the editor of the largest circulation national newspaper, the *Sunday Mail*, was unceremoniously removed by the company following his denunciation by the prime minister. The editor published a story about a large number of Zimbabwean students being sent back to Zimbabwe by Cuba following AIDS testing. His source for this, quoted in the story, was a highly placed government official. When the story was denied by government the editor bravely stuck to his guns, reprinting it and insisting that it was adequately sourced. His denunciation and removal have undoubtedly had a demoralizing effect on Zimbabwe's journalistic community, notwithstanding the fact that he has since been promoted to a well-paying managerial position in the same newspaper company.

The minister of information has on more than one occasion voiced strong condemnation of the *Financial Gazette*, and this has encouraged the view that he does not favor journalism that is critical of government. The *Financial Gazette*, however, has stayed firm against such criticism and has replied through its editorials in the strongest terms. It has not suffered in any way from doing so. Its circulation continues to grow. Its parent company still carries out government contracts, and the paper publishes government and parastatal advertising.

In general, although there are few restrictions on journalists, the flow of news and information is influenced by the prevailing atmosphere of caution among editorial staff, particularly those working on the state-owned electronic media or the national newspapers.

* * *

APART FROM THE incident involving Dr. Henry Muradzikwa, the *Sunday Mail* editor who paid a price for having the audacity to stand by his story on AIDS, 1987 also saw the expulsion of Jan Raath, a *Sunday Times* (London) correspondent; Goidwin Matatu, a correspondent for the *London Observer*, who had already been barred from reporting from neighboring Mozambique; the couple Peta Thornycroft and Peter Wellman, who worked for Australian Broadcasting and Associated Press, respectively; and the temporary arraignment of Independent Television News's (ITN) Tim Leach, along with Tommy Liddle and Paul Hughes of World Television News (WTN).

What makes the government so uneasy about the work of some newsmen is the fear that South Africa could easily infiltrate their ranks, and subsequently use them for intelligence gathering. Up to now, no journalist has ever been found to be in the employ of the South African administration. Those expelled have been charged with writing in the foreign press articles which the Zimbabwean government considers objectionable.

A major threat to professional journalism in Zimbabwe is the operational framework which some journalists bring here after having worked in other independent African states: 1) They practice excessive caution and self-censorship, depriving the media of some credibility. 2) As a holdover from the war years, when communiques were published without questioning, the media still carry a disproportionate number of stories that are government-centered. 3) Some journalists are political appointees who define newsworthyness in terms of their political support, at the expense of more objective information. 4) Many experienced and professional journalists go into the private sector to run house journals, and undertake public relations work.

A breath of fresh air in Zimbabwean journalism, however, was the publication in 1988 by the Bulawayo *Chronicle* of a front-page report of high scandal. A crusading editor, Geoff Nyarota, published the names of several cabinet members, security chiefs, and ranking bureaucrats who obtained cars and trucks from the government assembly plant and then sold them at a great profit. Cars are in great demand, so private buyers on the black market pay many times their original values. The defense minister, mentioned in the article, threatened to detain the editor. The *Chronicle*, interestingly, is owned by the government-controlled Mass Media Trust.

But the fresh air soon soured. Nyarota was removed as editor, threatened with arrest and lawsuits, and given a "promotion" to a public relations position. The minister of information denied he had had a role in the affair, but said "we do not believe" in investigative journalism "because it encroaches on the privacy of people." He added, "We cannot afford the wholesale publication of inaccurate news items or those that will tear up the fragile social and political fabric of our new society."

* * *

THIS INCIDENT, CONCLUDING eighty-eight country reports, epitomizes the all-too-prevalent reality: the courage of a young journalist, a brief moment of openness, and the sad return to the patronizing defense of censored, guided journalism.

9.

The Reagan/Press Conflict

TENSION IS THE name of the Washington game.

There, as in 167 capitals, a finger on a button every minute of every day mobilizes words that influence every citizen. That button, far more diversified than a nuclear trigger, may control a high-speed computer, a broadcast network or an aged typewriter. Such a button may be at the command of a journalist or a government official. Either wields word power. Often, the force of one is set against the other.

The nature of that struggle depends on how free is the journalist to seek and report information, and how restrained is the official by law, custom or political acumen. Official restraint permits a contest of ideas, a clash if need be between the journalist's words, and the government's. Paltry few of the 167 capitals permit a robust struggle between the official word, and the journalist's. In Washington, such struggle is the capital industry.

That spells tension: the ceaseless tussle over what the public should or need not know, depending upon whether the journalist or the official is making the judgment.

The need to know, as defined by the journalist, would open almost every government file, old or current, for reporting to the public. Officials would restrict sensitive information, and tend to hide the politically or diplomatically embarrassing. Yet there are sound reasons for classifying information that reveals truly vital security information, intelligence operatives or methods, diplomatic initiatives, discourse with allies, scientific or economic research of a strategic or proprietary nature and areas related to each of these concerns. President Reagan signed an executive order in 1982 favoring increased classification. In 1984, the federal government classified 19,607,736 documents, a 9 percent increase over the previous year and a 60 percent increase from 1973.[1] Of these, 350,000 documents got the "top secret" stamp signifying that "exceptionally" great harm would befall the nation if their contents were publicly revealed.

Every Washington administration at some time complains publicly about the "irresponsibility," the relentless prodding of the press. Walter Lippmann described the press-government relationship as a system of checks and balances by which officials may withhold information, and journalists may dig it out and publish it.

In times of stress or emergency, American leaders have drastically restricted speech and press. Fear of attack from outside or from subversion inside has dictated stringent controls in the name of national security. Sometimes particular groups were singled out as "threats": Catholics, Jews, Mormons, Blacks, southerners, northerners, German-Americans, Japanese-Americans, Communists, anarchists. These have been historic victims of American xenophobia. This history is a continuing warning of where unnecessary secrecy and censorship leads.

Far more prevalent, however, are the uses of state power to prevent disclosures of political acts deemed potentially embarrassing. Franklin Roosevelt was probably the most effective manager of news among modern presidents. Before the television era, he dominated radio broadcasting of his activities, and restricted attendance at press conferences by holding most in the small Oval Office. He ridiculed, not always good-naturedly, opposing journalists and their publishers. Lyndon Johnson called the press "the least guided, least inhibited segment of U.S. society." He preferred to broadcast directly to the people rather than run the risk of the press misinterpreting his words. When angered, Johnson withdrew White House credentials of some journalists, protested to their employers, and in some cases had FBI investigations instituted against reporters. Lippmann called Johnson "a pathologically secretive man."[2]

Richard Nixon gave this "for your eyes only" message to his assistant H. R. Haldeman (15 June 1971): "In view of the *New York Times*'s irresponsibility and recklessness in deliberately printing classified documents [the Pentagon papers] without regard to the national interest...Until further notice under no circumstances is anyone connected with the White House to give any interview to...the *New York Times* without my express permission. I want you to enforce this without, of course, showing them this memorandum."[3] Nixon's press secretary also phoned television stations to ask what views they had broadcast about presidential policies—a step some broadcasters may have considered a veiled threat. And Nixon's communications director (Kenneth Clawson) told a reporter,[4] "Newspapers are privately owned but we all have a piece of TV's ass and we're entitled to do something—although I'm not exactly sure what, if it offends us." Nixon frequently complained he was maltreated by the news media. One motivation for the illegal coverup of the Watergate break-in which led to the president's resignation was fear of press exploitation of the original, relatively minor criminal act.

The Reagan/Press Conflict

Every president expresses exasperation over leaks by executive and legislative officials. President John Kennedy said the only real secrets in Washington were those his underlings kept from him. He cancelled twenty-two White House subscriptions to the *New York Herald Tribune* because its editorial criticism displeased him. Three Kennedy officials threatened to sue newspapers. He also asked American publishers[5] for voluntary press censorship to preserve national security.

The eight Reagan years reveal expected government/press tension in this freest informational society ever, anywhere. It is the freest, yet there are still official efforts to restrict or influence the information flow. Ronald Reagan and his principal aides came to Washington believing that the most influential news media base their hard-news coverage, not only their commentary, on decided liberal biases. The principal adversaries were said to be the three national broadcast network news departments, ABC, CBS, NBC, the *New York Times*, *Washington Post*, and *Newsweek* magazine. Few doubted that most journalists in New York and Washington had liberal personal histories and voting records, and favored liberal social and political agendas. Several polls of "eastern-establishment" journalists added solid statistics to these long-time Reagan assumptions.[6]

There was, then, the obvious adversarial relationship between a press regarded as liberal and a conservative administration. A White House private memorandum, leaked in January 1987, said Reagan policies were reported by a "left-of-center working press which put its prejudices above its responsibilities to the public." The important question for our purpose is not whether the eastern press is biased in covering a conservative administration. A case can be made that during the first Reagan term the networks and the major newspapers provided highly favorable coverage for the president personally, and generally for his policies. Indeed, there was concern among some professional press monitors that the news media dropped critical investigative reporting, and climbed aboard the Reagan bandwagon. This was unflatteringly attributed to the press's own vote-counting. The polls showed deep public support for the "Great Communicator," and his programs were flowing miraculously through Congress.

The sense of the nation seemed to be: Give the new president a chance to deliver on his promises. And he and his advisers, in turn, believed they could not accomplish the policy changes they contemplated if their objectives, day to day, were clouded by the intercession of skeptical journalists or those believed to be biased. Said David Gergen, the Reagan communications director, "The government has to set the agenda" and "can't let the press set the agenda for it." To be specific, says David Shaw of the *Los Angeles Times*, news generated outside the "Cambridge-Manhattan-Georgetown axis" has little chance of gaining national attention.

Despite the preponderance of "liberal" journalists in the major media, however, the press (print and broadcast) was charged with having become a Reagan sounding board. "Journalists played dead for Reagan," wrote Mark Hertsgaard in January 1989. This critic had earlier published *On Bended Knee: The Press and the Reagan Presidency*. Such pusillanimity was also attributed to board-room control of the media: in 1987, twenty-seven large corporations controlled the major news organizations—ABC, CBS, NBC, the *New York Times*, the *Washington Post*, the *Wall Street Journal*, the *Los Angeles Times, Time*, and *Newsweek*—all Fortune 500 companies—an oligarchy of wealthy owners rather than a conspiracy of liberal journalists.[7] A more subtle, almost psychiatric explanation was given by Anthony Lewis, the *New York Times* columnist. He said journalists "found it upsetting to acknowledge to the public or to themselves that American leadership was" in the hands of a man with only "an anecdotal view of the world."

Yet the editorial and news policies of this group reveal potent anti-administration coverage, as well as support for the Reagan policies. While reporters certainly conveyed official positions, there was some—if not equal—coverage of skeptics and opponents. Inevitably, as during the liberal years of Roosevelt and Johnson, the ability of the White House to make news and manage reporting taxed even those journalists committed to balancing their reports.

Complicating the U.S. information picture is the growing interest of global media conglomerates in purchasing American electronic data bases, book and newspaper publishers as though they are merely cards in the monopoly board game. With the U.S. dollar low and foreign markets crowded, the big spenders are buying up U.S. information assets. The overseas conglomerate News Corporation bought Triangle Publications for $3. billion. Hachette SA, the French publisher, purchased more than $1 billion in U.S. publishing assets, including Grolier, Inc. and some thirteen popular magazines. British publishers were said to be bargaining for McGraw Hill, a major U.S. publisher. So far, the content of established publications seems to be little affected by the new managements. But the potential for new influences, particularly in foreign buy-outs, is always present.

In addition to the Reagan handlers' efforts to apply their spin to official hand-outs, were the administration's little-publicized efforts to contract the flow of official information. Presidents may be expected to use their pulpit to espouse policies, and criticize opponents of their policies—even journalistic opponents. Presidential power is not expected, however, to make robust debate more difficult. An official, especially a president, has the primary sources of information at his command. That is a clear advantage over any nongovernmental adversary. The question, then, is whether the Reagan administration attempted to convert its preconceived antipathy to "the news media" into institutional or policy changes that would obstruct or restrict the free flow of information.

In brief, where does the United States stand on the universal spectrum we have used in this book to estimate changes in seventy-four other countries in 1987 and 1988?

One quick answer: the U.S. still provides the freest journalistic system, but not without some incremental loss. That loss, while important to note, is not crucial. We are not on an irreversibly slippery slope. But retrogressive changes should be carefully monitored. Such steps, if repeated, may inhibit proper access to official information. That could reduce the government/press tension, but at too high a price. It would, as in the not-too-distant past, turn journalists and officials into enemies, not adversaries.

The ultra conservative ideologues in the Reagan administration dramatically enlarged their own creative "news" flows, and tightened controls over journalists' access to other information about the government's plans and activities.

Access restricted

Access to government information is a corridor to power. For eight years Reagan officials had the authority to reclassify information already available to the public, and restrict unclassified material. They did. In 1981 the new administration made drastic cuts in departmental budgets for reports of government programs. Valuable statistical programs were cut back. The stated goal was the "elimination of wasteful spending on government periodicals." Many public assistance pamphlets on health, education and farming issues were also closed down, along with analyses of U.S.-USSR arms and price comparisons. Funds for the National Archives were cut by 60 percent. The Office of Management and Budget wielded its hatchet for information control as well as dollar-saving objectives.

To control information at the source, the White House tried to force preclearance of all interviews on national security matters, but this was later dropped. The Department of Defense sought criminal penalties for government employees who leaked secrets to the press. Yet leaking—usually done by high-level officials—can have some value. It provides the public with more details about policies than the administration cares to reveal. Other departments required all employees to give written notice before being interviewed by journalists. The Office of Management and Budget sought to give to private corporations, some government publications and data bases which were then government property.

In April 1986 the White House adopted thirteen measures to increase information security, including prosecuting officials for leaks and limiting their access to classified documents. Rarely, however, are journalists subject to prosecution. In one case, they might have been. William Casey, then CIA director, carefully warned the *Washington Post* and *Newsweek* that they would violate the Communications Intelligence Statute in reporting U.S. intercepts of Libyan communications in connection with the trial of Ronald Pelton, indicted for selling secrets to the USSR. The U.S. Code, Section 798, Title 18, forbids publication of classified information and cryptographic intelligence. The *Post* delayed and edited its copy before publishing the account.

The FBI investigated news leaks from government sources concerning drug trafficking in Latin America. Clearly, said William Clark, as national security adviser, "The press has been doing its job—collecting information—better than the government has been doing its job—protecting national security information."

In a rare use of disinformation, a national security council officer revealed a campaign against Muammar Qadhafi. A November 1986 memorandum from the national security advisor gave new powers to limit release of government information and restrict access to security data by creating a new "sensitive" category. A member of the commission investigating the Iran-Contra affair said that the administration's "over-obsession with secrecy" was responsible for the "wrong policy" on Iran-Contra.

Freedom of Information cutback

Since 1967 the Freedom of Information Act (FoI) has enabled journalists and others to secure from federal government files a vast array of information previously controlled. The FoI is a "blessing for those who value a check on government snooping," wrote William Safire, the conservative columnist for the *New York Times*. Major news stories have been based on letters and documents released after formal application, and following procedures for which the recipient pays the government a fee. In some agencies, complying with journalists' FoI requests requires the hiring of special personnel.

The Reagan Administration sought to reduce the FoI flow. Early in 1981, the attorney general issued new guidelines stating that any information could be suppressed even if its release would not pose harm to the government or a third party. In October 1984, the President signed into law the bill excluding the CIA "operational files" from the FoI Act. An executive order gave agencies broad authority to classify information without a time limit. This reversed the trend begun by President Eisenhower of limiting the classifying of documents. The Defense Department announced directives to withhold from the public information that cannot be sold abroad under the Arms Export Control Act. This was intended to bar the dissemination abroad of sensitive technological information, but it may also suppress word about the effectiveness and cost of American military equipment. A bill passed in November 1984 authorizes the Defense Department to withhold technical information from disclosure under the FoI Act. In October 1986, Congress amended the FoI Act to allow agencies to withhold some information about law enforcement activities. In 1987 the president proposed amending the FoI Act to permit agencies to withhold information they deem harmful to their programs or commercial interests.

That year, too, the president issued an executive order requiring federal agencies to tell commercial businesses whenever information which they submitted is sought under the FoI Act. In February 1988, the president asked Congress for an exemption to the FoI Act to enable agencies to withhold information about science and technology where harm to the country's "economic competitiveness" might result.

Administrative restrictions

Editorial prerogatives can enhance the free flow of information; restrictions have the opposite effect.

Several times the administration announced that bureaucrats with access to classified information must take lie detector tests, if asked, when officials seek to trace news leaks to the press. The Justice Department removed restrictions on infiltrating the news media, political or academic groups, if needed for national security. In October 1983 a U.S. marshall posed as a news reporter in Athens, Georgia to secure information at an antinuclear protest meeting.

In January 1985 a freelance journalist's diary was seized at the Miami airport by an FBI agent who photographed the unpublished article, and then returned it. In July 1985 the Justice Department subpoenaed video and audio tapes and photographs made by the television networks, the Associated Press and news magazines during the hijacking of a TWA passenger jet and the hostage crisis in Beirut. Justice said a grand jury was considering indicting the hijackers, and would need the tapes. In September 1987 the FBI asked twenty New York librarians to create surveillance records on the use of the library by foreigners.

Court rulings

The Supreme Court supported Carter and Reagan administration demands that CIA agents must submit all their writing for review before unofficial publication; prior restraint is an effective tool. The Court also agreed in 1981 to a Carter-inspired suit cancelling the passports of U.S. citizens whose writings abroad endangered American security. In June 1982 Congress enacted the Agent Identities bill which prohibits publishing the name of an intelligence agent, even if the newspaper did not intend to harm the U.S. and even if the information posed no danger to national security.

In February 1983 the Justice Department required that films on pollution and nuclear disarmament, distributed by the film board of Canada, state that it was "political propaganda," under terms of the Foreign Agents Registration Act.

The courts provided mixed reviews for the press. In June 1988 the Supreme Court ruled that cities may not have "unbridled discretion" in deciding which newspaper publishers could put their coin-operated news racks on public property. Another Supreme Court action in May 1988 further bolstered the Court's opposition to prior restraint on free speech exercised occasionally by lower courts. At state and local levels, judges demanded to see interview notes (published and unpublished), and film (raw and edited for public use). Twenty-six states now have shield laws intended to protect the journalist from such demands by courts. But when a defendant's case hinges on the journalist providing background information, the judge backs the defendant about one-third of the time. Two state appellate courts—California and New York—ruled in the spring of 1988, however, that journalists' notes, photographs, films, tapes and other newsgathering information have special protection from court-ordered disclosure. The San Francisco court ruled in a capital murder case, and the New York court acted in a civil case. The New York decision broadened the shield by saying that even data not gathered in confidence should be protected from forced revelation in court. This ruling was based on the U.S. and New York constitutions, rather than legislated shield laws.

The New York ruling is important. It recognizes the special role of the press. "The autonomy of the press would be jeopardized," the court said, "if resort to its resource materials, by litigants seeking to utilize the news-gathering efforts of journalists for their private purposes, were routinely permitted." Journalists, therefore, should not be diverted from "journalistic efforts" by repeated subpoenas, the court added, unless it can be proved that the information is "critical to the litigant's claim," and is available nowhere else.

After the press had sought for years to protect its sources, two newspapers in Minnesota were fined $700,000 in July 1988 by a state court for identifying a man as the source of an article after they had promised him confidentiality. The papers were the Minneapolis *Star Tribune* and the St. Paul *Pioneer Press Dispatch*. The court ruled their pledge was a contract. One First Amendment expert said news organizations would weaken their claims to special protection against revealing sources under duress if they voluntarily broke their promises to keep sources anonymous. A California journalism school dean called the case bad journalism and bad law. The ruling will probably be appealed.

Libel suits diminished considerably in 1988. The Supreme Court gave the press one of its most significant victories in years. The Court unanimously ruled that Rev. Jerry Falwell could not collect damages for a satirical article about him in *Hustler* magazine. Falwell contended he suffered "emotional distress" when the article appeared. The ruling further strengthened the "actual malice" test for libel set in *New York Times v. Sullivan*. It held that public figures must prove that the medium knowingly carried false information. After *Falwell*, it would not be possible to avoid the legal burden of proving actual malice by claiming a report produced emotional distress.

In throwing out a $50 million suit by the Liberty Lobby against the *Wall Street Journal*, the Court declared the case "epitomizes one of the most troubling aspects of modern libel litigation: the use of libel complaint as a weapon to harass."

Military blackout

During the invasion of Grenada in October 1983 the military forces excluded U.S. reporters from covering the initial landing. Commented Secretary of State George Shultz, "These days, in the adversary journalism that's been developed, it seems as though the reporters are always against us. And when you're trying to conduct a military operation, you don't need that."

The U.S. distributed government films that provided favorable coverage of the landing, detained several U.S. reporters already on the island, and threatened to shoot others who tried to reach Grenada on their own. On the third day, American journalists were permitted on the island. By contrast, near the culmination of World War II, 180 Americans were accredited to Allied military headquarters for the vast D-Day operation against Europe. Only twenty U.S. reporters accompanied the assault forces on the beaches. Thirty-nine years later, about 369 journalists assembled to cover the fighting on the tiny island of Grenada.[8] Yet because of the three-day delay in permitting press coverage, the Reagan administration is forever tarred with having barred civilian reporting of the Grenada affair. There should, indeed, have been at least a civilian-reporter/picture pool accompanying the initial attack. Months later, the Defense Department created and tested an emergency pool for wartime coverage, and with journalists developed guidelines for that arrangement.

An American naval intelligence analyst who gave *Jane's Defence Weekly* three classified satellite photos of a Soviet aircraft carrier under construction, was convicted of espionage and other charges and sentenced to two years in prison. The Supreme Court late in 1988 refused to review the conviction. The *New York Times* called this act of leaking "a dangerously broad definition of espionage." It added, "coupled with the power to declare anything classified, the power to criminalize leaks becomes a loaded weapon against democracy."[9]

"Panoramic" security

The Reagan administration's adoption of a "panoramic definition" of national security led to a steady expansion of government controls on scientific information and communication. That is the finding of a 1988 paper by the Association of American Universities (AAU). The AAU notes that the Export Administration Act (initiated during the Carter administration) has been extended to scientific communication. This limits the access of foreigners to U.S. scientific meetings, and places controls on papers presented. While intended to prevent the loss of sensitive information, the action reduces intercourse among Americans, and tends to enlist scholars and librarians as monitors of government policies. The survey also noted that the administration expanded the security classification system, particularly in enforcing "prior restraint." There were also new provisions for blanket nondisclosure agreements which impose an "open-ended requirement" for official clearance to allow government employees to write about matters they dealt with while in federal service.

The historic U.S.-Canada trade agreement passed by the U.S. Congress contains one precedent-setting restriction on journalists. For the first time, the U.S. government has placed itself in the role of deciding who is, and is not, a journalist. Before a journalist may cross the border for work he must have a bachelor's degree and three years of experience in journalism. This is tantamount to state licensing of journalists. It was opposed by press-freedom groups, and an amendment of the agreement sought.

This is not an exhaustive list of administrative or legislative acts to restrict the flow of official information during the Reagan years. It does indicate, however, the trend toward restriction, and the counterbalancing factors which limited further restrictions. Attorney General Edwin Meese III said, shortly after assuming office, that stronger measures than those used might be sought. Any government decision to prosecute journalists for making secrets public would, he said, "depend on the circumstances of the case." He hoped that journalistic ethics would "prevent people who have obtained what is in effect stolen property, stolen information, from utilizing it in a way that would compromise or hurt the national interest."

Through his years as adviser to the president and running the Justice Department, Mr. Meese frequently expressed serious concern over leaks to the press. Lie detector tests to find officials who did the leaking were often proposed, but never as widely employed as threatened. To stop the flow at the source, government employees were threatened with prosecution, but that was rarely enforced. The attorney general nevertheless proposed reducing the amount of information that is classified and then working with journalists "to make sure that that information is not improperly disclosed." He said at the beginning of his tenure that he would avoid the temptation to close up sources of information.[10]

At the beginning of his administration, President Reagan said he was determined to redress the balance between "the media's right to know and the government's right to confidentiality." The media do not have the right to know *everything* (the limitations, while very few, are important). The government has no *constitutional* right to confidentiality. It needs limited privacy for the high-level formulation of policy, and maximum security for its most important defense secrets. Of these two forces, the government is formidable. It has police power and the authority provided by an electorate. The news media have neither. Yet when they act responsibly, without hidden agendas, and are perceived to move in the public interest, they acquire the symbolic strength of the body politic. That, after all, is the highest source of legitimacy in a democracy.

The tension, then, is between two legitimate forces, both acting on behalf of the same client. Since their real power is asymmetrical, however, the government must take special precautions not to overwhelm the news media. By unremitting pressure, the Reagan administration tried for eight years to alter the government/press relationship. By signals, outright threats, court action, and legislative acts the administration sought to tighten its own control of information. It reduced the volume of reports (partly a money-saving step), limited the number of information-dispensing officers, made access to information through the Freedom of Information Act somewhat more difficult and costly, and applied prior restraint in diverse ways. More often than not, the Congress and the Supreme Court agreed with administration proposals to limit the flows.

The executive orders on secrecy enable the government for the first time to classify private technical research. While this may inhibit foreign theft of U.S. technology, it also makes it harder for American scientists and technicians to share their creations and provide new applications or still more advanced processes and machines. The reclassification of documents, intended to reduce the number of secrets, can also remove from the files previously open material and classify it. "History," remarked one observer, can be treated as "a menace to national security."

And regarding future history, the president on 17 February 1984, ordered held in abeyance National Security Decision Directive 84 which requires all government employees with access to "sensitive compartmentalized information" to sign agreements that subject them to official censorship for their lifetimes. If NSDD-84 is made operative, an employee included must submit every piece of writing on "intelligence" before publication for government review. Almost any defense or security subject can be included in this definition. In October 1983 the Senate voted against lifetime censorship for government officials.

Yet, Reagan was the president who in 1988 removed many of the old restrictions on private use of U.S. picture-taking satellites. This will enable the press to secure finely detailed pictures from the spy-in-the-sky satellite. This could provide "live" coverage of military maneuvers, natural disasters, and other subjects now hidden behind national borders.

An ironic example of the failure of a policy of extended secrecy began soon after the Reagan administration was reelected in 1984. Late that November the Defense Department issued a directive stating that Pentagon officials must withhold from the public all "technical data" regarding "contractor performance evaluation," and tests of military hardware. These are the areas in which the Pentagon-contractor scandals broke in mid-1988. Perhaps the tightened restriction on such information kept corruption from public attention, or, at least, gave those practicing fraud, and worse, the sense that no one was watching.

The honeymoon between the press and the Reagan White House was longer than that usually granted presidents after their first election. By 1983, however, analysts could write that "the nation's reporters have written or said that Reagan is dumb, lazy, out of touch with reality, cheap, senile, ruining NATO, tearing up his own safety net, even violating his constitutional oath."[11]

Television covered the president's activities, but not always with the Reagan spin. During his June 1984 visit to Europe, for example, the pictures showed the president as statesman: at Buckingham Palace, with Margaret Thatcher, and saluting the flag. Leslie Stahl's words over the CBS picture, however, said, "When it comes to political oneupsmanship, Ronald Reagan is a master....have lunch at Buckingham Palace, meet with Britain's Prime Minister Margaret Thatcher, and pay tribute to the thousands who died on the beaches at Normandy. White House officials say this trip was not scheduled for political reasons, not designed to cast a shadow on the Democrats' final stretch, but as long as the president had to come for an economic summit, why not?" One analysis recorded that the president was shown and heard simultaneously for just 10 percent of the coverage of his European trip. The voice of a network correspondent was dubbed over his picture 26 percent of the time. That commentary usually contained U.S. domestic political references not necessarily favorable to the president.[12]

Other developments during the Reagan years tend to support the journalist and news media in relations with the government. After thirty-eight years of debate, the Federal Communications Commission in mid-1987 abolished the Fairness Doctrine for broadcasters. It required holders of broadcast licenses to air contrasting views on issues of public importance.

A similar effort to enforce the publishing of letters to newspapers was decisively rejected by the Supreme Court in the 1974 *Tornillo* case. Chief Justice Warren Burger declared then that "the choice of material to go into a newspaper...whether fair or unfair...constitutes the exercise of editorial control and judgment." He added, "It has yet to be demonstrated how governmental regulation of this crucial process can be exercised consistent with First Amendment guarantees of a free press." Radio and television have thus been second-class citizens, despite the First Amendment. The Fairness Doctrine was, at least, a toe in the door for government influence in programming. Yet no broadcaster has ever lost his license for violating the doctrine. Perhaps Clay Whitehead had the pragmatic perspective. As director of telecommunications policy in the Nixon White House he said, "The value of a sword of Damocles is that it hangs, not that it falls."

The FCC's ban on broadcasting "indecent" language during evening hours was struck down, July 1988, by a federal appeals court. The ban had been intended to shield children from "indecent" language. Thus, indecent but not obscene material is protected by the First Amendment.

Other changes in broadcasting, particularly television, served to enhance the function of the independent media in a democratic society. Early in the Reagan era, television cameras came to the floor of the House of Representatives, and later to the Senate. Such coverage added a new dimension to the politician's tie to his constituency. The cameras made possible gavel-to-gavel coverage of Congress by the new cable network C-Span. A loyal new viewing audience developed. This provided extensive coverage of committee hearings, as well as floor debates. The "town meeting" aspect enabled hundreds of thousands of viewers across the country to watch not only the highly touted Iran-Contra inquiry, but scores of nuts-and-bolts discussions of Nicaragua, arms control, the budget and other issues. Indeed, young congressmen took advantage of C-Span exposure to rise out of obscurity.

Still other changes in the electronic media suggest greater diversity in news coverage in years to come. The new Local Program Network and Conus both set up satellite interviews with government officials and other newsmakers for transmission to local stations around the country. This enables the newsmaker to reach several major publics with a personalized message, while permitting the local station to broadcast "exclusive" interviews not appearing on the major networks. This increases diversity of views and actors. During the Reagan era the White House briefing room steadily accommodated new radio and television stations arriving to cover Washington—directly, and independent of the networks. This reflected burgeoning efforts among independent broadcasters to cover national news with a local twist, despite network affiliations. In some cases, satellite transmissions sped this development. Perhaps most significant is the opportunity for greater diversity in news coverage. The three large radio-television news networks are, in the words of one academic observer, "the last remaining controlled format the president has."[13]

Ambiguity still plagued that important arm of U.S. informational broadcasting, the Public Broadcasting Service (PBS). On the one hand, PBS provides extraordinary educational and commentary programs almost never seen or heard on commercial television and radio. The MacNeil/Lehrer News Hour provides nightly the clearest, least-hurried analysis and projections of events. It is scrupulously balanced. PBS's nature, science and dramatic programs are unrivaled in commercial programming. On the other hand, PBS receives only about $214 million in federal support (including radio). When the public system was created, it was regarded by some as a potential threat to free enterprise. Now, PBS relies on major corporations to support much of its programming. Such support inevitably narrows the range of controversy even PBS can address. Federal funding has also been drastically reduced since 1982. The U.S. spends less on public service television than any major industrialized country. Canada's excellent programming costs about $25 per citizen. The U.S. spends $3.77 per person, of which only fifty-eight cents comes from federal expenditures. The ambiguity over PBS's political positioning continues. Both liberals and conservatives are dissatisfied with PBS documentaries and public affairs programming. Consequently, the basic questions are unresolved: Should a free society have a government-supported information network and, if so, should it be expected to carry programs that will satisfy and irritate different interest groups simultaneously? The answer to both questions, I believe, is affirmative.

Perhaps greater attention should be directed to other consequences of deregulation in the electronics industries. Cable companies, for example, will not be obliged to carry all local stations, though a cable license is tantamount to a monopoly. The cables, then, will make final decisions over the viewing habits of their customers. Some accommodation is necessary. Several cable companies could use the same wire and compete through diverse programming. That is more desirable than governmental regulation of cable programming.

When contemplating government influence over the broadcast media, whether through public ownership or too rigid governmental regulations, these words of Zechariah Chafee, an eminent First Amendment scholar, are pertinent:

> Whenever anybody is inclined to look to the government for help in making the mass media do what we desire of them, he had better ask himself one antiseptic question: 'Am I envisaging myself as

the official who is going to administer the policy which seems to me so good?' Justice Holmes remarked that, when socialism comes, he hoped he would be 'on the committee.' You and I are not going to be on the committee which is charged with making newspapers or radio scripts better written and more accurate and impartial. It is very easy to assure that splendid fellows in our crowd will exercise the large powers over the flow of facts and opinions which seem to us to save society, but that is an iridescent dream. We must be prepared to take our chances with the kind of politicians we particularly dislike, because that is what we may get.

Examining the recent restrictions on government information flows, a harsh critic writes, "Imagine a venerable republic, the hope of the world, where the habits of freedom are besieged, where self-government is assailed, where the vigilant are blinded, the well-informed gagged, the press hounded, the courts weakened, the government exalted, the electorate degraded, the Constitution mocked, and laws reduced to a sham..." [14]

Has America suffered that horrendous fate? If so, the cumulative effect would by now have resulted in stringent censorship. Yet the still-vigilant and opposition press does not seem to notice such a revolutionary change. Or if they do, they are uncharacteristically silent.

Does the incremental nature of the restrictions defy consistent observation, and adequate assessment of the cumulative danger? No. Press monitors and ample numbers of political adversaries challenged the Reagan administration.

Is the charge sheer hyperbole? No, it is an overstatement of the probabilities, a fear of the slippery slope. The worst possibilities, by far, have not materialized. The checks in the courts, the Congress, the professional integrity of the permanent bureaucracy, and most of all the latent good sense of that part of the public that watches, including the news media themselves, have provided some balance.

Though the Freedom of Information Act has been modified, it still provides the most effective citizen-spotlight on government anywhere in the world. Other countries only now are considering establishing an FoI.

Repeated threats to charge journalists for breaching the communications intelligence act never materialized. Indeed, journalists involved in the most prominent incident quietly conferred with intelligence and other officials before publication, and reached an arrangement. It would not be entirely satisfactory to the press or officials, however.

But that reflects the everlasting tension, in a free society, between the adversaries: government and the news media. In the Reagan years, as in some previous administrations, the news media have been threatened, cajoled, but not fundamentally hampered.

The tension, as it should, persists.

10.

Networks of Freedom

We hope that Prof. [Samuel Pierpont] Langley will not put his substantial greatness as a scientist in further peril by continuing to waste his time, and the money involved, in further airship experiments. Life is short, and he is capable of services to humanity incomparably greater than can be expected to result from trying to fly...[T]here are more useful employments.
—*New York Times* editorial, 10 December 1903

EVEN A GREAT newspaper can misread the future. So can corporate managers hired to invent the future, as well as make profitable inventions. Supercomputers, for example, can shape the future of global communications. They can also forecast weather, design aircraft, and conduct geological explorations. A Biblical concordance, an alphabetical referencing by word, can be created by a supercomputer in twenty seconds. It took eighteenth-century scholars seventeen years to complete one. Today's supercomputers can make billions of calculations a second. By 1992, a new version of this computer will be 1,000 times more powerful. The biggest breakthroughs may come in Japan, not the United States, the originator of these earlier miracles. Lowering of the American productivity curve is the reason—not quite the problem Samuel Langley faced in 1903. Professor Langley had placed a highpowered, lightweight engine in a craft intended to carry a man aloft. His early failures brought ridicule from the *Times* and others. In 1914, Langley's craft flew. It would be difficult to estimate the social, political, financial and military changes which the airplane and associated industries have since wrought. The federal expenditure of $50,000 in 1903 for a "man-carrying machine" must be one of the nation's most lucrative investments.

A quarter-century before Langley's brainwave, Alexander Graham Bell wrote that the telephone, Bell's invention two years earlier (1876), would enable "a man in one part of the country [to] communicate by word of mouth with another in a distant place."[1] That forecast seemed unlikely at the time. So did many other claims made for telephony a century before they were realized. The fiber optic telephone cable which went into service across the Atlantic in 1988 uses laser light to carry 40,000 voice and computer calls simultaneously. The telephone became the centerpiece for radio broadcasting, television, cable communications, satellites, and the newest systems being developed for the next century. Telephony changed the social structure of every country which embraced it. The telephone also influences the political systems, when permitted.

None of this was accidental. The telephone system was constructed as "universal service," said AT&T's annual report in 1909. The system would be "like the highway system...from everyone at every place to everyone at every other place." That is the forecast today for the Integrated Systems of Digital Networks (ISDN)—the linkages of diverse information systems worldwide—in the next century.

As early as 1879, a wireless phone was attempted, and radio telephony was predicted. The telephone was already regarded as changing the pattern of human settlement. The telephone favored the growth of skyscrapers (messenger boys would no longer need to crowd elevators). The development of suburbs and urban sprawl would be advanced. The telephone was even credited with abolishing loneliness, particularly for the farmer's wife. The telephone was said (1885) to change the law of evidence. It was foreseen (1906) as fostering national integration. And telephony was viewed as forever altering politics (1906). Most important for our projections—twenty-first century communications will depend largely on telephony—the telephone was regarded in the 1920s as a democratizing force, leading to the "spontaneous interaction of individuals." This was and remains the bane of authoritarians and totalitarians. In international affairs, as early as 1922 at the Washington Conference, the telephone was employed as an instrument of international diplomacy, and its widening use predicted. By 1931, President Herbert Hoover tells us in his memoirs, he used the phone daily to bypass the State Department and speak to his ambassadors in London, Paris, Berlin and Vienna.

In all, the dreams of the distant future, the twenty-first century, were cast in the glimmerings of the nineteenth, and the same basic instrument—telephony—is at the core.

The coming of ISDN

Now the world is moving toward a new age in which the inventions of communicators and the plans of today's bold dreamers will be tested.

The dream of Jean Monnet in 1952 for a United States of Europe will be partially realized forty years later. The European Economic Community (EEC) is scheduled in 1992 to drop trade, monetary, and especially communications barriers for 320 million Europeans. The truly common market is the most significant initiative since the Treaty of Rome created the market more than thirty years ago.

With trade barriers down, not only goods, services, capital and human skills will flow freely. All kinds of information will move via computers and telecommunications at a faster, more voluminous rate than ever. Information networks will be the life-generating arteries of the single European market and community. The sale of transborder telecommunications goods and services is expected to trigger a "supply-side shock." About half of the annual European equipment market is represented by purchases of terminal equipment, valued at 17.5 billion ECU per year. Only 40 percent of these contracts is likely to be competitively bid before 1992. The Europeans learned from the Americans' good-natured losses. When the U.S. opened its markets to foreign competition in the early 1980s, the American trade balance steadily deteriorated. After AT&T was dismantled, foreign manufacturers were allowed to enter the U.S. telephone market without assuring Americans equal access to foreign buyers. Foreign phone sellers "gouged" the U.S. market. The Europeans seem not likely to allow the U.S. a similar opportunity. Every EEC country that has made significant studies in telecom liberalization has had to design a new employment policy. Thus, as ISDN—Integrated Systems of Digital Networks—develops it must be seen by citizens of each nation as personally productive for them, and for their country.

Great jockeying for access to the European community by Japan and the United States began in the 1980s. The EEC tried late in 1988 to end fears of protectionism. The U.S. commerce secretary welcomed the reassurance but said this single market is "increasingly under siege," and "backsliding, narrow national interests, and protectionism may grow as 1992 approaches."[2] The national telecommunications administrations (PTTs) will continue to operate the network infrastructure and voice telephone service, but all other services—the major data, news, banking and other value-added services—will be unrestricted, left to the competitive market. Significantly, terminal equipment may be placed in the hands of the consumers without charge. This equipment may operate within and between the twelve EEC states. States are required to ensure that such equipment will work on the network.

Not only is a revolution in trade occurring, but so is the inauguration of information linkages that will generate revolutionary political power. Those networks can alter the work habits of millions of Europeans. The European Commission estimates that by the year 2000 fully 60 percent of Europe's work force will depend on information technology for their livelihood. More than that, however, cooperative and competitive companies, and friendly and not-so-friendly nations will be linked by electronic switches. People living outside Europe will want access on fair terms not only to the market, but to the information base as well. Inevitably, this freeing of information barriers leads to the liberalizing of political structures. This may affect people far beyond the EEC, for the European model will be based on the individual citizen's access to the European telecommunications network, particularly the Value Added Network Services (VANS), and the Integrated Services of Digital Networks (ISDN).

VANS will provide (1) an *electronic data interchange* (such as U.S. banks and retailers already use) to move information securely from one computer to another, (2) *electronic mail* which transmits large documents with more flexibility, (3) *managed data services* which will enable corporate users to choose an information supplier offering diverse services, and (4) *voice mail* networks which carry messages of one or many to one or many.

ISDN, commonly used, provides many diverse electronic "services." I use ISDN to mean worldwide integrated "systems." This "seamless telecommunications highway," says IBM, "is of fundamental economic importance" and without it "the dream of a united Europe will never be realized."

But what about the rest of the world? How will the U.S. and the developing countries—to cite two ends of the developmental spectrum—gain access to Europe's post-'92 information system? The technical tie-ins, at the outset, are restrictive. Certain computers cannot "talk" to those of another manufacturer. Software of one manufacturer is useless in the computer of a competitor. This has international implications. Computers designed in one country may be impotent across the border. Exclusivity is breaking down, however. Manufacturers, software developers and especially users are pressing for common, open standards of computer development. This holds out the prospect that one day every computer in the world will be able to speak to every other, and software written for one can be used on all.

Open standards are thus the ultimate liberalizing force in computerization. There already are work group systems functioning on this basis. AT&T is advertising "open architecture" already available to link its systems to others. IBM estimates that by the mid-1990s there could be one billion computers around the world all connected into one network or another, just as telephones are connected today into a worldwide telecommunications network. Several American and European manufacturers have established the Open Software Foundation dedicated to creating open standards for all software manufacturing. Standardization, or open standards for all computers, is a further step toward the freeing of the information flow. The movement toward open standards, as toward computerization and networking, is, I believe, irreversible. And that has political—indeed, citizen-empowering—potentialities.

The Gorbachev-led revolution in the Soviet Union, and the summit with "friend" Reagan recognized that Marxist economic policies may provide theoretical analysis of limited value, but utter economic disaster if pursued as national policy. The historic Soviet Conference of 1988 thus signalled the end, not of the Cold War but of the costly competition in unusable nuclear megatonnage, and the transference of the struggle from the arena of military stalemate to the economic battleground. There, the Soviet Union is admittedly second-rate among superpowers, and falling rapidly behind the looming Western European community as it points toward total economic Eurofication in 1992.

Macroeconomics is the new game. It leaves the military chips mainly in place. One still cannot be a major player without militarily defensible interests and borders. But the struggle for economic security and national development will henceforth be determined largely by the variety and quality of a nation's information systems.

The world market for all telecommunication services is growing by 12 percent a year. Within that general expansion telephony and telex are growing by 5 percent a year and new value added services at 20 percent annually.[3] Information power, as never before, will determine the quality of life in every country in the world. The new electronic linkages will greatly magnify the sensate aspects of culture, and require new human patterns for dealing with the technology. Such adjustment may be most difficult for those in the less-industrialized world—citizens of poor countries and poor citizens of richer nations—who may have little experience with older communications technologies. Information power, more than in the past, opens new opportunities for the realization of each person's potentials—provided political power within nations wills it.

During a speech in Australia in 1980 I mentioned the vast humanizing influence which communications technology can one day provide. Technology already here, I said, could produce in five years a wristwatch-like telephone by which anyone in the world could phone anyone else, anywhere. I absentmindedly touched my wristwatch with my other hand and then added my point: despite available technology this magnificent opportunity for human contact will be indefinitely delayed because there is not the political will, in most of the world, to permit unlimited, unmonitored conversations across national borders. Yet the telephone is, and is likely to remain, the principal democratizer of the information age. Information power will determine not only the quality of life, but the level of human freedom as well. The politics of communication, however, supersedes other factors. Freedom, therefore, will remain limited indefinitely in much of the world.

An Australian press photographer in my audience seemed to have missed this point. When my speech ended, he said, "Let me take a picture of you with that wrist telephone."

It's not here yet, but it's coming.

So, too, is ISDN—a far more revolutionary development, and one that poses a crucial challenge to the United States in the early years of the next century. The Integrated Systems of Digital Networks will link news, scientific, educational, financial, meteorlogical, entertainment, personal communications, and countless other systems. No matter where on earth they originate, these networks will be immediately accessed in homes, offices, schools or farms by men and women without technological or political hindrance. The wrist-telephone is a small but not incidental aspect of the linkage. ISDN envisions, perhaps three or four decades off, the first serious assemblage of these connections. This is the beginning of the ultimate in Marshall McLuhan's world-village concept. He may have been a century ahead of his time when he used the term, and he may not have truly foreseen the full implications of ISDN as it now evolves.

Before one can understand the magnificent possibilities of ISDN, it is essential to consider its negative potentials as well. For as the era of ISDN approaches it will be necessary to negotiate the framework of the network, and particularly who or what will stand astride the control points. For the power of information to influence every aspect of humankind will reach its zenith with ISDN. The linkage, therefore, must be constructed with maximum fairness to all the potential users. It must advance individual freedoms, not restrict or replace them. It must acknowledge national sovereignty, and not trample cultural independence. It must also encourage personal and corporate initiative, and not hamper proper recompense for further innovation and technological application of new science and advanced education.

Kenneth Boulding describes information and knowledge as "that which reduces uncertainty." The technologies of information, however, begin by creating uncertainties—about the impact of the new system on people, institutions and their culture—and then resolving those uncertainties. The new communications systems will bring constant change to the home and office. Ideologies are likely to be minimized in the sea of pluralism made possible by diverse information sources. Nationalism, too, may be restrained. Continuity, the hallmark of times past, will give way to continuous obsolescence. Adjustment will not be easy, nor will the retention of treasured traditions and cultures. Yet a place must be retained for them. That is a tall order for diplomacy and negotiation. No inventor can predict the real impact of his creation. Henry Ford did not envision gridlock, the growth of suburbs, and juvenile car-fixation. Eli Whitney could not foresee his cotton gin inflaming the Civil War.

ISDN will require the best scientific and technological brains, as well as the wisest social analysts, supported by the most astute negotiators and arbitrators. No less than human freedom and creativity are at stake far into the future.

ISDN will provide the "continuous, evolving process" leading to "a new world information and communication order." This process was called for in Unesco after a decade of bitter debate. At Unesco the "new order" was demanded by Third World countries from the information-rich Western nations. It is likely, however, that after the framework of a global ISDN is put in place, most of the clamor for its expansion will target domestic controllers of information flows. In rich and poor countries alike, citizens without access to computer terminals—the nucleus of ISDN linkages—will clamor for inclusion. In the most free countries, this will spark pleas for fundamental economic and political readjustments. In the least free places, such struggle will become increasingly possible, but only as some political power is permitted the poor and the disenfranchised. By that time, the low cost of ISDN technology will no longer make economics a major excuse for denying universal linkage. The political will to share information power will be essential.

Consider where we are now, and examine the challenges already here, and the ones that are inevitably ahead. Technologies have merged. Telephones and other communicating instruments have acquired the memories and processing skills of computers. New connectives—microwave, copper wire, cable, satellite, and glass optical fiber—greatly extend the lines to and from computers. Fiber as thin as hair can carry many times more messages simultaneously than copper wire, at far *less* cost. Data can be sent swiftly around the globe. Computerized telephone exchanges automatically switch calls over great distance. The same channels, also using satellites, convey digitized information to and from computers a world away. Data bases on other continents are tapped. Business is conducted over facsimile, telex, telefax, telephone-conference systems and electronic mail. They employ television, radio, and satellites, as well as land lines. International banking and airline reservation systems could not continue without the linkups of many of the instantaneous communications systems. Within a few years, television newspersons sent abroad on urgent assignments will routinely use a portable satellite uplink enabling them speedily to get time on a transponder, and send live pictures a continent away. A *Washington Post* journalist aboard the 1988 presidential-campaign plane carried a laptop computer and modem, cassette recorder, cellular phone, portable printer, and a watch with an alarm and memory for telephone numbers.

This has been called the "technetronic" society.[4] Suddenly, distance is no longer a deterrent to communication and understanding. Every town and city—no matter how small or large, no matter how remote on land—is equidistant from every other place. We are all 22,300 miles and two seconds away by satellites. In Algiers, I have reached a distant phone in southern Algeria when it was impossible to complete a call within the capital city. The country used a satellite for some domestic phone links. When a U.S. officer during the fighting in Grenada could not reach his superiors by military communications, he used a credit card at a Grenadian public phone and immediately talked to the Pentagon. Technology, thus, is compressing time and space to link all humankind in that "global village."

The fusion of computer and communications technology, moreover, is drawing America (and, later, other countries) into a massive information-based economy. This creates social changes—improvements in the quality of life—which have domestic and international implications. Most important, I believe, the political forces—the power centers—being energized by these changes can either support or undermine democratic countries, or raise crucial challenges to authoritarian regimes.

America is preeminent in virtually every field of communication: news reporting and news distribution, video- and data-processing and transmission, manufacturing of macrocommunications hardware, international communications channelling, commercial and governmental satellite transmission, remote-sensing.

Yet the same information technologies that carry news, information and data rapidly around the world can alter employment patterns. American companies are moving entire business operations abroad where salary scales are far lower. Telecommunications and computers link those operations to home offices in the U.S. as though they are in the next room. McGraw Hill Inc., the New York publisher, for example, processes subscription renewals for its magazines in Galway, Ireland. American Airlines has its data base in the Dominican Republic and does all its ticketing, anywhere in the world, via the computer links. Some 2,000 people are now employed at the Jamaica teleport, and the industry is growing in the Dominican Republic as well as Barbados. This can be an employment boon to Third World countries. They are leapfrogging a generation of technology and moving straight to the latest systems—without investing in the costly intermediate communications technologies.

For forty years American policy has been based on commitment to the principle that the free flow of information strengthens political liberties and democracy and, further, that free enterprise in an open and competitive market is the most productive and efficient means to foster innovation, meet the requirements of citizens, and stimulate economic growth. But America has stood virtually alone in holding both these convictions.

No other country has the equivalent of America's First Amendment which separates private from public ownership—but especially, control —of communications and information power. Most other democratic governments own or control the broadcast services, and operate their telephone and mail services—the PTTs (postal, telephone and telegraph). The three-quarters of the world regarded by Freedom House as least free still more clearly reflects governmental control of the information and communications services. Communications, thus, can be a "bearer and destroyer of values and behavior patterns," a Pakistani scholar warns. "The emphasis remains on the introduction of the technology, while it is the message that subsequently produces the changes," she says.[5]

As a bearer of values and behavior patterns, news and information technology strongly affects nearly all aspects of our lives, our work, play, family relationships and physical health. A Congressional study ascribes to twentieth century technology: "power, scale, precision, invasiveness, pervasiveness, persistence, and imperfection."[6] Information technology, in all its varied forms, is the central agent of vast social and scientific change.

As never before, we can gather, store, combine and analyze information about people. That can invade the privacy of American citizens. Sometimes the subjects are in other countries. Foreign nations, increasingly, pass laws to restrict or monitor cross-border call-ups of their nationals' personal information stored in computer banks abroad. Electronic surveillance also dramatically shrinks the locations and activities in which one has a recognized expectation of privacy. Most important for power centers, information can challenge the traditional sources of authority and institutions built on that authority.

Developing countries, consequently, officially abhor—even while they import—American news flows and Hollywood films and videos. Soviet bloc communications dub such American dominance "cultural" or "information imperialism," and cry that national or cultural sovereignty is threatened by this "one-way" flow of information and data. This gives rise to increasing challenges to American communications and information systems from all the rest of the world—the Western European and Japanese communications competitors, the Third World en bloc, and of course, the Marxist adversaries.

Challenge to America
In other eras, such challenges could be regarded as normal. Indeed, those who profit from market forces should, at least theoretically, be expected to accept defeat or at least unfavorable treatment, from the actions of the market. But not if the market is rigged against one or more of the players, and especially not if decisions being made lead to the ISDN—the network of networks—which can be so inclusive as to affect all internal market forces in the next century. To be denied fair treatment in the ISDN can be the most drastic, almost permanent penalty in the nascent Age of ISDN.

America must face that possibility—now. We are widely being painted as the big bully on the block. And because of our First Amendment we are seriously hampered in responding as a nation. For the vast communications structure of the United States is mostly privately owned and domestically competitive, sometimes within one system (telephones; AP versus UPI, for example); sometimes system against system (newspapers and magazines against one another; radio/TV networks against cable; IBM against the data field); and with further competition among the carriers, domestically and internationally. Marxist analysts, nevertheless, state that "massive growth of information flows...which enable the companies to transact their global business and further integrate the internationalization of capital" is accomplished by "the management of the world economy by a few thousand transnational corporations, two-thirds of which are owned by U.S. private interests."[7]

Yet the anti-trust laws of the U.S. remain a significant deterrent to real monopolization. As a consequence, editors and managers of AP and UPI hesitate to discuss mutual problems lest they be accused of rigging the news market for financial gain. What appears to be a game of giants in the communications field represents far less monopolization of resources and word-control than is apparent in *all* authoritarian and totalitarian countries, and even most democratic states which run PTTs. Indeed, the vast American communications systems represent fragmentation and division, rather than centralized control—either by private or public power-wielders. As for information sovereignty—the United States is the world's largest *importer* of information and cultural materials. We are also rapidly and deeply influenced by the languages, customs, foods and other cultural exports from the Third World. We are subject to major corporate buy-outs from Japan, the Arab states, and other countries. We are also the largest debtor nation in the world—an importer of credit in exchange for dollars. And, too, at a still undetermined cost to America itself, we have expanded to global size the arena of commerce and politics. We have, in effect, greatly enlarged the scope of the transnational corporation (TNC) which diminishes the ability of any nation, the U.S. included, to retain sovereignty over the TNC's disposition of jobs and other production and economic policies.

The capacity to transform information into action—more information, and faster more widespread action—is a stupendous feat. Yet it is a daily, almost unnoticed occurrence. Examine the technology deployed twenty-four hours a day by the Associated Press, the cooperative of U.S. newspapers and broadcasters, the largest news dispenser in the world. Computers transmit millions of electronic images to satellites for nearly instantaneous radioing back to earth at distant places. About 80 percent of the news gathered by the AP's global reporting staff is fed into the GTE Spacenet III in geostationary orbit (synchronized with the movement of the earth) 22,300 miles above the equator. Transponders take radio signals and retransmit on another frequency to some 4,000 AP earth stations across North America. Impulses taken from these signals are fed directly to the news and photo equipment of AP's clients.

Such linkage enabled the *Wall Street Journal, New York Times* and the *Toronto Globe and Mail* to become national newspapers. Japanese newspapers had gone to the satellite and national publishing earlier. The satellite provides cheaper transmission rates than land lines used to charge. Reuters bureaus in seventy countries serve clients in 120 countries. The *Wall Street Journal* supplies more than two million copies daily through eighteen earth stations at publishing points across the U.S. *USA Today*, which is a product of satellite journalism, color pages and all, bounces its complete pages off American satellite's ASC-1 facility. Transmissions are received at thirty-three printing sites. Different satellites are used for *USA Today's* same-day printing in Zurich, Hong Kong and Singapore.

And President Reagan's SDI can be described, in Secretary of State George Shultz's view, "as a gigantic information processing system." Perhaps the Strategic Defense Initiative's multi-billion-dollar investment may not provide the projected giant nuclear shield, but it is almost certain to produce significant technological innovations that may add new industries and technologies to America's economy.

Even now, a word spoken or leaked in Washington can, within minutes, send ripples through stock markets in London and Tokyo, and gold shares moving in Ghana and South Africa.

Secretary Shultz says the U.S. "does not advocate free trade because we are adept at pioneering technologies," particularly information technologies; "we are adept at them *because* the dedication to freedom is intrinsic to our political culture."[8] That has led us to pioneer change, at home and abroad, in the agricultural, then the industrial phases of our development, and now in the age of information.

Opponents see it quite the reverse. We sought free markets and the free flow of information when the British or French dominated Africa, Asia and Latin America. We continue to support free flow today when American communications are preeminent. But will we support the freedom of Western Europe and Japan to make sweetheart deals that favor other than American suppliers and carriers? Will we yield without a struggle to Third World news and information purveyors that favor other national or multinational agencies, and put U.S. news flows at economic and political disadvantage? All of this is contemplated in the name of cultural sovereignty in defense against presumed cultural imperialism.

Whether the Age of ISDN comes sooner or later, however, American information power will be seriously challenged even more than in the recent past.

History of warnings

For a decade, House and Senate committees of Congress have quietly examined the threats to American communications power, but with little practical effect. A "new order" of international communications is coming "whether we like it or not," the Senate Committee on Foreign Relations was told in November 1977 by George Kroloff and Scott Cohen.[9] It could be "the driving force," they said, "for the 'new international economic order' called for by the Less Developed Nations (LDCs)." The U.S. was "totally unprepared" for discussions of the issues looming the next two years, the authors added. At its worst, they speculated, a "new world information order" could mean state control of media entering or leaving a nation; eliminating Radio Free Europe/Radio Liberty, and drastically reducing the Voice of America; a dramatic loss in the amount of information about the world available to the U.S. government and all Americans; rigid cutbacks on the information sectors of the U.S. economy, including computer manufacturing; loss of parts of the electronic spectrum used by U.S. space satellites or reduction in vital intelligence, and loss of frequencies used by U.S. radar and electronic-based weaponry; higher phone bills for overseas calls; and operations abroad of U.S. firms becoming uneconomic. At its best, wrote Kroloff and Cohen, the "new order" could mean a much greater flow of news, movies, television programs and other information; closer relations and better understanding among societies and individuals in the world; less chance of war; greater chance of increased wealth, health and worldwide well-being; and a friendly rather than hostile climate for international business and investment. The paper focused on the issues immediately facing American negotiators at several international communications meetings in 1978 and 1979. But the long-term challenges remain. And the U.S. is still ill-prepared, if not "totally unprepared," as it was twelve years earlier.

The best or worst potentialities of a new information order may depend in large measure on how the U.S. responds to demands of 100 developing nations for enlarged communication infrastructures, and the response of other developed countries to the vast opportunities presented both by these Third World demands, and the unlimited progress in worldwide information linkages.

The House Committee on Government Operations[10] in December 1980 examined the barriers to the flow of information, the effect of these hurdles, and the U.S. response. The report stated that several nations have created economic barriers: tariffs and discriminatory pricing, inconsistent technical standards favoring their citizens, restrictions or denial of market entry. Some have created serious political barriers: privacy protection, national sovereignty rules, external control of domestic activities, charges of cultural erosion, and national security technology-transfer protectionism.

In the face of these developments, said the House investigators, "the U.S. government is unprepared." It had developed neither comprehensive plans or policies nor a coherent strategy for responding to the policies of other nations which may damage U.S. interests. Moreover, said the report, "the U.S. government does not even have the organizational structure to develop such policies, coordinate its actions, and effectively protect U.S. interests." The assessment turned still more gloomy. "Even if the existing organization had worked well in the past it is incapable of providing the coordination of government action and the development of policy necessary to anticipate and respond to the rapid change in international communications and the growth of barriers to the flow of information." Specifically, "the current (1980) organization is based on a dangerously obsolete compartmentalization of government policy-making. The world has changed and continues to change with increasing speed," the analysts told this nation which created much of the information age. "Yet the U.S. government holds to an organizational structure, the essential weaknesses of which have been recognized for more than a decade." More than the stock values of communications-oriented corporations were involved, the report declared. "Barriers to the international flow of information threaten injury to fundamental democratic freedoms and individual rights throughout the world."

The report recommended the establishment for five years of a cabinet-level Council on International Communications and Information. It would develop and implement a uniform, consistent and comprehensive U.S. policy in response to the barriers to international communications and information. The council would receive problems brought to it from the private communications sector, and an advisory committee—public and private—would provide guidance to the council. The report also suggested that a Bureau of International Communications and Information administered by an assistant secretary be created in the State Department. The report would also strengthen the negotiating capability on telecommunications and information products in the office of the U.S. Trade Representation. An office of international communications was also urged for the FCC.

As appropriate as were all of these recommendations—and not all were followed—the report reflects one fundamental failure. Apart from serving a very useful educational function, the report adopted a mainly reactive stance. It did not grapple with the basic challenges to American communications power; it dealt mainly with the symptoms: the "barriers," not the realities of changing power structures, reflected in the evolution toward the ultimate linkage through ISDN.

Congress and the executive inched slowly toward that realization. In the early years of the Reagan administration the Senate Committee on

Commerce, Science and Transportation sought long-range goals for American policy in the field of international telecommunications and information. Again, the investigators found "disturbing trends," and a "gradual erosion of the U.S. position in the telecommunications flow, and associated high-technology markets." Again, too, the report concluded that the U.S. government's "dispersal of responsibility with lack of policy authority at the highest levels of our government have prevented the United States from responding effectively and quickly to [the] escalatory challenge to its defense, economic, and political interests." Note, again, the emphasis on "response" to challenges, rather than a call for positive, creative action programs. "If we wait another ten years before we counter these adverse trends," the Senate analysts warned in March 1983, "the likely impact on U.S. defense capabilities, on employment and economic growth and on freedom itself will be catastrophic."

The report listed twenty-two international conferences in the recent past at which other countries received approval for a wide variety of controls over communications flows and products. The report also noted what it regarded as the "politicization" of the International Telecommunications Union (ITU), the oldest intergovernmental agency in the world, and the United Nations Educational, Scientific, and Cultural Organization (Unesco). The report declared "There is a great international strength in U.S. ideas, exchanges, and free enterprise." It added ominously, "There is great weakness and danger in complacency and indecision."

Such reports appear every decade or so, though more frequently in the recent past. In 1946 a Senate committee concluded it was "obvious" that "government departments were not of one mind with respect to the policy which should be laid down by this government to govern American international communications." In 1951, President Truman's Communications Policy Board stated it "once more encountered dispersion, confusion, gaps, and deficiencies in the product and performance of those agencies charged with telecommunications policy responsibilities." The U.S. was also said to lack a "satisfactory means of determining policy as a basis for negotiations with other nations." A 1968 Presidential Task Force on Communications Policy concluded that policy "has evolved as a patchwork of limited, largely ad hoc responses to specific issues, rather than a cohesive framework for planning." The 1980 House report reiterated this concern, and the 1983 Senate analysis picked up the thread. All of this flies in the face of the perception abroad—in the Soviet bloc and the Third World in particular—that American preeminence in information power is cleverly orchestrated from some oligopolistic center.

The fact is, little else informs broad U.S. communications policy beyond the 1945 commitment pressed by the U.S. at Unesco: "the free flow of information." The variety and applications of communications technologies soon exploded, the number of nation-states tripled, and information became a pervasive factor in everyone's life. Today, an American undersecretary of state must acknowledge that "information" is not a simple commodity in trade. "It is a consumer good, a capital good, an intellectual value, and an element in the individual's enjoyment of his fundamental civil and human rights."[11]

There have been glimmers of concern for several decades that America's preeminence in information-gathering may be critically challenged. The target is massive. The United States now houses the world's largest collection of knowledge. Though seldom recognized, this vast array of information is in the tradition of the great libraries of human history. And the modem that links the widely diverse data bases is communications. Wilbur Schramm called communications the crossroads where all the disciplines meet.

Keeping America's communications crossroads free of domestic gridlock or foreign blockage, has been an all-too-occasional concern. The Communications Act of 1934 established legislative standards for radio broadcasting, and created the Federal Communications Commission (FCC). But the act created to regulate telephones and radios could not address the vast technological changes ahead.

The appearance of the Soviet Sputnik, and a new American communications satellite energized the Kennedy administration. AT&T's Telstar became the first satellite able to handle incoming and outgoing messages in space. The Communications Satellite Act of 1962 planned an American-managed international satellite system. It meshed public and private sector agencies in the highly successful Intelsat global system through Comsat as the single corporate instrument. This American conception now has a phenomenal list of 115 national co-owners. Intelsat provides great communications advantages to all its users. But it was a product of reaction to the anticipated Soviet threat to produce a satellite system linked directly to the USSR. Intelsat is *not* tied to the U.S., as the Soviet satellite system, Intersputnik, now serving the Soviet bloc, is closely linked to Moscow. Indeed, messages from one Eastern European country to another must go through Moscow center.

The Comsat legislation temporarily aroused interest in other aspects of U.S. communications policy. A commission headed by Eugene Rostow, a Yale law professor, took a first stab in 1968 at laying out a national communications policy. The task force called communications a crucial issue for America and proposed some actions to support communications development. The report was filed and forgotten, though each subsequent task force has generally followed the Rostow agenda.

A year later the Nixon administration seemed to take heed, and created the Office of Telecommunications Policy (OTP) in the White House. It was supposed to coordinate government policy and planning—a need stressed by every communications specialist before and since. But OTP was soon accused of pressing a political attack on the news media, and in 1977 President Carter abolished the agency. Some of OTP's functions were given to a new National Telecommunications and Information Administration (NTIA) in the Department of Commerce. Yet other federal agencies were major users or influencers of communications. The Defense Department uses half the radio frequencies assigned to the government. The State Department represents the U.S. government at many international communications conferences at which crucial frequency and other communications matters are negotiated. And the U.S. Information Agency, through the Voice of America and varied print and broadcast programs, is a primary user of open communications channels.

It was not until 1976, when Vice-President Nelson Rockefeller reported to President Ford on national information policy, that the necessary recommendation was made: that "the United States set as a goal the development of a coordinated National Information Policy."[12]

Congressional studies in the 1980s were to flesh out the growing threats to American information services abroad, and call repeatedly for greater coordination of U.S. actions based upon a realistic national communications policy. Coordinators were named in several federal agencies during the Reagan era—and a coordinator of coordinators. Not until the National Security Council commissioned a major study in 1983 did communications policy and coordination get the high-level attention that was long overdue.

What is U.S. policy?

Fundamental American communications policy hinges on the First Amendment's separation of state power and information power: "Congress shall make no law...abridging the freedom...of the press." That separation assures that the primary means of America's daily flow of news and information be free of governmental influence or control. "The press" of the Founding Fathers now covers radio, film, television, magazines and, it should be assumed, the newer technologies, cable TV, casettes and others. The rights embodied in that amendment reflect the high priority given "the press" by the authors of the Constitution.

That "preferred place," as Justice Rutledge called it, is being widely broadened as new communications technologies add new channels, new forms of communicating, and new linkages with traditional components of "the press." These new technologies change not only the forms of communicating but the power relationship among citizens, and between individuals and the state. Satellites, computers and digital transmissions are changing the ways that ideas, facts and polemics are formulated and conveyed. As with the arrival of each new technology—the telegraph, telephone, radio and television—the mode affects who can communicate to whom, and at what cost. And with each new communication technology governmental regulation in some form soon superseded the First Amendment guarantee. The rationale: telegraph and telephone companies were common carriers. They simply moved other people's messages from one place to another. The question now is whether cable television networks, for example, should be regulated as common carriers, rather than regarded as editors and producers deserving of wider freedom under the First Amendment.

One Congressional study in January 1988[13] declared:

> Taken together, advances in computers and telecommunications may change the concept of "the press" from one in which one organization publishes for many to one in which many share information amongst themselves. With these changes will come new First Amendment challenges to the power of the government to regulate access to and ownership of communications media. New technologies, such as electronic publishing, may not fit easily into old models of regulation, and First Amendment distinctions between the rights of print publishers, broadcasters, and common carriers will become increasingly difficult to justify.

The new technologies, particularly the ability to gather and retrieve information about an individual, may raise questions of privacy protection and possible limitations on the freedom to disseminate new forms of individual "speech." The decentralization of the editorial function in computerization also recasts the issue of liability under the First Amendment.

In 1983, Ithiel de Sola Pool properly called his last seminal work *Technologies of Freedom*. He examined the new modes of communications and concluded they would not necessarily centralize control of the vastly heightened power of information. On the contrary, the new technologies would be engines of freedom, Professor Pool predicted. He took the next prescient step, and urged that the new technologies be regarded as "the press," and not common carriers subject to regulatory action by the government. Said Dr. Pool:

> Networked computers will be the printing presses of the twenty-first century. If they are not free of public control, the continued application of constitutional immunities to the nonelectronic [press] ...may become no more than a quaint archaism, a sort of Hyde Park Corner where a few eccentrics can gather while the major policy debates take place elsewhere.

Electronic publishing covers many services: electronic mail, electronic bulletin boards (a private or publicly accessible computer listing), teletext (visual broadcast of pages of information), videotex (a two-way flow from the user to the computer which stores news files, law cases or other specialized services), home-shopping and banking publications.

By 1988 there were 3,699 publicly accessible on-line data bases in the U.S. established by 900 producers and 300 distributors. This is 3.2 times the number of bases available in 1982. That number had tripled since 1979. There are more than 6,000 nodes available on the Arpanet information system alone.

By 1987 there were 2,546,000 subscribers to U.S. data banks. This was a 19 percent growth in one year. Subscribers included 900,000 for general subjects; 620,000, financial; 350,000, science, technical and professional; 170,000, airlines; 150,000, credit data; and 25,000 news organizations. This is a $6.2 billion industry—and counting.

Users may range from high school students to opera buffs, graduate engineers, lawyers, journalists and Wall Street investors. Each may want categories of information that range from the most current—today's news, market quotations, entertainment schedules, or the newest journals; to longer-term information such as trends, scientific theories, private or public regulations; or longest-term data such as history, all-time market statistics, technical formulas, back-issue publications, legal case histories and entertainment classics.

Electronic publishing using common-carrier networks—telephone, cable or satellite systems—should not be feared as potentially restrictive. Commercial users of these networks may well become more competitive than present-day newspapers, magazines or television proprietors. Government controls are inappropriate if electronic publishing is treated as "the press" with First Amendment protection, and without regulatory intervention influencing the content of messages or free access to the networks. Since it is far less expensive than print, electronic publishing may well become far more diverse in content, and consequently much more competitive.

Home users may choose to receive information from cable systems with one hundred different channels, and ISDN links to thousands of on-line data banks—not to mention videodisks and videocassettes. Systems will also compete with other systems: digital terminals, cellular radios, and cable networks will vie for customers with the local telephone service.

Newspapers have already tried to head off such competition from electronic publishing which the papers regard as a threatening new industry. The American Newspaper Publishers Association (ANPA) persuaded Congress in 1981 during the break-up of AT&T, to stipulate that the telephone companies "may not provide...mass media service, or mass media products." That would cut electronic publishing of phone directories, among other services. The act was later bolstered by Judge Harold H. Green who ruled in the AT&T divestiture case.

"To get the prohibition on electronic yellow pages adopted," noted Professor Pool, "the press raised the spectre of AT&T becoming the nation's monopoly publisher. In the name of pluralism, the press persuaded the government to designate who may and may not publish. One might have expected the press in its own interest to advocate loudly the constitutional right of AT&T, like anyone else, to publish whatever it wishes." But Dr. Pool recalls, "when radio was young, the press in both Great Britain and America sought to hamstring it by censorship." The basic protection: AT&T is a common carrier and must open its system to all comers.

In Europe or wherever PTTs serve as the actual monopoly carrier, it would be appropriate to prevent those government-owned networks from acquiring a monopoly in electronic publishing as well. But the First Amendment creates a totally different relationship: to maintain the freedom of the new electronic publishing media it is essential, therefore, that it be required to interconnect with any other system, whether or not it is competitive (and it is likely to become so).

The biggest users of electronic publishing are most likely to be not homes, but large business firms. They will probably create the most elaborate competitive systems of their own. They can save millions of dollars annually by leasing transmission lines and creating dedicated networks for their special purposes. The global news services and the largest Third World news agency already lease satellite, cable and telephone lines. In some cases they charge their customers for the use of the dedicated lines, or simply add the charge to the proportionate cost of selling the news service.

These are decentralizing factors in the larger scheme of ISDN linkages. Indeed, such subnetworks will increasingly defy centralization and monopolization of the content of the news, data and information flow. Such decentralization by big users will greatly reduce the cost of storing information, and make much more information accessible to more inquirers. The very act of large users processing new banks of information will thus have a decentralizing effect. When such data banks are linked to satellites, as they already can be at low cost, the interactions become global. That introduces new ideas, new relationships, new supports which can inspire salutary human connections across great distances.

Such significant human as well as national interconnections rest in the capability of the electronic systems to interconnect. That requires universally conforming technical standards. And, as the French government realized early in the 1980s, this requires a directory that facilitates use of the system. The French made a citywide test of substituting electronic terminals for printed telephone directories. They replaced the books with terminals given free to users. The test worked, and electronic networks are growing rapidly.

Clearly, electronic technology advances human freedom. As with the long-delayed wristwatch telephone, only political will is required to speed the fullest use of ISDN.

An early test of American political acumen will be the treatment eventually accorded cable television. It was regarded at its beginning as the discerning viewer's answer to the "vast wasteland" of commercial television programming. There is, indeed, far greater diversity on cable, though it, too, accommodates large-audience tastes. Cable proprietors (probably more than cable producers or editors) determine which local or regional channels, as well as diverse special programming, they will provide to their viewers. These special offerings may include sports, arts, pornography, public-service or other subjects—far more, and more regularly than the old-line channels provide. These special offerings are similar to neighborhood newspapers, particular-interest magazines, or radio talk shows conducted by a controversial interlocutor. All appeal to small segments of a large audience, and all have an equal right to be seen or heard. All cable programming should be protected by the First Amendment. Cable should not be regulated on the old fallacy that broadcasting, in any form, is a scarce medium, unlike print publishing. There are far more limitations on entering the newspaper market today than adding to diversity on cable. Yet no one would seriously recommend abrogating the First Amendment in the field of newspaper publishing.

Print publishing will continue, as Prof. Pool suggests, but electronic publishing will probably predominate. New forms of publishing are likely to develop. Individuals will be able to select those blocks of information they seek, and then reassemble them in new forms, perhaps sharing the result through computer conferencing. Material received via electronic mail may be telecast widely, recorded on magnetic tapes, stored in a new database, and accessed later in edited forms by widely separated audiences. If these greatly enhanced capabilities are not to be stifled, the freeing umbrella of the First Amendment had better be raised over these systems.

Domestic concerns

Due allowance can be made for the restraints of libel law. A new body of law will probably be needed to establish liability for each successive use of questionable material. There must be due recognition of proprietary rights, when applicable. This may apply particularly in the field of journalism. It will be necessary to determine at which point information secured by a reporter and edited for print or broadcast journalism moves into the domain of public ownership of that news and information.

Press freedom could be restricted, perhaps unnecessarily, if society fails to recompense private owners of the physical plant for the public's right to share the content of the media flow. It may not be easy to reconcile both sets of interests. The individual will also need protection from the use or misuse of new technologies. It will be much easier to gather information about a person from a variety of databases. Individual privacy may be threatened by such revelations.

Security will also be affected by greatly enhanced reportorial capabilities linked to clever analysis and the combining of separate data sources. Remote sensing satellites are already able to observe weapons emplacements, troop movements, and a vast array of underground resources and crop conditions. Greatly magnified pictures made by remote sensing will soon be made available routinely for purchase by the press. This has important implications for nations concerned over protecting their physical and man-made secrets. The prospect of owning a "mediasat" dedicated to journalistic coverage via remote sensing is enticing. But the mediasat would need more than high-resolution pictures (now available). Two satellites would be needed simultaneously to cite news and transmit it to ground stations all around the world. It is a costly operation, and may have to wait until several nations provide the funding. They are almost certain to set restrictive rules for photographing sensitive incidents or regions.

Remote sensing also touches on personal privacy concerns. For satellites may photograph private property, and acts of a personal nature, behind enclosures formerly unseen by the public. Even the protection of private records may present a First Amendment problem. The government's removal of personal records from electronic files may be challenged on the ground that press freedom is being violated. But the free press/fair trial conflict of rights is not new in America.

The inevitably wide use of wristwatch pagers may also present problems. The pagers may function over great distances. The flip side will be the instrument's ability to track persons or vehicles with considerable accuracy—not always to the advantage of the subject.

The electronic press will need to change its modes of editorial control and liability. When the press is responsible for creating the content of its messages, it is liable for any actionable statement. Still unresolved is whether the electronic press is liable for words or pictures provided by linkage with a second database, or an individual simply permitted to speak as a subscriber or participant in a bulletin board conference. When the linkage becomes global, involving the international press, the issues of press freedom are still more complex.

Unlike other countries, the U.S. cannot have one grand information policy. The vast reservoirs of information held by the federal government are governed by such statutes as the Federal Records Act, the Federal Reports Act, the Privacy Act of 1974, the Freedom of Information Act, and the Sunshine Act. All of these statutes provide mechanisms for the public and the news media to secure information developed inside the government. The Sunshine Act requires federal commissions and regulatory bodies to meet in public and share with anyone requesting them, materials relating to government meetings. Legal policies govern the right to secure information whether it is in the public or private domain. The policy presumption favors openness, except for limitations based on overriding public policy concerns. The policy also favors the free-market system, wherever possible, for distributing information products. The contents of most American information flows are fully controlled by private operators. Information carriers are subject to regulation. But U.S. policy so far fails to address the responsibilities of private compilers and purveyors of information, the full protection of personal information in public and private sources, the new role of information in international trade, and the sharing of information and technologies with the developing world. Rules will be needed to assure the free flow of teletext and videotex, which will soon provide one- and two-way information systems linking the home to the transmitter. Videotex, with its sampling of viewer attitudes, may be subject to manipulation, if left unregulated. Choices or purchases made electronically should be protected from political or commercial exploiters of such information.

To speak of regulation, however, reverses the trend of the early 1980s. Deregulation has become the policy. Regulation is neither automatically an ungodly restraint of the free market, nor a guaranteed problem-solver for issues that loom larger or more complex than private enterprise can cope with. Deregulation, indeed, does not necessarily decrease government intervention. It may increase it, as the judicial fine-tuning of the AT&T divestiture demonstrates. Yet, clearly, the day has passed when the telephone monopoly could be defended as requiring great economies of scale. Now, with the diverse services implicit in ISDN looming, there appears to be ample opportunity for competitive systems, and competition within those systems. A wide variety of textual, pictorial, duplicating, digital processing and household services will be accessible over the telephone lines. Eventually, if not immediately, competition will set in, and deregulation may prove its value.

Neither the free market nor regulation is a scientifically assured method of grappling with the hazards of information systems which have great influence—negative as well as positive—on the lives of millions of people. Private competition can rig the market and short-circuit its "freeness."

Regulation is subject to political or industry influences, or less-than-expert tinkering which damages the market system.

The present system, a middle road between laissez-faire and regulation, is not always a viable track. It can lead to a chronic gridlock: no action or only feeble action, because opposing interests clash and produce weak solutions. Political gridlock is increasingly the bane of the federal government in all essential fields. Private interests, focusing effectively on one or more of the three branches of government, can delay or short-circuit almost any proposed governmental act. And within the government, one of scores of small Congressional subcommittees, or middle-range executive-branch bureaucrats can hold hostage almost any legislation or governmental action. So, too, in the regulatory agencies.

Perhaps the answer lies in enlarging state governing powers, but doing so by coopting widely recognized private citizens as regulators. This corporatist model was used successfully to keep New York City from going bankrupt in the 1970s. The Municipal Assistance Corporation (MAC) co-opted business, labor and government leaders to produce reforms in the government as well as adequate financing to avoid bankruptcy. The architect of MAC was Felix Rohatyn, who now recommends a national corporatist strategy in place of what he regards as the failure of market-based deregulation. Says Rohatyn, "Deregulation, as with most things in life, has to be done in moderation; it has been carried too far." He cites the deregulated airlines and financial markets. Rohatyn concludes that "The free market is not always right; it surely is not always fair," he adds. "It should not be turned into a religion."[14] Rohatyn urges the creation of a commission to provide a "coherent frame for debate of the economic alternatives." He would include representative political leaders from both major parties, and others from the fields of business, labor and the university.

Whatever steps a commission recommends, it will not be easy to protect the citizen from the negative potentials of the vast information array and, especially in the Age of ISDN, the network of networks. The dilemma remains: how to rein the power of the state and the power of the private information systems for the broadest benefit of the greatest number of citizens, and how to do so in the international as well as the domestic arena—without the threat of gridlock among the domestic monitors, or more adversarial controls and power plays from the world outside. The fear of the slippery slope—that some governmental regulations must lead to the destruction of First Amendment rights—should be dispelled. Consensually approved regulations should be fully defined in advance and reasonably enforced.

Before turning to the impact on the international scene of open American domestic information flows, however regulated, it is useful to consider the new steps needed to defend United States security, as well as share the data. The computer networks worldwide reverberated long after 2 November 1988 when a U.S. graduate student injected an electronic "worm" into the 60,000 machine system. A worm is a relatively benign code. This one reproduced wildly, and was readily detected. A computer "virus" is more dangerous. It insinuates itself into existing programs and forces them to act out the virus's orders. The 1988 attack was timely warning, however, that computer mischief can be very damaging.

With greater dependence on computer and communications systems for business, personal affairs and government, there must be greater reliance on the confidentiality and integrity of the information processes. Military and intelligence services will provide their own security controls for the communications hardware and software they employ. But there is increasing blurring of distinctions between industrial, scientific and technological data evolving from the private sector, and applications of this information in classified uses. How, then, can openness and free-market forces coexist with secret activities and controls on sensitive information? Some commercial databases, while not classified here as secret, may include clues to innovations which could compromise U.S. security interests if developed abroad, or placed in the hands of foreign competitors. The practice of commercial piracy is not new. Whole economies have been built abroad by infringing U.S. patents and copyrights. A small innovation in larger hardware can be replicated, or used to develop a next-stage innovation. Some new communications elements magnify the fear of unfriendly transfers. Microwave radios and cellular telephones increase the vulnerability. Optical fibers, however, decrease this threat. Underground fiber systems are harder to tap than broadcast signals. Digital communications may add to the problem. Decentralization of computing functions increases the vulnerability of computers and allied systems, especially for use by those not authorized.

To counter these threats, secret cryptographic designs are employed by some U.S. companies and civilian government agencies. But they raise the possibility of restricting private use of the computer or allowing the National Security Agency to eavesdrop on corporate communications. As in the matter of regulating information flows for non-security purposes, excessive accommodation to one interest or another—private sector or defense/intelligence concerns—could damage overall national interests, and not the least, personal freedoms of citizens. Here, as in the nonsecurity field, a formula for weighing competing interests is needed. That formula, supported by fundamental information policies, is essential for American negotiators to carry into the international arena where adversaries challenge American information power.

International information policy

Before examining the challenges to U.S. information power, present and future, it is useful to acknowledge the steps recently taken to restrict the flow of information deemed detrimental to U.S. national interests. The FCC decided in 1986 to publish in the Federal Register thereafter only summaries of actions it would take. The public is thus less informed about governmental policies, and less able to comment. Controls, meanwhile, have increased on the export of scientific information and goods, and limits on the participation of foreign experts in the American economy. An NSC directive in October 1986 sought to restrict unclassified information affecting "government or government-derived economic, human, financial, industrial, agricultural, technological, or law enforcement information." This broad sweep could necessitate monitoring all U.S. computerized databases and information networks. Congress objected and the White House withdrew the notice the following March. But the basic policy—NSDD 145—remains. This calls for "a comprehensive and coordinated approach" to restricting foreign access to all telecommunications and automated communication systems. The premise: "information, even if unclassified in isolation, often can reveal sensitive information when taken in the aggregate."

And so it may. But under the same premise, restricting foreign access to "all" automated information systems is certain to inspire retaliatory measures abroad, and limit the full use of information processes at home. There can be high costs.

The U.S. has demonstrated that it can negotiate with other countries over sensitive information issues. We do so at the International Telecommunications Union (ITU) in approving new radio-spectrum and satellite assignments. We support the OECD privacy guidelines as "a true landmark of international cooperation and resolution"—the words in 1983 of Diana Lady Dougan, then Coordinator for International Communication and Information Policy in the State Department. In subsequent remarks she revealed other aspects of U.S. policy affecting the world community:

> [T]he world community, developed and developing countries alike, has entered a critical period of decisions on the structure and operations of the national and international information and communication systems. The decisions made and policies advanced in the next few years will chart the economic frontier for decades to come.
>
> We are deeply concerned that human rights not be lost in the hum and whir of the computer age. The right to privacy, the right to know, and the right to free expression must be protected.
>
> We recognize that "free flow of information" is subject to widely varying definitions. Nevertheless, the United States will always insist the burden of proof is on those who claim a restriction is

necessary. If the purpose of such restrictions is the protection of the rights of individuals, or the rights of property, or the denial of military technology to potential adversaries, those concerns should be stated candidly and clearly.

We cannot accept such broad generalizations as the "protection of cultural integrity" to be a sufficient justification for information control, particularly as these are too often only a guise for economic protectionism or censorship of the press.

We recognize, at the same time, that those who transmit information, and who themselves seek protection of such rights, must respect the rights of others as well.

We consider the free flow of information internationally as an extension of our domestic democratic traditions. Our laws and regulations are designed to encourage maximum access to information and minimize its abuse with minimal government involvement.

...The essential need for personal contact and human esteem ensure that we will remain the masters of our machines. We have seen technical wonders fail in the market place because they were the stuff of novelty rather than need....But we should not deny the entrepreneur the freedom to take such risks...It is not enough for our governments to work with each other for political consensus on the issues of international telecommunications. We must also explore with our private industries the practical consequences of political decisions...[T]he OECD, the ITU, the European Community and other such organizations must remain sensitive to the issue of public control versus private enterprise and remain committed to a constructive working relationship with the private sector.

...Fears of a communications world dominated by a few major powers—a "mainframe cartel"—have proved unfounded. Computing power has been almost universally dispersed, rather than closely centralized, with the advent of remarkably small yet remarkably sophisticated data processing units.

The SIG report

While Ambassador Dougan was addressing the OECD in London in November 1983, a Senior Interagency Working Group (SIG) under her general direction was preparing the most important examination of U.S. communications assistance policies ever undertaken. The report was commissioned by the National Security Council (NSC), and delivered to the White House just after Thanksgiving Day 1984. The NSC commission paper was classified secret, but the SIG report is unclassified.[15] This seminal fifty-page paper places the issue of communications development and communications assistance to the Third World formally on the White House agenda.

Ironically, the report was delivered within hours of the State Department's formal announcement that the United States would withdraw from

Unesco on 1 January 1985. That withdrawal culminated years of bitter debates over Unesco's communications programs. The heart of those debates was the demand of developing countries for communications infrastructure and, as well, better news reporting in the West of their events and personalities. There were other communication issues involved: Some participating countries wanted to control the content of press coverage, international as well as domestic. And there were noncommunications complaints—financial, administrative, personality—that led to the U.S. withdrawal from Unesco. But the irony remains: Just as the U.S. was withdrawing from Unesco over Third World demands for communications assistance, the NSC received the SIG report recommending—for the first time—that "international communications development be explicitly recognized as a strategic priority on the [U.S.] foreign-affairs agenda."

This conclusion (among others) is particularly significant. It not only recognizes the needs of the developing countries, but regards the assuaging of their demands as a crucial American strategic concern at a time when not only Unesco but the U.N. generally, and many other intergovernmental organizations, and the Third World itself, were being trashed by the Administration in the news media and on public platforms. It was also the time when an assistant secretary of state was engaged in the cynical charade of promising financial support for alternative communications and other programs traditionally conducted at Unesco. These programs presumably would be financed thereafter through other agencies. After numerous press conferences announcing the alternatives, and the letting of contracts to study potential carriers, nothing came of the publicized alternatives—as, indeed, observers had predicted all along.

Yet, the NSC-commissioned SIG study proceeded quietly, and given the heated climate over Unesco, was greeted at the White House with equal silence.

The SIG report deserves further attention. It stated unequivocally that helping Third World countries improve their communications capabilities is crucially important to the U.S., and to world peace. "Information is a basic resource without which full participation in today's world is impossible," SIG said. "It is highly destabilizing to allow the world to remain separated into two groups of countries: a small group that is information rich, and a large group that is information poor." The report acknowledged that "weak links" in global communications are in Africa, Asia and Latin America, and in such strategic regions "vital" to U.S. interests as the Caribbean. SIG noted that the developing countries had stated their need under the rubric of "a new world information and communication order," pressing their case at Unesco and the ITU. SIG further acknowledged that despite Third World pleas for aid, "the United States has resisted LDC demands for large-scale direct transfers of funds and technology." However, the White House was told, "it is in our political and economic interests to recognize the seriousness of the North/South communications imbalance" and the need to take practical steps to correct the problem!

SIG, moreover, recognized the implications of the unanimous Third World demand: Unless the U.S. and its allies respond sympathetically, "there will be an increasing disposition on the part of the LDCs, pushed by a major Soviet effort, to close off sources of news and information." The result "will be the development of institutions and mind sets antithetical to Western values and interests." If given "significant...U.S. thinking," the report stated, "the developing nations [are] likely to favor our views." This is hardly the case—certainly not on the basis of "thinking" alone. By 1984, the time for theoretical debates alone had passed. Western journalists and governments, quite properly, displayed increasing irritability over the prolonged "information order" discussions at Unesco. The alternative, since the struggle exploded at Nairobi in 1976, was meeting some of the negotiable demands of the developing countries. Providing communications technologies and training was not only negotiable but long overdue.

The American ambassador at Unesco's general conference in 1976 pledged a sizable grant to assist Third World communicators. This last-minute pledge defused the rhetorical attacks for "a new order." The policy began of promising infrastructure in return for cooling the ideological debate over the content of news and information flow. Not a single U.S. dollar followed the pledge at Nairobi, however, neither in 1977 nor through 1978. At the end of 1978 the U.S. ambassador again pledged a large grant. And again it was never provided. By 1980, with the rhetoric heating up again, the U.S. floated the idea of creating a new, somewhat independent body within Unesco to fund solely Third World communications projects. This was called the International Program for the Development of Communications (IPDC). From its inception to 1988, the U.S. has not contributed one dollar to IPDC, but has made "grants in trust"—bilateral gifts for communications, for which IPDC credits the U.S. as a donor. In eight years, the U.S. has contributed $773,260 for such aid, most of it spent inside the U.S. The U.S. also supported the concept of ITU's Center for Telecommunications Development. America promised much and provided little.

The SIG report examined past U.S. assistance to developing-country communications. SIG identified sixty programs administered by twenty-six agencies. Most of the funds went to the poorest LDCs. Aid was mainly for communications used to improve housing and health or meet other "basic needs." For years, the U.S. Agency for International Development (AID) was limited by Congress to providing assistance only for "basic needs." Communications infrastructure was not mandated. SIG told the White House, however, that "communications is, in fact, a development priority in its own right."

The report also noted reticence in the government competing with the U.S. private sector, which sought customers in the economically middle-range, newly industrializing countries. Yet, SIG noted, "U.S. exporters are losing out to competitors in these countries, in part because the European and Japanese governments have developed sophisticated export promotion programs, including tie-ins with development assistance support."

Clearly, in the long as well as the short term, American interests, public and private, would be served by having U.S. telecommunications and other information systems become the standard instruments and channels in many Third World countries. West German foundations, tied to political parties, are very active in supplying communications assistance—expertise as well as funds—to countries in Africa, Asia and Central and South America. The Japanese have returned in force to Asia and the Pacific basin. They offer communications this time. The French and the British have long-standing communications enclaves in their former colonial outposts in Africa and the Caribbean.

With ISDN—the networking of networks—likely in the mid-twenty-first century, U.S. preeminence cannot be sustained without substantial linkage to the Third World. Early emplacement of U.S. telecommunications systems is more likely to assure the American objective of "free flow of information" than would a latter-day accommodation of U.S. technology to systems long since in place. The SIG report regards new assistance to poor countries as supporting U.S. political interests and trade, as well as being essential for economic growth of Third World countries. Finally, said SIG, "Such an initiative would once again give momentum to the continuing U.S. leadership role in the creation of an integrated worldwide communications network [ISDN], a fundamental part of U.S. international communications policy for two decades."

U.S. policy was summed up in 1983 by Secretary Shultz.[16] Communications development, he said, "is fundamental to the advancement of democratic institutions throughout the world...[It] reflects the strategic contributions communications and computers make to the expansion of opportunities for worldwide trade and investment...(and) recognizes communications as an important catalyst for growth in developing countries."

The Senior Interagency Working Group made these recommendations:

1. The National Security Council (NSC) should designate international communications development as a "strategic priority on the [U.S.] foreign affairs agenda."

2. To reflect the new NSC mandate, U.S. assistance for Third World

communications should be increased, and such development assistance given visible identity.

3. Such aid to middle-income developing countries should be given priority.

4. "Significant" new communications projects should be considered for countries already receiving substantial U.S. aid in other fields (e.g. Egypt, Pakistan, Philippines and southern African countries).

5. The State Department communications coordinator should coordinate all the communications development programs in other agencies.

6. The federal government should examine ways to assist U.S. private industry to invest in Third World communications development, and serve as a broker between U.S. communications equipment suppliers and Third World nations.

7. Inform the global and regional development banks of the new NSC policy, and ask each to review its own assistance for communications activities.

This agenda is admirable. The full report is frank and balanced. It cites American geopolitical and economic interests clearly. It relates those interests to broadening the global flow of information, and ultimately strengthening democratic societies in the Third World. The report also acknowledges American failures. It notes that the U.S. helped create the IPDC and the U.S. Telecommunications Training Institute (USTTI), a private technical training center. "Yet," SIG added, "neither organization has received a sustained financial commitment from the U.S. [government], leaving the U.S. open to suspicions of bad faith from the LDCs."

Moreover, said SIG, the U.S. observance of 1983 as World Communications Year generated some "fanfare" but little in the way of concrete assistance resulted. This writer was a member of that U.S. committee and can vouch for this negative assessment. Further, added SIG, "Our major bilateral assistance agency, AID, has not extensively addressed the issue of communications development." Communications equipment, and related goods and services, have been given low priority in allocating resources for basic human-needs programs.

More than the judgment of bad faith is at stake. Continued failure to help developing-country communications systems can produce confrontation. Dr. Sarath Amunugama, associate secretary-general of Worldview International Foundation, believes that "a confrontational attitude between the developing and developed nations on information would impose a strain on global networks. This could be a serious threat to *developed* countries."[17] Dr. Amunugama was formerly director of the IPDC at Unesco.

The reality is bleak. Internal telephone service, radio and television are disastrously poor in almost all lesser developed countries. Newspapers and books serve only a small urban elite, with news and coverage of questionable value. Computerized information links are few. Personnel trained in the uses of these media are rare.

For many years it was stated by authoritarian rulers that centralized control of economic development would improve the quality of life and increase individual freedoms in Third World countries. Three or four decades after independence in many African nations, however, living standards have not substantially improved, and political and economic freedom are still a dream. Many rulers still argue they cannot grant freedom until living standards improve. And for that, they say they need communications systems. Increasingly, communications development is regarded as an engine of development. But it can also be an empowering machine. If chained to central authority, communications can be restrictive. If placed in the hands of citizens, communications can convey not only new ideas—political and social, as well as developmental—but provide the choices among ideas which may ultimately free the individual.

The telephone is, indeed, a major instrument of human development. An ITU study in Kenya calculated a cost-benefit ratio of 115 to 1 for telephone investment. An MIT study in rural Egypt produced a 35 to 1 return. The United Nations has set for the year 2,000 the goal of a telephone within easy reach of every person on earth. That is a significant step toward the ISDN.

* * *

WHILE THE GOVERNMENT was slowly acknowledging the erosion of U.S. communications power the private sector made the first appraisal of assistance already being given Third World communicators by the news media, universities, and foundations. The World Press Freedom Committee released the first lists at Talloires, France, in 1983. The tally showed training programs in seventy countries. The list was updated in 1988 by the newly created Center for Foreign Journalists in Washington. Its register describes more than 213 programs in eighty countries providing education, training, fellowship and intern programs for journalists worldwide.

For many years, U.S. wire services (AP, UPI), newspapers large and small in all regions, and, to a lesser extent, broadcasters, have welcomed foreign journalists for short visits and, occasionally, service for some months as interns. The U.S. Information Agency (USIA) runs a year-round program bringing foreign journalists to the U.S. for a month or more of observation. Several private programs bring journalists to the U.S. for intensive training, and several send U.S. journalists overseas for six months to conduct seminars in Third World newsrooms. All of these are useful projects. They introduce visiting journalists to American values as well as technology and journalism.

But they address only a small part of a very large problem: the need to share not only journalistic skills, but communications technology (and then the skills). That can be provided only by a sustained, major-scale program. And that will require funding such as large U.S. corporations and the U.S. government can provide, or make available on long-term, low-cost loans.

The largest U.S. training program for men and women working in Third World broadcast and telephone systems has been operated since 1982 by the U.S. Telecommunication and Training Institute. The USTTI is a nonprofit organization funded in 1988 by $1.8 million from private contributors and $300,000 from the U.S. Agency for International Development (AID). Some 1,400 Third World professionals from 108 countries were trained over seven years in 127 diverse courses. It is an advance for the NSC to have acknowledged the scope of the communications needed in the Third World, and America's crucial interest in finally joining the assisting nations at more than a symbolic level.

It is also an advance for the U.S. private sector to have begun to assume a share of the burden. But it is not clear that the U.S. private sector can make a major impact without conducting training programs on site in the Third World (as the British, French and West Germans do), and without providing, by gifts or loans, the basic technologies that are needed now.

Concentration on technology-sharing, rather than content-changing, would be a major advance in the communications field. The election of Federico Mayor as director-general of Unesco and his early pledges for "uninhibited" flows of information, suggest that he believes "new information order" rhetoric should be traded for technology-sharing to benefit Third World countries. Western journalists and governments are slowly accepting this trade-off. Americans, so far, are the slowest to respond.

More than Third World rhetoric should change, however. The prevalent concept of "communications sovereignty," as linked to national sovereignty, should be discarded. The Bulgarian delegate to the Outer Space Committee of the United Nations stated in March 1982 that the draft principles concerning direct broadcasting (into countries via satellite) "should be subordinated to the general recognition of the principle of state sovereignty and strict respect for sovereign States and non-interference in their internal affairs."[18] Concern for the sovereignty of the state is not relevant in a discussion of communications. A state's sovereignty signifies the control of its own borders and the authority over its citizens. Sovereignty is not threatened by communications, since a government's power cannot be abrogated by messages alone. A government that feels threatened by transmission of news, information, and cultural products must be fearful of its own citizens, and probably uncertain of its own stability.

Cultural protection

Cultural "dependence," however, is a valid consideration for many Third World countries, although their protests are often directed to the wrong address. The same minister of information who complains that the American television serial, "Dynasty," destroys traditional values, may have approved the scheduling because it is available at far lower cost and with more striking interior decor than locally produced programs. Yet, cultural independence—"cultural sovereignty," if you will—may be overwhelmed by the powerful new media.

The protection of traditional cultures should, indeed, be the concern not only of indigenous peoples and their governments, but the world generally. In this communicating age, technologies have the power to overwhelm weakly defended cultures. That problem must be addressed in every country. All of us face the homogenization which mass media encourage. The potential value of the universal linkage of all peoples to the ISDN, a worldwide system, will be the possibilities of expanded choices. This will mean that all cultures should find a place in ISDN systems. For that to happen, however, cultures must survive until ISDN becomes a reality. For the survival of cultures sustained now by small groups, or in small, poor countries, it will be necessary to protect the indigenous culture from extinction. That process is not advanced by raising the spectre of "cultural sovereignty," or suggesting that mass media are guilty of cultural genocide.

Perhaps a starting point in cultural protectionism would be to reexamine the concept of international direct television broadcasting. In heated debates at the U.N., starting in the late 1960s and continuing for ten years, the countries were unable to reach a consensus on a direct-broadcast resolution. Soviets insisted that each country must "bear international responsibility for all national activities connected with the use of artificial earth satellites for the purposes of direct television broadcasting, irrespective of whether such broadcasting is carried out by governmental agencies or by nongovernmental agencies..." This formula in almost the exact words was used at Unesco by the USSR to apply to all forms of international news and information exchanges. It would mean that the U.S. government would be responsible for the content of news or views conveyed by the National Broadcasting Company (for direct television transmission) or the Associated Press (for news service abroad). This is inconceivable under the U.S. First Amendment. The Soviets also demanded that television transmission across borders would be illegal without the express consent of the receiving state. The Soviets would also ban certain subjects from transborder broadcasts. In the ensuing negotiation in 1980 a compromise seemed possible. State responsibility and content regulation would be eliminated from the draft resolution, but prior consent retained. The entire exercise, however, foundered. In the interest of broad agreement on the uses of the new technologies and the protection of cultures, however, "prior consent" for certain categories of cross-border telecasts seems not a high price to pay for expanded communications linkages and intercultural exchanges. Such negotiations in the past have been conducted for the U.S. by an ever-changing body of federal personnel, sometimes assisted by prominent private citizens with particular expertise. Often, these citizens were brought onto the stage just prior to an international conference, too late to be properly briefed or to operate in tandem with the U. S. professionals. Sometimes these private citizens and the professionals are no match for representatives of other countries who may have spent many years in relevant specialized roles. Soviet delegates, for example, are often far better prepared for technical discussions. And Third World spokesmen, too, have a far better background in the history, legends and rhetoric of these debates.

It should be clear now, as the report commissioned by the National Security Council demonstrates, that international communications are the crossroads to trade, jobs, and political influence for the United States. Indeed, the fastest growing area for U.S. trade is with the developing countries. The largest segment there could be telecommunications equipment and servicing. No less important, the attitudes of the developing countries—two-thirds of the world—count. Communications diplomacy will be increasingly important. More support for this is needed in the State Department, and in the Department of Commerce and other agencies. The Bureau for International Telecommunications Policy has been created at State, and a senior interagency group coordinates policy. But, as Rep. Dante B. Fascell recommended in 1985, "In communications and information policy we need policy integration, not just policy coordination." Policy integration can only flow at the insistence of the president, Rep. Fascell added. He called for creation of a deputy assistant adviser to the president for national security affairs and the establishment in the State Department of a viable career track for officers specializing in communications and information.

International regulation
Every country is a victim of what some call "cultural imperialism" when applied to Western culture. And every country practices it. No country is so isolated that some news, information, fashions or other cultural products cannot permeate its borders. And all nations, to some degree, export their own ideas.

Long before "cultural imperialism" was coined, cultural exchanges were encouraged. "Getting to know you," was not only a popular song, but a liberal sociological objective in Western societies. And sociologists regarded acculturation as a normal process of accommodation to a new environment. Indeed, it was also assumed that the process of acculturation did not obliterate the arriving culture or those bringing their traditions to a new land. On the contrary, the new arrivals were expected to contribute something of their past to the present and future cultural motif of their adoptive society. In brief, the newcomer and the new culture were not totally assimilated, but integrated. The differences meant that cultural integration and cultural exchange were positive rather than negative attributes.

Using the political pejorative "imperialism" as descriptive of culture pollutes the discourse. Such rhetoric introduces harsh political elements as well as terminology. The flow of news and information, the sale of film and cassettes, and the movement of data are recast as villainous Trojan horses insinuated behind the borders of an unsuspecting society. Who is the victim, who the victimizer?

Rock music is as authentically American as any cultural form. It combines Black cries of troubles and hope, with Southern white country music. That linguistic and musical form known initially to only a minority of Americans, has swept the globe, especially its youth. Rock has been called "the most universal means of communication we now have, instantly traversing language barriers in a way that is barely understood by the sociologists and linguists."[19] The 1986 Live Aid television benefit for famine victims in Ethiopia was said to have assembled the largest global audience. Yet it was rock music and its stars that attracted that huge audience. Those who decry rock, either as a musical form, or as a threat to traditional music or indigenous culture in general, could say that the ultimate form of cultural imperialism had arrived. Never have so many succumbed to so few, in response to music. Was this not a Euro-American intrusion into cultural traditions the world over? If it was, it did not remain so for long.

In December 1987, Sony, the Japanese conglomerate, purchased CBS Records for $2 billion in cash. The sale places in Japanese hands recordings of CBS's best-selling rock stars, Michael Jackson and Bruce Springsteen. More than that, CBS Records controls a catalogue of American music recorded over fifty years. This major segment of America's cultural heritage is now also controlled in Tokyo. Perhaps more than music led to the Sony purchase. Sony had been developing a digital audio tape (DAT) which CBS had opposed by building an alternative system, and hoping to have its product become the U.S. standard. Now, Sony is in control, as are the Australians who in 1989 bought the Hollywood film conglomerate, Metro-Goldwyn-Mayer/United Artists. Who is the victim, who the imperialist?

Critics have long attacked the U.S. advertising industry for generating consumer desires, and imposing American products overseas. The MacBride Commission of Unesco criticized the role of advertising and commercialization. "The advertising field is often influenced, if not controlled, by branches of international companies," said the report. The advertising field, said to be mainly American-owned, was derided. By 1988, however, three of the seven largest American advertising agencies, including J. Walter Thompson, had been sold to British companies. American investment abroad is still larger than foreign direct investment in the U.S., but the gap is closing. Who is the victim?

RAI, the Italian radio network, introduced in 1985 a new direct broadcast from Rome to New York City. The programs are directed to those of Italian descent in the New York area. In the first month audiences of 700,000 listened, Italo-Americans, even the third generation, said to be "looking back to their roots."

Increasingly, now, foreign journalists are appearing on U.S. television screens—especially on Cable News Network (CNN). Local correspondents of some seventy countries have appeared on CNN. Traditional U.S. journalism avoided putting foreign journalists on the air. It was feared they would either yield objectivity to their fear of governmental reprisal, or use the freedom of American air to criticize their own government. Either way, the journalistic role would change. Given the high cost of stationing U.S. correspondents abroad, hiring foreigners might become routine —again, breaking down a tradition.

The reverse is also happening. American television programs, particularly the most popular entertainment shows, are being purchased at an increasing rate in Europe. This provided Hollywood and independent producers with life-saving income during the protracted writers' strike of 1987-88. The privatization of television stations from Spain to Finland created a great demand for new programs. So U.S. films, often decried, were again highly marketable.

Indeed, in the present globalization of information of all kinds, the marketer rather than the creator of the product determines its distribution. Profitability, rather political or cultural objectives, is the likely determinant.

Turkey provides an unusual reversal of "cultural imperialism." Ownership of VCRs in Turkey, as elsewhere, has mushroomed. (VCR penetration in Oceania has been placed at 46 percent of the population; in some countries of the Middle East and Asia, 85 to 100 percent). By 1985, 2.5 million VCRs were estimated to be in use in Turkey. But Turkey also harbored video pirates. Illegal videos were rented in retail shops all over the country. One network of video distributors has 4,000 to 10,000 estimated outlets.[20] Foreign producers distributed film to theaters, but stopped attempting to market VCR products. In mid-1987, the Turkish government passed a video and recording law providing penalties for video piracy. If the law succeeds in curbing piracy, says one observer, "it will pave the way for restoring the economic aspect of media imperialism." While piracy continued, who was the victim, and who the victimizer?

In Brazil, Chile, Colombia and Mexico, entrepreneurs produce television soap operas called *telenovelas*. Regarded as an authentic Latin cultural form, the *telenovelas* have brought a sharp decline in the purchase of U.S. television programs. Moreover, U.S. commercial television channels now import 38 percent of their programs. In 1973, it was one percent. About half of the imported programs now come from Latin America, the rest from the UK and Europe. No victims, no victimizers?

Among the most blatant examples of domination are the PTTs (post, telegraph and telephone systems) owned by governments, except the United States. They not only control the communications channels but wield price-fixing power as well. PTT rates are generally set not by market standards, but as a means of manipulating long distance rates. The PTTs favor local systems and discriminate against the mostly American multinational operators. There are also rules governing access to the systems that favor local people. Other international, Japanese, or European standards also discriminate against U.S. communications. Who is the victim?

The answer seems clear. Everyone is at risk, and everyone may profit from expanding communications systems inside countries and across borders. Yet the cry gets ever louder to regulate communications flows of all kinds—domestic and foreign.

The Council of Europe's ministerial conference in 1987 recommended the preparation of a West European Mass Media Convention. The OECD is grappling with similar regulations. A Marxist scholar from East Germany comments that even "Western experts predict that the wave of 'deregulation,' which was introduced by the Reagan administration, will be followed by a new wave of 're-regulation' because they realize that in the absence of the protection function of international law, national cultures, national industries and even national security are at stake." The author, Wolfgang Kleinwachter, appropriately calls his paper, "The Interrelationship Between the Introduction of New Communication Technologies and the Need for International Regulations."[21] He argues that international regulation is increasing, and must be stepped up. He cites the chief regulators, ITU and the Outer Space Committee of the U.N. He then adds as potential regulators, however, agencies that are not rule-setters but run only forums for discussion of communications and other issues: Unesco, the U.N. Committee on Information, the U.N. Center for Transnational Cooperation, and the U.N. Commission for Human Rights. These latter do not create legally binding covenants. To suggest they should, is to reveal the objective of controlling the content of communications, not merely assuring the free movement of the flow. Many problems created by new communications technologies require professional examination and even-handed, consensual resolution in an international body. Most would not require legally binding standard setting.

A strong case can be made for the widest possible openness in information flows. There is support for the East German's forecast that the U.S. may veer toward communication protectionism. With the transference of major U.S. industries to other countries, particularly in Asia, a clamor for protectionism has arisen (so far for protection of remaining U.S. textiles and womenswear industries, but not for the radio and television industries which have nearly departed). To place under wraps many new technological or scientific developments, as the Reagan administration attempted in the 1980s, is to hamper domestic as well as foreign progress. Karl Deutsch puts the point dramatically:

> Imagine if the Germans in 1875 had classified the internal combustion engine. The French then in retaliation would have classified the gearbox, and Goodyear would have classified the vulcanization of rubber. The automobile industry would have had a very hard time. In other words, pieces of knowledge put together are more useful than when kept separate.

The fear on all sides is that the new communications technologies lead to centralization. For awhile this seemed to be so. In the brief era of the giant computers it seemed that outside of the U.S. few would be able to compete. Everyone else's data would be fed into the maw of this huge machine, and Americans would control the world's data flow. It did not happen, and that fear should have evaporated in the age of the microcomputers. Not only miniaturization is the result, but incredibly lowered costs as well. Computerization is widely decentralized. A small law firm with Lexis can equal the library of the largest Wall Street partnership. An understaffed newspaper can secure through Nexis the back-copy file of the *New York Times*.

Most important, when ISDN arrives—the universal hookup—there will be no scarcity of information, no information-have-nots. This equalizer will permit access from everywhere to everything. There will be little need then for cross-border regulation; nor would it be needed now, if nations were to adopt the principle of openness. One can see glimmers of ISDN in today's electronic geography. Economics, following technology, is changing the center of information. The 115-nation satellite system of Intelsat—an American creation, and a product of U.S. management—has reduced the need for "gateway cities" in the United States, Europe and Japan. In the past, communications to the rest of the world moved through a few cities and were channeled outward from there. This produced delays, and sometimes errors. Every Intelsat city is now a "gateway city," with access to the world via satellite.

So it can be in the Age of ISDN.

How would ISDN work?
In the U.S. and especially abroad, work is proceeding to create Integrated Services of Digital Networks (ISDN). ISDN is growing both conceptually and operationally in the minds of scientists, engineers, communicators, and political leaders in many countries. ISDN is a massive, almost limitless, puzzle whose electronic parts must be fitted in place carefully, piece by piece, with awareness that new machines and new linkages may be added to the network at any time. It is unlikely that at some one moment ISDN will be inaugurated, full-blown. More likely, systems will be linked as national and international agreement is achieved, and all components are found to be compatible. Because the transmission and reception of signals are digital, the sending and receiving devices may be computer terminals of quite different manufacture that would have proved incompatible without digital linkage. Today's variety of computers, which cannot now talk to one another, would be able to do so in an ISDN linkup.

Through ISDN, two network users may not only speak to one another and see themselves as they converse, but exchange video pictures and text while editing them on the same screen. Together or separately they can call up today's news or century-old history, program a telephone to take or refuse calls or bills, receive a burglary warning from a distant point, market, settle accounts, make travel reservations or pursue other activities.[22]

Compatibility is the key to ISDN. First, domestic communications systems must be married to one another through the telephonic integrated digital network. That must then be made compatible with the international linkages. The variety of services will probably include digitalized voice, facsimile and graphics, video, telemetry, videotex, software transfer, electronic mail, database access, computers, and terminals for banking and other services.

The U.S. system is bound to be the most complex, and most difficult to accommodate in a single network. There are technical reasons for this, but there are also multiplicity of common carriers, and large number of deregulated, private systems that must be linked to one another and to public networks. There is also a multiplicity of international carriers operating in the U.S. There has been no public discussion (as of April 1989) of the cost of ISDN, and its domestic and global potentials.

The private character of much of the U.S. communications system will provide economic and political problems. The crucial controversy may arise again: Should the ISDN be treated as a public carrier, and regulated as a monopoly, or freed as part of "the press" under the First Amendment? There should be the maximum freedom on the premise that the user will have the right to choose most aspects of the service and, where applicable, the content. Favoring certain linkages over others can affect the competitiveness of one private operator over another. Then, too, equipment manufacturers must be prepared to redesign production and, if necessary, change standards to make products compatible with both the domestic and international requirements. There is the possibility that without adequate coordination of both the private and public sectors—or in the face of unrestrained private competitiveness—some users may be isolated when left with incompatible ISDN elements. Not only American producers should be affected. Will the PTTs allow their service and product needs to be met by private suppliers? If not, will protectionist standards freeze out American private industry? Countless other known and still unknowable questions remain. Yet the move toward ISDN is going forward in many countries.

The U.S. has begun to make some technical preparations. Common carriers are adding switches to receive digitally-coded messages from any sender. This is like fitting a 220-volt plug on a 110-volt shaver so it can be used in Europe as well as America. Many more such adjustments will be needed in the decades ahead. It is easier for the Europeans to mesh ISDN because they have state-owned networks. France has gained a jump by establishing a new digital telephone network. All of France may be linked to ISDN by 1990. State-owned telephone companies such as British Telecommunication PLC, Japan's Nippon Telegraph and Telephone Company, and West Germany's Deutsche Bundespost have operated ISDN technical and market tests since 1984. Some have pilot networks wired with optical fiber and tied to satellite transmission. The twelve European Economic Community members planned to coordinate ISDN efforts starting in 1988.

It is urgent for the United States to remain an active partner in negotiations over ISDN. Negotiations proceed at Geneva under the umbrella of the ITU's International Telegraph and Telephone Consultative Committee (CCITT). Unusual cooperation has so far been demonstrated at CCITT in the early development of ISDN. But considerable communications diplomacy for the United States lies ahead.

The interests of U.S. manufacturers, service providers and network operators should be protected if private enterprise is to be viably competitive in the communications field. The inclination of many ITU members is to regulate private information systems as though they were public entities. The U.S. will try to keep nonpublic services free from international regulatory control, and continue to assure the voluntary character of CCITT standards. But no less important is the need to assure the freeness of the flow of information, worldwide, in the ISDN age. The system should not be constrained by those who fear universal person-to-person access to information, as they now fear such access to information by their own citizens.

The 1988 negotiations over public/private issues at the World Administrative Telephone and Telegraph Conference (WATTC) were the first since 1973 under the ITU umbrella. Some 112 countries set rules for international telecommunications through the next decade. Despite earlier fears, private networks will not be strictly regulated, and may be freed of all controls. Government-run telecommunications will continue to control international services available to the public. This represents a compromise between developing countries, which wanted greater regulation, and the major private carriers favoring the status quo. The United States, however, was a "loser" at WATTC, says G. Russell Pipe, an American telecommunications specialist.[23] The U.S. was "isolated and outmaneuvered," he adds, in the most difficult conference in twenty-five years. Other countries accepted compromises needed to complete the trade-offs; the U. S. did not. The issue of whether economic harm to Third World countries ought to be included in the convention was left unresolved, and placed on the agenda for the plenipotentiary conference of the full ITU set for May 1989. The U.S. wanted to remove any reference to economic harm, and voted against the compromise package at WATTC. Pipe noted that users of telecommunications services should be "delighted" at the outcome: "Member countries are obliged to ensure that administrations (governments) and recognized private operating agencies" maintain "a minimum of quality of service corresponding to the CCITT recommendations."

These include access to the international network whose service is sought, international services available for dedicated use (restricted for a particular user), reasonable access by the public to telecommunications, and the capability for linking different services to facilitate international communications. Pipe concluded that the sovereignty of states was recognized, as was the possibility of "special mutual arrangements" with public and private systems. Finally, Pipe noted, WATTC supported the claim by trade experts that telecommunications services are indeed traded, thus opening the way for further attention by the General Agreement on Tariffs and Trade (GATT).

Whether at GATT or elsewhere, neither protectionism nor blaming the lack of American "competitiveness" can fully explain the loss of U.S. industrial markets, including those in electronics. Certain major U.S. public and private policies have contributed to reduced U.S. preeminence. The government accepted a massive burden for defense and military development and deployment; our market-competitive allies did not spend commensurately on arms-related programs. Our private manufacturers opted to move subsidiaries overseas to low-cost areas, or purchase abroad products for the U.S. market, even phasing out whole lines of U.S. manufactured goods. U.S. labor productivity declined in key industries. American investors encouraged foreign buy-outs of U.S. firms, which reduced the competitiveness of American products. For all of this, American entrepreneurs lost market shares in production, but gained shares in foreign investment and profitability.

The consequences of this—massive federal deficits and the transference of skills and jobs overseas to name only two—will hamstring America for a long time. Those repercussions should be kept separate from the great, new challenge coming from the ISDN linkages. These we should address with a clear understanding of the consequences—as we did not when the public and private policy decisions of the previous three decades were undermining American "competitiveness."

We have a new opportunity. By late 1985, thirty different national systems indicated they would join ISDN. Some, including the U.S., plan public-service trials of ISDN by 1990. Only a few developing countries responded affirmatively. If this continues, the gap between the developed and less developed worlds would widen. To avoid that, and to ensure an American technological presence in most of the world, the U.S. should heed the advice of the 1984 SIG report to the White House, and begin now to help developing countries join the ISDN.

Protectionism—not the biggest threat

Protectionism—for textiles, automobiles, steel—is still debated in U.S. politics. Protectionism—directed strongly against the American communications industries—is already here, and growing. But decreased U.S. productivity, not foreign protectionism, is the greater threat. The U.S. had better understand that its preeminent power in all fields of communications flows is eroding. Not one VCR is made in the U.S.—the country that invented and first marketed it. Few electronic consumer products and television sets are produced here.

One imminent development, high definition television (HDTV), is just coming on the horizon. It will have a broad impact on many other parts of the electronics industry. The U.S. market for HDTV by 1997 may be $20 billion a year. The world market by 1991 may be $100 billion. The larger, earlier market overseas reflects the ten-year lag in American initiative in this field. The Japanese began developing HDTV in 1970, including extensive psychophysical testing. Seven years later, the U.S. began studying HDTV. The first demonstration in North America was in 1981, a decade after the Japanese began. European companies have also been working on HDTV, in competition with Japan. The EEC will try to establish its system as the world standard. That may be decided in 1990.

The signs of U.S. industrial decline have been visible since the 1970s. An ominous photograph dominated the *New York Times*'s business section on New Year's Day 1989. It showed a Japanese engineer consulting with two Americans at the National Steel plant in Ecoise, Mich. National Steel, America's sixth largest steel producer, is now owned 50 percent by Japan's second largest steel manufacturer. Still, the U.S. plant is not doing well; Japanese control may be enlarged. Some early steps are being taken to regain productivity. Not until 1994, however, may American automobiles close the quality gap with their Japanese and European rivals.[24]

Virtually all U.S. electronics are now produced in Japan and Europe. Americans are less able than in the past to link next-generation inventions with effective marketing skills. This is symbolized in the present race to see which nation will control the high-temperature superconductors that will revolutionize computers and electricity generation. Patent battles are being fought now, though marketing of the products is still years off. The first three companies securing the most U.S. patents in 1987 were Japanese. America's annual share of U.S. patents has dropped from 73 to 54 percent in the past fifteen years. In just three years, the U.S. market share in electronics has dropped from 50 to 40 percent.

By substituting short-term dividends in Wall Street for long-term, costly research—among related considerations—Americans have missed a major learning curve in the newest communications, as well as smokestack, technologies. Being a generation behind, a ten-year gap, makes catching up very difficult, if not impossible. Ironically, catching up requires more, not less, interfacing between the U.S. government and private entrepreneurs. In the 1980s, defense production lured many creative scientists and favored their noncommercial productivity. Senator Bennett Johnson estimates that 69 percent of federal research and development funds went for defense in 1988. That helped speed America's decline in nondefense productivity—even in consumer electronics that may ultimately have had some defense application. Even when we produce new ideas, other countries develop them faster in commercial production. Apart from defense contractors, American businessmen have usually fought to be free of government involvement in their industries. American communicators, especially, hoisted the First Amendment as a shield to avoid any contact with government. To join the global communications network now forming for the early twenty-first century, U.S. governmental cooperation and some regulation will almost certainly be necessary. Americans—journalists, data processors, and other information users and marketers—must recognize that private/public sector cooperation in the vast field of communications can be achieved without loss of personal freedoms; indeed, such cooperation will assure the development of communications technology under democratic rules rather than authoritarian controls elsewhere.

Public/private interfacing is one of the reasons for Japan's phenomenally successful reindustrialization after the nearly total loss of productivity during World War II. Not only did Japan have the "advantage" of a devastated industrial base on which to build the most modern plant (while Americans struggled with older, less efficient models), but Japan traditionally created a governmental and private-sector consensus for development. No less important were other complex legal, cultural, debt-to-equity ratios, and other financial patterns which assure Japanese employees and managers lifetime positions and the prospect of long-term advantages.

While it may be too late for the U. S. to recover the present consumer electronics market, it is still possible for American inventors and entrepreneurs to leap-frog over the ultimate field of High Definition Television. Indeed, HDTV may be symbolic of America's inability thus far to regain some preeminence in communications. HDTV is not just an improved entertainment system, though it *is* that. Images will be seen with far greater clarity. Many industries will be affected. The medical catscan will be greatly improved as picture grain is reduced and the high-resolution picture reveals cancer far earlier.

Perhaps even greater values can be secured by leapfrogging soon-to-be "old fashioned" HDTV and using instead the systems of ISDN. HDTV relies on the traditional analog signal. It varies in voltage, and sound level and brightness of picture. The signal is "analogous" to the information it conveys. The ISDN systems use the digital language of computers. There, all sound, pictures or data are based on a two-number code. It is more precise and clearer—most appropriate eventually for application to high definition television. Ultimately, too, digital TV will be carried directly to homes and offices by underground fiber optical cables.

This leapfrogging of HDTV, Japanese style, will require computer chips of significantly higher quality, and research of a new order of competence and inventiveness. Americans are doing research in the field but they no longer stay with such long-term development. Great investments are required. Shareholders become impatient. And researchers, deprived of funding for one stage in technological development, find it difficult to return at a higher level of investigation. Years of losses or low financial returns may precede the breakthroughs which ISDN-associated HDTV will require. There is still no lack of U.S. technical capacity or human skill though they can be insufficient if the experimenters continue to miss the experience being accumulated abroad.

The solution lies in some U.S. government cooperation with private industry. The "millions" which the Pentagon says it has allotted to assist HDTV development is hardly adequate, and probably the wrong way to cooperate. Several billions of dollars—a steady flow of funds over several years—will be needed for the development of HDTV. Most of this need not come in government outlays, particularly at a time of a record national deficit. Private funding would probably be available if the government were to assure a consistent policy for many years of tax freedom for the development of HDTV. Long-term, low-interest loans might also be arranged for this purpose. To make such an offer, and make it politically credible in an era of governmental as well as industrial shortsightedness, would require a major educational campaign. Citizens generally, and their officials as well, would need to be persuaded that a basic national interest is at stake in the development of the next generation's communications technology. By sounding a call to support the national interest, leaders would be taking several unprecedented steps. They would signal that

1. domestic and international communications will soon be the most crucial element in sustaining American democracy;

2. domination of communications systems elsewhere may deter the free flow of information of all kinds, and in some places overseas enhance restrictions;

3. Americans, finally, will accept long-term objectives and the responsibility to see them through to development and fulfillment—an advantage for sound governance beyond the field of communications.

Government funding alone will not accomplish these goals, even if the money could be found (highly unlikely), nor will persuasion alone. In the 1970s the Pentagon failed to persuade private firms to develop very high speed integrated circuits for its semi-conductor program in chip technologies. Meanwhile, the Japanese made progress in a large-scale circuit and prepared it for commercial application.

Chips are opening whole new industries. John Diebold points out that "new communications paths are opening between and among people, machines, buildings, and a host of inanimate objects. Buildings will 'talk' to other buildings or with trucks; home appliances will communicate with other appliances; vehicles will talk to roads and to other vehicles; all without human intervention...Increasing amounts of information technology is imbedded in products through chips...For example, as electronic road maps begin to appear on computer screens in automobiles, it will create a new advertising vehicle for restaurants, hotels and filling stations."

Diebold cites the quickening of the pace of the development of information marketing in Europe in a recent twelve-month period:

Merger of the telecommunications manufacturing interests of Plessey and General Electric Co. in Britain to create the GPT deal, a joint venture with sales of 1.2 billion British pounds a year.

Moves toward manufacturing integration, such as the large amalgamation of Alcatel and ITT, and the Ericsson takeover of CGCT in France.

Creation of three international consortia to set up pan-European digital cellular car phone networks in the 1990s, followed by an agreement among European governments on a common standard for car phones.

A collaborative research and marketing agreement between STC in the UK and Northern Telecom of Canada.

A breakthrough by the UK's Cable and Wireless in Japan, where it will have a significant stake as one of the new international telephone service operators.

Establishment of the first cross-border European electronic data interchange services, by which suppliers and customers can exchange orders and invoices throughout the region...

Acceleration of a digital cordless telephone industry.

Establishment of a new European Telecommunications Standards Institute to streamline standards.[25]

The U.S. is also falling behind in a crucial X-ray technology that will manufacture computer chips in the mid-1990s. Consumer electronics as well as military technology are at stake. Semiconductors (chips) are the basic elements of computers. At present, light is focused to etch circuits onto wafers made of silicon. Advanced chips can store one million bits of information. Ultimately, a single chip may accommodate sixteen million bits. (A bit is a single digit which in combination forms a word or number.) By using X-ray lithography, such a chip about the size of a fingernail could store some 1,000 typewritten pages. Japan has set up a joint government-industry program that will have $1 billion to pursue X-ray lithography in semiconducting. This is regarded as essential to the next generation of computer development. In the U.S., only IBM is working on such a program. Current chip-making technology is obsolescent. Failure here, says Erich Bloch, director of the National Science Foundation, "would be like at the beginning of the twentieth century not being involved in the design and construction of internal-combustion engines."

The hard-core problem is the same in the development of high-temperature superconductivity (HTS—total loss of resistance to electricity—a door opening to vast, new uses from ultrafast computers to magnetically levitated trains). The Office of Technology Assessment of the U.S. Congress puts the HTS problem clearly:

> The pattern already evident in HTS resembles that in other troubled American industries. The United States dominates in science, and

American companies often make the major technological breakthroughs. But as industry moves along the learning curve, the competition quickly comes down to engineering and manufacturing, where Japanese firms excel and U.S. firms continue to lag.

The Federal Government's traditional, mission-based approach to supporting technology development worked well for several decades. But the world has changed. Military technologies have grown steadily more specialized, the defense sector more isolated from the rest of the economy. If [Department of Defense research and development] funding was ever a cornucopia for U.S. industry, it is no longer.

Measures designed to stimulate commercial innovation indirectly—through tax policy, or relaxed anti-trust enforcement—have not proved to be an adequate response to new international competition. Although such policies have been in place since the early 1980s, U.S. industry invests proportionately less in R&D than West German and Japanese firms. Indeed, the gap has widened.[26]

Private enterprise in the U.S. would accept the risk if it is made ultimately profitable to do so, and research is sustained by government assistance in the development stages. Other stop-gap measures won't do. Washington has said it will fashion regulations to restrict the number of lines on HDTV to an American-friendly scale. But foreign manufacturers would soon modify their scale to comply, and enter the U.S. HDTV market. The Japanese plan to begin broadcasting on their HDTV system around 1991. Inevitably, there will be one standard worldwide, as there should be, for all new information systems using computers, television, telephones, etc. That era will be based upon the globally installed Integrated Systems of Digital Networks: ISDN. The first linkages, the networks of networks, are already being installed on a small scale, without fanfare.

For in communications, technology leads policy. The Soviet Union is reluctantly pressing glasnost (wider publicity and discussion of management objectives) in order to increase the nation's productivity and distribution system. Government officials in both East and West Germany have modified their broadcast receiver rules because radio and television flows from across their borders were picked up by listeners regardless of official restrictions.

John Diebold says the productivity challenge stems from "the now endemic short-term thinking of United States business management, which is motivated by the need to show quarterly profits, as well as the simplistic and often inappropriate application of accounting practices taught by our venerable business schools." Diebold advises further, "We cannot expect to excel in HDTV, or superconductors, or much longer in biotechnology, aerospace or any number of other key technologies, until we focus on the basics: insuring at least a minimally educated work force, encouraging risk taking and investment, rather than consumption, and taking the steps necessary to refocus public and private sectors on long-term social objectives."[27]

* * *

CHINA IS PRODUCING glass optical fibers in association with Japan and the UK, and is holding talks with West Germany, the UK, the Netherlands and Japan for introducing terminal manufacturing technology. By 2000, China is expected to have produced 300,000 kilometers of optical fibers for home consumption and export. Optical fibers are the natural channels for ISDN networking. Meanwhile, in 1988, a fierce interagency battle engulfed the State Department and the Transportation Department over allowing the export of U.S. communications satellites for launching on Chinese rockets. Transportation objected to the transfer because it feared American companies would not only lose launching business to the low-cost Chinese satellites but would also transfer technological expertise. In free trade, nations win some and lose some. The big factor is the freeness of the international system.

By 1992, the European Economic Community (EEC)—twelve West European allies of the United States—will complete their economic integration. The EEC will become the largest production and marketing unit in the world, accounting for 40 per cent of the world's trade, and rivaling the U.S. in size and influence. U.S. exports to the EEC declined by nearly $7 billion from 1980 to 1986. It is likely the ISDN, digital telephony providing widely diverse integrated services, will link many of the twelve countries. The United States may be testing some domestic linkage by then, but will not be as fully tied to the EEC's linkage as the West Europeans themselves.

The Europeans, moreover, are looking beyond 1992 to the next century when ISDN will make possible a host of new communications and information technologies. By April 1988, some 700 universities and research institutes in Europe applied for a share of the EEC's $740 million fund for information-technology research. While U.S. R & D spending was still high ($120 billion for all purposes in 1987), it was dropping. The call for research bids was the second phase of the Esprit program under which the EEC contributes to basic research in information technology. The first phase, starting in 1984, brought together some 3,000 researchers from more than 400 research organizations in all twelve countries. In 1985, French President Francois Mitterand regarded a European research coordinating agency as an EEC alternative to America's Strategic Defense Initiative (SDI). Europe would not depend on the priorities of U.S. research, but would develop Euro supercomputers, artificial intelligence, and advanced laser technology. The program has already produced significant new technology, including the most powerful computer chip on the market—the T800 floating point transputer.

Half of the new applications to Esprit are related to projects in computer and support technologies. One third is for the development of information processing systems, and much of the remainder for research in microelectronics and peripheral technologies.[28] The political purpose is no less significant: joint research with national barriers removed among EEC countries produces common technical norms and the creation of a powerful domestic market. This is translatable into economic and political power outside Europe. While the British were somewhat hesitant, the West Germans welcomed the lessened dependence on U.S. technology. Esprit is a civilian program, but its technology is bound to have military applications, just as SDI will have civilian uses.

The French have developed an information tool, Minitel, which may become the computer-terminal prototype for all of Europe. This prototype resulted from a deliberate political decision made to gain a lock on the technology in Europe. This system was designed, at the outset, to replace printed phone directories with an electronic directory service. Minitel terminals are used to secure telecommunications services known as videotex. The terminal has forty columns, a keyboard, a small two-color screen, and a telephone. The French began by issuing three to four million free minitels. This service handled forty-seven million calls in December 1987. Minitels also provide some 7,000 databases and other services across France. Says one observer, "The minitel is changing the face of France" and it may "also change that of much of the rest of Europe."

Protectionism is already rampant in Western Europe. It has placed quotas on more than 1,000 products, and continues to create new restrictions. While after 1992 the twelve nations will continue competing with one another in telecommunications through their PTTs, proposed laws would make it more difficult for the U.S. and other nonmembers of EEC to compete in the new market. It appears that trade barriers will be lowered for members, and raised for others. Such policies, as in Japan, invite U.S. retaliation, to everyone's loss. That loss will be especially great in the field of ISDN, and information-sharing.[29]

* * *

SALE OF INFORMATION as a product is not the only motive for communications protectionism. The power of information, and the urge to gain and protect that power is a major objective of the new transnational regulator, especially in the field of data-processing (ADP). The efficiency of ADP is itself a threat to the bureaucracies, and to the trade unions. Individuals regard ADP as a loss of control of their personal information, especially when personal data are transferred from one place to another—and particularly when the information is accessed in another country. This occurs frequently when multinational corporations are headquartered abroad. Such a situation is regarded as a human rights issue.

The U.S. is the target of much of the new EEC legislation precisely because many of the multinational databases are in the United States. Personnel records of European employees are, therefore, subjected to examination in Cleveland or San Francisco. The EEC, the Canadians, and the Japanese take legal steps to regulate ADP border crossings. They tie that to preferential rates and tariffs for information services and products, in order to foster domestic information industries. The developing countries, however, press for greater information capabilities and also target U.S. preeminence.

As a consequence, the trend is to insist that data files remain in the country of origin rather than be transmitted across national borders. This reduces efficiency, and adds to the cost of doing business. There is the need, however, to separate information flows so that news and nonpersonal data can, indeed, be swiftly moved across borders, while privacy is assured for personal files. The technology can readily accommodate this separation. In the early negotiations at GATT and elsewhere, "information" was not protected as a product or activity subject to fair trade and principles of fair competition. Information, according to one observer, is "in a no-man's land to be blocked, controlled, taxed, tariffed, or used as a bargaining chip by governments."[30] There is a hazard here for news reporting and commentary. Increasingly, particularly in Third World discussions, news is mingled with "information" and treated as a commodity. Under such perception, journalism could readily be subjected to cross-border controls, as are ADP and related data flows.

Some sample regulations:

Sweden was the first to restrict the flow of information to protect its citizens' privacy. One of the early anomalies was the negative effect of the law on the fire department of Malmo. It had compiled a building-by-building file of fire hazards and occupancy. When a call came, the fire fighters could learn in seconds about the troubled building and its occupants from a printout at Malmo—through a computer in *Cleveland, Ohio*. Later, Swedish law prevented the use of the U.S. data base for securing information about the Swedish population. West Germany's federal information-protection act halted what is defined in the law as "improper input, access, communication, transport and manipulation of stored data." This vaguely worded law could open many corporate data flows to charges of impropriety. France penalized violators up to

$400,000 and demanded prison terms for up to five years for recording or transmitting data defined only as "sensitive." The United Kingdom requires the post office to be able to read transmitted information. This could reveal proprietary business information now sent in confidential commercial codes.

Nine East European countries have also planned an information-exchange system that uses high technology. It proposes to accelerate scientific and technological progress among the communist nations. The program initiated in 1985 would develop (1) a new supercomputer capable of handling 10,000 million operations per second; (2) wider-use industrial computers; (3) a uniform East-bloc information-exchange program; (4) fiber optic technology for exchanging information; and (5) a new satellite-exchange system for television and computer use. The new Communist-bloc system is being directed from Moscow. It is unlikely, however, that individuals will receive access to the new system. It will probably be available only to centralized research institutes and other government organizations. Yet the program does advance the Gorbachev drive for technological progress.[31]

At home, Gorbachev plans to double the USSR's telephone system by the early 1990s, introduce computers and databases at all levels of the Soviet economy, and train youth, starting with secondary school students, to handle computers. For most of this, the Soviets are likely to build mainly on established Western expertise. The challenge, nevertheless, will be the philosophical/political implication of Western technology: the empowering of the individual, and access to new and ideologically divergent ideas and data. The threat to Soviet leadership is nothing less than the diminishing of central control. For example, an explosion in telephone usage will make more difficult the traditional Soviet monitoring of calls. Late in 1988, the USSR announced that it plans to increase by 20 percent the amount of money spent by government in 1989 to support fundamental research. The money would support promising research in high-temperature superconductivity and information technology.[32]

For its part, the United States has put in place some limitations on the transfer of telecommunications equipment, if not the messages themselves. The National Security Agency, which has the technology to monitor the transmission of data from the U.S., has been examining the software export issue. In this effort, the U.S. has repeatedly asked European PTTs to cooperate in detecting and checking unwarranted data flow. This effort to deal with software smuggling by wire is part of efforts by the Defense and Commerce Departments to restrict some computer software from export. Software can readily be exported over telephone lines by using inexpensive modems. A personal computer in Washington could "export" to Europe during a short phone call a computerized design program that would be helpful to a weapons designer.[33]

Similarly, in 1987, the Reagan administration gave India approval to purchase an American supercomputer, the first ever permitted to go to a developing country. But permission was not given for India to buy the state-of-the-art supercomputer the Indians wanted. This was held back because it is capable of tapping the U.S. global code network and deciphering top secrets. The Indians have security ties with the Soviet Union. The Indians presumably promised not to use the supercomputer's secret-tapping potential, but U.S. intelligence agencies opted not to rely on the pledge.[34] In mid-1988, India and the Soviet Union signed a protocol allowing India to purchase Soviet-made supercomputers. The protocol also provides for Indo-Soviet collaboration in the manufacture of consumer electronics, radio navigation, optic fiber systems, and television and radio broadcasting equipment. Electronics thus adds another chip to geopolitics.[35]

America's Third World market is hardly based on the sale of supercomputers: the lowly telephone is still the prime export instrument. The phone is still regarded as an elite system in all but a few industrialized countries, and telephone services are virtually nonexistent for most citizens of developing countries. Yet the phone line—especially in the Age of ISDN— is the potential link of the outlying tribal, provincial and rural areas to the urban centers and especially central authority. To construct phone systems in the nineties may require expenditures of nearly $800 billion for a billion phones—twice the number predicted for the 1980s. Third World nations invested $8 billion in telecommunications in 1983. Soon after, the ITU's Maitland Commission recommended the creation of a global center for telecommunications development, and expenditures of $12 billion a year for that purpose. Competition from European, Japanese and other Asian manufacturers is increasing, but U.S. equipment firms still may service much of the new network. To do so in the years prior to the initiation of ISDN, when long-term services are to be set in place, is particularly important. A reminder of the competition is the case of Saudi Arabia and its contract for more than $4 billion in the late 1970s. All the major U.S. firms bid for the contract. The contract went, however, to a group composed of Ericsson, Phillips of Holland and Northern Telecom of Canada.

In the computer field, two-thirds of all international communications originates in the United States. U.S. electronic equipment manufacturers still dominate the world market, but new competitors enter the field each year—including Third World nations who regard a computer industry as a 747 on the runway: a mark of national pride.

Not only will marketing competition increase, but the politicization of the information flow is conceivable. Wilson P. Dizard Jr. points out that British telecommunications workers in 1978 voted not to handle any international traffic to South Africa to protest apartheid policy there. The ban lasted a short time. But, says Dizard, "it is possible to imagine a similar ban against traffic to [the U.S.] because of the way it voted on a certain issue in the United Nations, or because of some other political event."[36] He cites Harvard specialist Anthony Oettinger who regards access to information as a major policy challenge for the United States in the future. He notes this chronology: in the '50s U.S. strategic interests focused on the real estate needed for military bases; in the '60s the emphasis became global expansion of U.S. industry and banking. The major concern now is shifting to threats to information and communications access. This is the connective between an increasingly dependent U.S. society and other nations.

As more than 50 percent of the American gross national product is now service oriented, and a large part of that is information-servicing, the country is increasingly vulnerable to the fiber-optic-thin ties to other nations. Two European specialists, Michael Palmer and Jeremy Tunstall, regard the 1986 breakup of the American firm, International Telephone and Telegraph (ITT), as "important a landmark in the evolution of world telecommunications as the breakup of AT&T." By the 1990s, they say, "we could see European telecommunications greatly strengthened around a French-German axis, and with U.S. influences in Europe much weakened."[37]

It is easy to fashion an adversarial scenario involving the Third World and the information-rich West. Yet information wealth is not like oil or mineral resources—used up once they are extracted. Information is meant to be shared. The more it is shared, the more valuable it may become. The developing country's future may best lie in joining the information linkages, and adding to the great pool of knowledge and experience. That will require cooperation rather than resistance, an acceptance of those Western modes that encourage diversity and openness, and particularly the freedom of the individual.

Western governments have the great responsibility and opportunity to assist telecommunications and radio development in the Third World, and perhaps head off nasty confrontations if the communications gap becomes deeper and wider. Telecommunications development in the Third World has importance beyond empowering the individual. Expanding the flow of news within developing nations, between Third World countries, and between them and the West, is largely a function of low-cost telecommunications. The national PTTs frequently apply high rates to news delivery, restricting the flow, while simultaneously blaming Western carriers and news organizations for being restrictive. A great deal of Third

World investment has gone into radio broadcasting rather than telecommunications. Radios provide the central authority with the quickest, cheapest channel to the entire population. It is mainly a one-way flow, the kind Third World spokesmen decry when referring to the Western dominance of the news flow.

Small is revolutionary

For years, this writer has favored Third World concentration on the smaller communications technologies: radio, yes, but also the telephone, casettes, and small printing facilities such as the copier, telefax, and the mimeograph. With ISDN decades off in most places, it is essential to work now for greater basic literacy, and the small achievements which must precede the bolder promises made by high technology and eventual ISDN linkage. Indeed, there are serious political pitfalls in Third World countries if the new communication technologies are perceived to fail, as have past development efforts in most of Africa and part of Asia and Latin America. The post-colonial rulers of many Third World countries provided centralized control—including a tight grip on information power—as the key to social and economic development.

Neither authoritarianism nor totalitarianism has worked in post-colonial Africa any more than in Marxist Europe and Euro-Asia. The failures have produced frustrations that are slowly being acknowledged and addressed. Great promises, which may be difficult to fulfill, should not be floated anew. The ISDN will almost certainly have an early impact on the industrialized countries. It may not, at first, be as widely or as deeply felt in the Third World. It may, in fact, appear that the information gap has been increased. But once Third World countries have joined the network, new possibilities for information and services will expand over their *present* potential, if not equal to advantages to citizens in the industrialized world.

There already are significant Third World models. More than seventy countries use computer packet switching systems to improve the quality of life of their citizens. These switches, enabling one source of information to tap the data resources of another system, cost only $30,000. At the Oswaldo Cruz Institute in Rio de Janeiro, for example, Brazilians use microcomputers linked to laboratories in other countries to fight chagas, a deadly disease which afflicts eight million Latin Americans. A Caribbean cooperative computer network helps farmers in the Dominican Republic diversify their crops. In Mauritius, a small sweater factory has become a major exporter because it is linked by computer to design services in New York.

There should be no doubt, however, of the political implications of even the "small" information machines. Xerography made possible in the

U.S. the speedy copying of volumes of data hitherto kept secret or slowly accessible. Xerography led to the Freedom of Information Act, and that enhanced investigative reporting. While the scale of such development may be different in other countries, no regime—authoritarian of the left or the right—will ever again be "closed" once such information machines become available to the people.

To reduce the disparity between developing and developed countries, the U.S. should heed the SIG report to the 1984 White House, and recognize the mutual advantages to sharing communications technology with the Third World. At the height of the debates at Unesco challenging American dominance of the news and information flow, William Harley, a consultant to the State Department, proposed the deployment in U.S. missions in Third World countries of satellite-fed computer circuits. They would enable local citizens to access a variety of American databases. News and information would be offered without charge, as selected by the Third World user. Some arrangement would have been needed to recompense the American suppliers of the information, but that was a minor consideration. After some study, the idea was dropped. An important opportunity had been missed to demonstrate the sharing of information, and provide concrete rather than rhetorical answers to Third World challenges. It is still time to consider sharing communications technology as well as information. There could be a natural, mutually profitable trade-off. In return for the gift of information and technology, Third World countries would guarantee the free flow of information within and between countries. Will inequality remain? Yes. But there will also be considerable progress. And there seems to be no other way to speed Third World progress faster. To emphasize the continuing disparity rather than the real improvements would be a disservice to everyone in all worlds. It must be recognized: technology is not magic, nor all human leaders divine.

Everywhere—in the industrialized West or East, or in the developing countries—the problems and opportunities provided by the ISDN and general information technologies should be widely examined and understood. The twenty-first century will be the first in which all people everywhere must now relate intimately to a global society (no longer simply to family, village, tribe, city or nation). Individuals everywhere, then, need to know how they can be empowered by the new technologies.

Hazards of freedom

Deregulation of broadcasting in the 1980s opened the way for greater diversity, particularly since commercial operators could now broadcast in some countries where only state-owned radios were previously heard. Deregulation, or privatization, of radio and/or television has already begun in Argentina, Brazil, France, Spain, New Zealand and Malaysia. But deregulation also makes possible the purchase by a single owner of many radio or television outlets in one or more countries. Indeed, this concentration of ownership in broadcasting, as in newspaper publishing, tends to defeat the possibilities for diversity which broadcast deregulation provided. The International Federation of Journalists (IFJ), the grouping of journalist trade unions, regards the concentration of ownership as a "threat" requiring a "coordinated campaign" to attack the problem before a single European market is created by the EEC in 1992. The IFJ also "deplored" the decision by the European Parliament to permit commercial sponsorship of programs in transfrontier broadcasting throughout the European community hereafter. The IFJ said "the impartiality of professional journalism can only exist independent of political and commercial pressure."[38]

On the contrary, the mixed broadcasting system—both state- and commercially funded—is the *only* solution for democratic European countries. The mixture provides diversity and prevents the domination of the system by either component. This, again, properly defies the slippery slopers who insist that even one step down the road to state broadcasting spells authoritarianism, or that a move toward commercial ownership spells mindless profit-making. To be sure, both possibilities must be avoided. A public-cum-commercial mix is an appropriate arrangement. It is not likely that professional journalists, left to run the show without "political and commercial pressures" (as the IFJ asks), would be any less influenced by political or economic issues.

Other aspects of new information power raise warning signals, if not cries of danger. For as with every new right or privilege there is an accompanying restriction. Home computers with two-way, interactive services provide many different functions. The QUBE system available for several years to residents of Columbus, Ohio, carries thirty channels including one with talk-back capabilities. Consumers may use this system to purchase and charge merchandise, do personal banking, and respond instantly to political polling or game-show questions. The danger arises from the capability of the system to report to the originating source—the store, bank or polling place—the personal choices or attitudes of the citizen making the electronic selection. This can be an unwarranted invasion of privacy. If instantaneous political polling or voting is undertaken, a serious flaw could develop in America's system of representative government. The Congress is elected to serve as a deliberative and consensus-forming body. It is not, at its best, merely the messenger of its constituencies. Citizens at home hardly have the time or the resources to research the intricacies of legislative bills. They certainly cannot engage in the process of compromising diverse views. Yet the QUBE system of instant talk-back puts a premium on both speed and superficiality, hardly the best features of a deliberative legislature.

We witnessed the hazards of automated, instantaneous decision-making at the New York Stock Exchange on Black Monday, 19 October 1987. The big-money managers had programmed their computers to cry "sell" whenever certain economic conditions loomed. The automated responses thrust millions of shares onto the market and immediately depressed stock prices. The strengths of the market were buried ever more deeply under the cascading sell orders triggered mainly by the big-money computers.

And, the cry of "cultural imperialism" long leveled at Americans by developing countries may be applied to the European purchasers of American television programming. In 1987-1988, for example, a British commercial television franchise agreed to buy MTM Enterprise, the U.S. variety producer, for $325 million. An Australian investment company now controls more than 50 percent of Hollywood's Hal Roach Studios. London Weekend Television owns Silverbach Lazarus Group, a U.S. program distributor. NBC searched the European field to barter its programming for an investment in a broadcast or cable service. Showtime/The Movie Channel was considered for purchase by France's Canal Plus. And MGM/UA, a traditional Hollywood consortium, considered acquisition by either a Japanese or a Dutch buyer.

The field is fluid. Major communications companies are bought and sold as chips in a board game. There are no obvious villains, mainly entrepreneurs. The culture products are treated as merchandise. Indeed, in many cases they are little more than that. Significantly, however, it becomes increasingly meaningless to charge one country, always the United States in the past, with homogenizing the international culture market. U.S. films still flow around the world, but at a lessened pace. And, increasingly, the home office for U.S. culture products is Tokyo, Paris or London.

Networks of freedom—or *1984*?

George Orwell regarded the nascent communications revolution as a counterrevolution—the final domination of evil power-wielders over the helpless masses. Orwell observed at first hand in Spain in the 1930s how forces professing to liberate the masses betrayed and murdered them. *1984*, the fictional forecast of the eventual victory of statist evil, gave information power the place of highest dishonor.

Are we shaping the instruments today for the eventual centralization of each nation's information power in the hands of political controllers, and, through them, the universalization of information controls at the whim of a dangerously small elite? These serious questions should be indelibly placed before every technician, politician, and citizen whose decisions influence the development of communications technologies. Everyone in every country, directly or indirectly, is part of the process by which the Age of ISDN is being incrementally established. One transnational owner of newspapers, magazines and television says that by the end of this century the world's media will be dominated by six large corporations. His will be one.[39] Other threats:

• Three-quarters of the world's nations already strongly influence print and broadcast services, and most operate the carriers of other communications.

• Cross-border information flows can compete with indigenous cultures, and lead to cultural homogenization.

• New technologies, highly miniaturized, can eavesdrop on private intimacies and public secrets—from near or far.

The list can be extended. Technologies, for the most part, are neutral. They can kill or cure; as the laser beam. But the designers and controllers of communications technologies are *not* neutral. Today, many are attracted by the power that information provides or enhances:

• Some would use information to construct a more open, freer society.
• Some want a more tightly restricted polity.
• Some see information mainly as a profit-making vehicle.
• Some regard information as the key to their nation's economic development.
• Some seek several of the above simultaneously.

Information technologies and their applications are not predestined. At every stage, the social and political forces of each society determine the assignment of research skills or technological deployment. Next, the responses to research and deployment generate new social and political mediation. This occurs in closed as well as open societies, in widely varying ways. But if information technology has any inherent inevitability, it is in its nature as a driving power that satisfies human instincts: *curiosity* about life in the neighborhood, and on the planet at large; *participation* in the society of humans; *wonderment* over cosmic matters; and, perhaps most of all, *personalization*, how I, the individual, can use information to make the most of my years.

Information technology then, is not simply glass fiber linking terminals via satellites to another fiber connective. Real people operate both terminals. Their needs and aspirations are human, not automatonic. To be sure, social, economic or political forces will influence the kind of information that will flow through the ISDN systems. So, too, however, the mediational forces of societies will influence this process and the content of the flows.

One can imagine a kind of technological predestination serving as a balance wheel. As new technologies are put on line, in closed societies and in free societies strongly influenced by giant entrepreneurs, the "small" technologies—copiers, telefax, telephones, audio-cassettes, VCRs, transistors, radios—generate alternative channels of communication. Indians in 1988, for example, were offered "Newstrack," popular video news-documentary cassettes produced by nonpartisan news magazines, intended to break the government's TV monopoly. These alternative publishers and broadcasters become increasingly more difficult to monitor from central authority. Indeed, they take on a life of their own. They generate a recharging effect. One "small" system feeds another. The political and social climate is changed. The alternative media move onto the mainline. Whether that is strongly controlled by centralized political force (as in closed societies) or influenced by dominant economic power (as in free market nations), the alternative systems take their place at the terminals. Third World countries that complain bitterly that their voices cannot be heard adequately on global news services, will appear on the keyboard. Dissident movements will find some channel for their views. Young entrepreneurs without funds to compete with establishment press and broadcasters can use the "small" technologies to test their public appeal. The networking of networks —ISDN—promises great diversity of content, facility to link and interact, around the corner and around the world, in real time, and with little political or economic opposition.

To do all of this, ISDN must first surmount serious obstacles. The economies of the world are dominated increasingly by a few industrialized states (the U.S., Japan, and the unified EEC). Their information power is essential to their continuing as pre-eminent world-class countries. This is clearly understood in Japan and the EEC, but not widely enough in the United States. Fortunately, for the future of the human race, all these nations are clearly oriented to basic democratic institutions. There should be minimal *cognitive* dissonance—but there will be some. The inclination to centralize, to use surveillance, to retain secrets, to manage news is present in all governments, everywhere. But the mediational factors of political and social traditions, not to mention constitutional and common law commitments, should assure the continuing rapid development of ISDN as a freedom-advancing factor in the preeminent information societies.

To seek to enter that charmed circle, the closed societies of the Communist bloc and the authoritarian developing countries will have to yield significant political, ideological and idea-control prerogatives. As they do, the technological predestination of even the "small" communications media will impose its own, nonideological reality. One step will follow another. The fax machine will empower the private user to turn the telephone into a publishing mechanism. The VCR will enable the citizen to see and hear ideas and modes of behavior from distant and different places. Interactions become unpredictable. Freedom takes on new meaning. Many things, hitherto only dreamed, become possible, not least, the democratization of politics, as well as technologies.

Indeed, all countries, democratic and authoritarian alike, will find secrecy rapidly disappearing. A small sign is already apparent in Sweden. A group of journalists there created a satellite surveillance system to watch both American and Soviet activities in space. The Swedes' Space Media Network sells what it sees to press and television clients. The network first detected the Chernobyl disaster in 1986, and thereby made inevitable the Soviets' reporting of the event.

But much can and must occur before such a sanguine time is reached. And the pitfalls are harsh and real. The period of danger may be just ahead. Indeed, we may already have entered such a time. All of the negative forces are still massed in strength. They may not yet fully realize the challenge to them from the Age of ISDN. The sophistication and determination of the information-controllers have never been greater. Chapter VI describes the many ways in which they impose their rules. Chapter VIII depicts in detail how most countries during 1987 and 1988 acted out their will to hamper human exchanges, and crucially subvert the flow of information. This is reality. One should not expect the censors and their governments to yield voluntarily the power of governance or information power. Indeed, one may expect to see in many places the further restricting of print and broadcast freedoms as local political needs dictate. The roller coaster-like censoring, relieving and recensoring in Nicaragua and South Africa may be models for other fearful authoritarians. Eventually, I believe, even they must yield to the impulsion of information power. For, in denying its broad use a nation will soon seem to choose darkness in an age of light, followed by social and developmental reversion to preindustrial lifestyles.

That is not to suggest there will be a simplistic dichotomy between the democrats and the anti-democrats merely because they employ information power differently. Mass communication, by definition, does not easily enhance the individualism of the citizen. But mass communication tied to the discrete potentials of the computer terminal enables the individual to choose among hundreds of channels and modes of information, and also "talk back." That is, indeed, a democratizing aspect, the new massification of idea flows. In the development of the new information technologies, however forceful may be the grand entrepreneurs of print and broadcast, and however itchy are the political fingers on the regulatory mechanisms, the greatest potential of ISDN—the enhancement of individual freedom of choice, the empowering of the citizen through new information systems—will require democratic political institutions to sustain the information networks that free choice will have inspired.

Now is the time for American negotiators—private and public—to cease cries of European protectionism, and thoughts of counterbalancing that with bilateral U.S.-Canadian or -Asian "community building", and instead, bargain persuasively with the EEC to open the European process to all, and forestall discrimination of any.

Bargaining alone is not enough. It must be bolstered in the U.S. by more nationally funded research in support of graduate education and lifelong training. Reports of the Office of Technology Assessment and the National Academy of Engineering reveal that the majority of graduate students in U.S. engineering departments are foreign citizens. Only 41 percent of the small number of Ph.D's are native-born Americans. Most foreign students are required to return to their homelands. And, increasingly, U.S. companies contract some of their engineering to Korea, India and other places where these students reside. The NAE indicates, moreover, that the half-life of an engineer's skills is 2.5 years in software engineering, five years in electrical engineering, and 7.5 years in mechanical engineering. There is, consequently, the need for lifelong learning. One specialist observes that "were the federal government to devote several hundred million dollars annually to graduate fellowships in the physical sciences and engineering, a substantial change" would occur in our national competitiveness.[40] The United States must relentlessly plan and build the networks of freedom—human as well as technological.

11.

Don't Fear the Slippery Slope, and Other Recommendations

THE U.S. JOURNALIST'S stock-in-trade is skepticism, sometimes leading to cynicism. That skepticism is directed to most leaders of establishments, particularly government. Beyond skepticism, however, is utter rejection of any governmental relationship with journalism except that of supplying information. Limiting that supply for any reason, increasing the supply for governmental reasons, or influencing the content of information is rejected by newspersons as the first step on the slippery slope to authoritarianism, or worse.

I have earlier described scores of ways in which governments wield information power. Some of that power is deployed without subtlety. In much of the world, journalists are murdered, arrested, harassed, deprived of information, fired from their jobs, expelled from countries, and treated as second-class citizens—just because they seek the facts and practice journalism.

I have only briefly referred to the social contract that journalists have implicitly made with their fellow citizens. That contract, despite the vagaries of the domestic government, calls for truth-seeking, honest and balanced reporting, and avoidance of favoritism whatever the source. Most governments are not as permissive as this suggests. All governments want to manage the news. A few are restrained most of the time by constitutional constraints. One moral of Chapter VII, the reports from seventy-four countries, is the fact that *all* governments, to greater or lesser degree, want to use official power to influence the flow of information. Political influences on newspapers are felt in most-free and least-free states; far more so in the latter. There is commercial influence in both groups, though stronger in the least-free. Licensing of journalists also influences the content of their reportage. Licensing is bound to affect coverage when the right to withdraw a credential is in official hands. Our survey showed that licensing influences press content in 12 percent of the most-free countries, and 50 percent of the least-free. There is obviously no equivalence between the freeness of information flows in the most-free and least-free countries. Indeed, the distinctions are varied and extensive, and must be acknowledged and addressed. Acknowledging these crucial distinctions between the free and unfree should lead us, however, to recognize that the hallmark of a free society is not primarily its tradition, statutes or fixed policies. A principal element of free governance is the viability of the democratic process: the give and take of competing interests, but always with limits set by a sense of civic responsibility, and a live system of balances.

It is here that the fallacy of the slippery slope enters. The fallacy, in logic, arises from errors of perception, judgment or interpretation—fallacies of induction. The slippery slope is to argumentation as the snow-covered peak is to the skier poised on high. Once underway, there is no stopping, runs the argument. Having set out on a certain course or policy, one must inevitably encounter the worst-case possibility.

Journalists and their information colleagues often display hypersensitivity over their prerogatives and freedoms. It is difficult for me to acknowledge that, after having published for some years the Journalism Morbidity Table (Appendix A)—the record of horrendous maltreatment of journalists in scores of countries. It is important, however, to avoid exaggerating presumed threats by democratic government, for such governance is fragile, and extravagant complaints can weaken the political structure supporting all citizen-freedoms.

Limiting the number of broadcast outlets a newspaper may own in a single area, some argue, erodes First Amendment freedoms, and invites further governmental interference in press content. Yet that restriction protects the larger citizenry from monopolistic control of mass media in a limited geographic area. Publishers, on the other hand, welcome federal government bailouts of financially troubled newspapers, and accept a federal judge's 1988 intercession permitting a joint operating agreement between two ailing papers (the Detroit *News* and the Detroit *Free Press*) despite anti-trust implications.

One journalist, who later in 1988 became president of NBC News, is suing the federal government because Congress prohibits the U.S. Information Agency from releasing to Americans the texts of USIA broadcasts intended for foreign audiences. This represents an ominous loss of press freedom, the complainant charges. The law was passed, however, to prevent some administration from using federal funds and facilities to propagandize the American people. The American polity has sufficiently matured so that it can receive USIA scripts, and still be on guard for abuse of the government-supported information flow.

In cases such as these, and in limitations on the press in courts of law—fair trial versus free press, a distinct clash of basic rights—the rule of reason must prevail. A judge's demand that a journalist provide evidence essential to a court proceeding is not an unwarranted breach of the First Amendment. Shield laws for journalists, while they are often important protections, are no more absolute than any other right in a democratic society. Neither the individual journalist, nor his medium—certainly not society as a whole—has necessarily started down the slippery slope because a journalist loses in the balancing of individual rights and social responsibilities.

There are dangers to poising on any slope. One learns to master the terrain, no matter how slippery. The keys to such mastery are specificity, rationality and reasonableness. Any law or regulation should clearly limit the objective of the act, and restrict its application just as precisely. The formulation of the act or regulation should involve rational procedures, designed by reasonable individuals. The administration of the act should be placed in the hands of reasonable people, with an opportunity for fair appeal. Thus, the dangers of the slippery slope become manageable. To act out of "worst-case" fears leads to impotence. Every policy or law can be abused or become destructive. One can always argue the inevitability of extreme consequences.

And yet reversibility also is possible. The Fairness Doctrine was removed by the FCC in 1988 after decades of less than robust political coverage by broadcasters. The doctrine required broadcasters who air controversial news programs or commercials to give opponents free time to respond. This provided some balance of political views, yet it never led to government control of private programming (as some had feared). The U.S. has ended monopoly protection for the vast AT&T system, and deregulated routes and prices on domestic airlines (for better or worse). Similarly, France, Spain, Jamaica and many other countries are selling government radio and television outlets to private owners—a clear change of course, away from government control. Moving in the opposite direction, Scandinavian countries now provide subsidies to newspapers, among the freest in the world.

United States history is replete with slippery slopes in the field of communications and information. The Morrill Act helped create the land-grant colleges and the extension services which provide vast quantities of information. Federal funding made possible radio development after World War I. Congress provided money for the first Morse telegraph experiments, and the first U.S. telephone network. The U.S. Public Broadcasting System is a limited experiment in federal broadcasting, but far from a dominant force in radio or television, and no threat to commercial broadcasting. Indeed, conservative critics of National Public Radio argue that federal funds frequently support anti-administration programming —quite the opposite of fears that U.S. public radio would inevitably favor the party in power.

There are, indeed, worrisome acts by government. As Chapter IX indicates, the Reagan administration reclassified much information already in the public domain, useful statistical and information services have been reduced, executive orders have sought to restrict scientific exchanges. Secrecy agreements and testing have been applied to many officials. Access to some government information has been made more difficult. Such restrictions tend to run in cycles, followed by periods of greater openness, some abuse, and then tighter controls. The key is frank public discussion of the restraints. Crying havoc is counterproductive.

There is distinct need now for cooperation between the federal government and private entrepreneurs in the further development of ISDN, pointing toward the massive networking of the next century. The slippery slope should not be feared in such a relationship, even if government must set some rules assuring competition and diversity. A free society is built on everlasting tensions: the clash of rights, and the mediating of diverse interests. Such tension is good. Fear that leads to policy gridlock is not constructive.

The world is full of slippery slopes, especially in the field of communications. With every new information system comes increased power over the word, and the capability of exploiting that power for improper purposes. The very magnification of information potentialities and power, therefore, requires broad agreement between government and the governed, and between governments in the international arena. Regulation, now more than ever, may be necessary in international linkages, but not for the purpose of controlling the content of messages. Quite the opposite. Regulation will be necessary to assure fair access to communications technologies—to the full potentialities of ISDN; and no more. In mid-1988, twelve member-countries of Asiavision (AVN) held a second trial of "live" TV program exchanges from many countries to many countries. It is called "hot switching." Technical quality was high, and more than 200 news reports were exchanged over ten days. Some controversial reports from Asia and Europe were aired. The experiment, intended to test technical feasibility, also inspired hope that cross-border news flows may become less intimidating to sensitive governments. Regulation for control's sake, or for maltreatment of journalists or other communicators, and their instruments, is no more implicit in the new information systems than in the current communications media. They are designed to be engines of freedom, not control.

Some recommendations flow from our reports and observations:

I. Policy formulation

It is essential, in the Age of ISDN, to end the avoidance of major U.S. policy making and agenda setting in the communications field. Constitutional ways must be found to coordinate some activities of public and private communications and information institutions in America. This should lead to the formulation of policies that U.S. representatives can carry to communications negotiations with other countries. For such negotiations, a full-time cadre of skilled communications diplomats should be retained on a career track in the State and Commerce Departments. High-level officers and technicians of private communications and information companies should serve, if not full-time, certainly on a continuing basis, as participants in U.S. policy formulation affecting the private industries. To formalize these functions:

1. The White House should lead an educational campaign to restore American preeminence in electronics, particularly the crucial information sectors, by stressing long-term planning, research and marketing, bolstered by tax, loan and other inducements, and reversing the short-term profit syndrome. Restoring the investment tax-credit (removed in 1986) would be a notable start. Expanding credit for research and development is no less important. The president should signal his concern for longer-term development of superconductors—the key to twenty-first century electronics—by forming a consortium of universities, private industry and government as recommended by the Committee to Advise the President on High Temperature Superconductivity.

2. The White House also should promptly adopt the 1984 recommendation of the Senior Interagency Working Group (SIG). It urged that international communications development, particularly in the Third World, be explicitly recognized as a strategic priority on the U.S. foreign-affairs agenda.

3. To assure continuity of that policy, once adopted, a deputy in international communications should be assigned to the office of the National Advisor to the President for National Security.

4. To advise the NSC, State, Commerce, FCC and other related agencies, a Presidential Commission of Private- and Public-Sector Communications Specialists should be created. The private representatives would be drawn from two major political parties, labor, management and the universities—all with special expertise in communications and information. The primary mandate of the commission would be to monitor the overall integration of U.S. policies in communications and information, particularly with regard to policies needed to deploy American instruments and assure fair access to all linkages—domestic and international—in the Age of ISDN.

5. The commission's early task should be to act on the SIG report's

recommendation that massive assistance be given by private and public agencies to Third World communications and information systems. U.S. policy here will have to avoid the slippery slope syndrome. Some communications aid will go to news outlets in less-free countries, where mainly—in most places, *only*—government-run news systems are permitted. Even there, however, the advantage of U.S. association outweighs insistence that the foreign government first build democratic or commercial institutions. The U.S. should nevertheless urge such recipient countries to open their government-controlled news and information systems to diverse reporting and viewpoints. Senegal and many Caribbean states already do so.

6. The Age of ISDN will enhance all flows of information. Among the most important is the exchange of new ideas in science and technology. The information revolution is a product of the application of science and technology. But the revolution continues. Indeed, one-third to one-half of all increases in the U.S. gross national product are attributed to advances in science and technology. While the U.S. is still the leader in basic research, other countries now apply it more productively to the marketplace. More is needed than pleas for greater American competitiveness or protectionism. The office of science adviser to the president should be strengthened, and supported by a blue-ribbon committee of scientists and technologists. They should have ready access to the president's science adviser, who should also be responsible for stimulating the transfer of basic research to useful technological applications. That will require the setting down of estimates and an agenda for long-term uses of science and its applications.

7. Electronic publication creates problems which the federal government should address. Laws governing public access to government information (the Freedom of Information Act and others) should be modified to permit citizens access to unclassified electronic records. Electronic technology makes such access more complex. New processes are changing the distinctions between reports, data bases, publications, etc. Increasingly, federal information appears in these forms. Present statutes on releasing information are outdated. This new technology also raises questions about government services competing with commercial electronic publishers over distribution of value-added information. It is essential for Congress promptly to resolve these and related issues.

II. Regulatory changes

1. For purposes of U.S. regulation and enforcement, the new communications and information systems such as cable television, videotex and others should be regarded as "the press," and covered by the First Amendment, and not seen as common carriers, without First Amendment protection.

In return, these new "electronic publishing" technologies should exhibit editorial responsibility to the public, providing diverse news and entertainment, under competitive conditions. Deregulation cannot be achieved immediately, but the political/philosophical principle of deregulation should be adopted, and worked toward. No country has ever achieved full access for all its citizens to any medium of communications or information. The U.S. comes closest. By adopting the policy of ultimate universal linkage, tied to the First Amendment, we will have chosen the appropriate goal.

2. Regulatory protocols may be needed for linkages with new international communications systems. These new rules should not be regarded as retrogressive. New regulations should not inhibit or control the content of messages (except to assure protection of personal privacy and freedom from libel, and avoidance of compromising national security). Such regulations are the price to be paid for access to worlds of new information and services. Here, U.S. communications diplomacy will be most tested in securing fair access to all new systems, and participation in the manufacture and deployment of the technologies. The U.S. should not be frozen out of communications manufacturing, deployment or access because of economic competition, or geopolitical or security considerations.

3. Conversely, the concentration of media ownership, particularly across national borders, presents a new threat to media diversity. It is difficult to limit ownership and still adhere to First Amendment rights of free expression. But—again, that slippery slope—a line must be carefully drawn to prevent the near-monopolization of one or more information industries. The problem is far more urgent in the UK and France than in the U.S., but it is, nevertheless, a development to be watched.

4. There will be actual competition—it's already surfacing—between U.S. newspapers and telephone carriers for control of information transmitted over phone lines. The Bell operating companies intend to go into the information business, linking information providers with users. Newspapers see this as a threat. They seek government regulation to prevent the carriers—"monopolies," the publishers call them—from entering the information business themselves. The publishers call this an "electronic threat." Television was so regarded at first, but the U.S. print press is thriving. Competition among all the purveyors is desirable. Newspapers have their natural advantage—completeness, when attempted; diversity; ease of recall; etc.—and electronic media have different natural advantages—speed; greater selection of topics; etc. The First Amendment should protect both media from all but strictly monopolistic challenges, and normative considerations of libel and national security. What is needed with telephonic as well as cable systems is open network architecture. This

would open the same "monopoly" loop to competing information companies. There would be no need to set in place parallel telephone or cable lines. Indeed, competing phone or cable companies could lease lines from competitors which, in turn, would charge information providers for the use of the lines.

III. Universal access to communications/information

1. Telecommunications—particularly the telephone—is the key to progress in the twenty-first century. ISDN—the network of networks—will be the magnifier of telephone-line services. It is essential, therefore, that everyone, everywhere, have access to a telephone. For Americans, the need is no less urgent than for others. It is essential for the U.S. communications industries to coordinate their activities so that no segment of the U.S. public is left out of domestic linkages, and to assure that European or Japanese protectionism does not restrict U.S. linkage through international ISDN. The U.S. should take the initiative in communications diplomacy, not simply be reactive to Euro-Japanese political and technological developments.

2. To engage fully in the international debate over the development aspects of communications, particularly in the Third World, the United States should be a better prepared, active participant in the old-line international regulatory agencies, as well as in the intellectual forum provided by Unesco. This will mean returning to the United Nations Educational, Scientific & Cultural Organization from which the U.S. withdrew in 1985. While some reforms have been completed, and a new director-general has indicated his intention to strengthen Third World communications capabilities without harming press freedom, it should be recognized that the developing-country majority will always have a strong voice at Unesco.

The forum, therefore, provides a legitimate expression of reality in a widely diverse world. Third World nations should recognize that past colonialism is not responsible for all present failures. They should, instead, give freedom a chance. Neither can the U.S. escape reality, although it can shut its eyes to it. That, we have done for several years. It is time to return to Unesco, restore its universality, and help it concentrate on the many constructive programs created there. This may mean eliminating some intensely controversial programs in communications and information. Many fundamental concerns and potentialities raised by the new communications technologies are valid subjects for examination at Unesco. However fruitful such discussions, they should not be regarded as regulatory or standard-setting, and certainly not elements of new international law.

3. Unesco itself should become a massive electronic "switching" center.

It should provide access to bibliographic and data bases in science, education, culture and the other sectors within its competence. Research done outside Unesco could be acquired through this system. Unesco itself would engage in far less original research and conferencing, and avoid divisive debates on a "new information order." Unesco should provide the electronic structure and content bases so that Third World investigators can pragmatically use the vast new information resources rather than emphasize political debates on information theory. Unesco already has made a start: It is developing software for desk-top publishing of small papers in rural Africa. A type font has been devised to harmonize scores of African alphabets.

4. The U.N. Development Program (UNDP), through Unesco, should examine ways to speed the linkage of *all* Third World countries to ISDN. This should lead to the greater coordination and rationalization of basic telecommunications linkages among all developing countries. The survey should emphasize assuring individual citizens access to the new technologies. This will mean examining objectively the present obstacles to the free flow of information, and the linkages needed to bring diverse communications to all developing countries. There should be particular emphasis on the need for journalists, especially indigenous newspersons, to have access to information from varied sources—official and unofficial.

IV. Domestic linkage

1. The ISDN should become the world's first demonstration of large-scale power sharing. The domestic linkage will provide citizens with hitherto undreamed-of banks of news, information and services. This should be a democratizing force since information leads to empowerment. Internationally, the ISDN linkage should demonstrate concern, even compassion, for the interests of poor states and their citizens. The ISDN will require a degree of pragmatism and consensual arrangements that bring all nations into a peaceful, coordinated conglomerate—at least for purposes of creating and sustaining informational association. It would seem difficult to break that linkage thereafter for political or military considerations.

It will be essential, in tying U.S. domestic to international systems, to avoid political or corporate gridlocks. Countless legislative, regulatory-agency, and executive-branch restraints, on the one hand, and diverse corporate hedging on the other, can delay and hamper American access to overseas systems. A group representation of private communications/information industries should prod government, as well as one another. For this limited purpose, they should be freed of anti-trust prosecution.

Education and training seminars on ISDN should be held throughout the U.S. by secondary schools, colleges, and the communications industry. These should be designed for the users of the services as well as the technicians and managers.

2. The Reagan administration tightened many information sources in the federal government, and closed down others (as Chapter IX indicates). Some reduction in government permissiveness was expected. Presidential elections in 1980 and 1984 overwhelmingly supported the candidates who promised a tougher policy in information as in other aspects of governance. Tension between press and officials is still the name of the Washington game. After information restrictions set in place during the Reagan years, there should be continuing public *reexamination* of whether to:

(a) restore some accessibility to government documents through the Freedom of Information Act;

(b) restore some publications and statistical sources;

(c) restore the restriction (removed by the Reagan people) against infiltration of news media by the Justice Department;

(d) end FBI requests that librarians provide surveillance of use of their facilities by certain foreigners;

(e) assure that at least a civilian press pool will accompany any U.S. combat abroad;

(f) remove the ban on presentation of certain scientific papers if foreigners are present;

(g) abrogate National Security Decision Directive 84 (NSDD-84) which subjects officials to lifetime silence on "intelligence" without prior review.

3. Cable television systems should be required to become competitive —within the cable field. More than one system could use a single network of wired installation. A regulatory agency could insist that the cost of installation and maintenance is properly shared. Cable-TV systems could then compete for public purchase of diverse programming.

4. New technology should not be used for surveillance not already authorized by the courts. Remote sensing should not be permitted to invade personal privacy.

5. Libel laws should be reexamined to make certain that the new linkages and services do not relay actionable statements from one medium to another, thereby magnifying the libel and spreading the liability.

6. Property rights should be protected as proprietary news and information is moved from one system to another.

7. Personal records should be kept confidential, particularly as private information is pieced together from diverse systems.

V. Stop the growing censorship

We have listed, in Chapter VI, eighty-two different tactics, procedures and subterfuges which governments use to control or influence independent news media. The list ranges from outright murder of journalists to relentless pressures on their institutions. Censorship in all its forms primarily targets the masses of citizens. Such controls diminish the citizen's (a) understanding of, and participation in his/her own society and the world beyond, (b) his/her opportunity to mature rather than become an automaton in an authoritarian society.

All governments try to influence the independent press. Some do it within the laws of democratic states (and suffer if caught breaking the laws); some do it autocratically, or by sheer force or monopoly power. The place of a country in this censorship spectrum—shown atop each country section in Chapter VIII—defines the nature of the society itself, and the degree of freedom of its citizens. A press is no freer than the citizens who depend upon it. And when autocrats take over a government, the first casualty usually is the radio station; next, the independent newspaper.

Since by our estimates barely 23 percent of all countries have a free print and broadcast press, these recommendations have broad application:

1. Licensing of journalists by governments is rapidly increasing. Twelve Latin American countries now license journalists. South Africa threatened for years to introduce journalist licensing, and in 1988 announced and then postponed the equivalent of licensing. ASEAN countries license journalists. The procedure in Latin countries does more than regulate who shall or shall not practice journalism (and, by strong implication, who shall be barred from the field because of writing that displeases the government). The *colegio* or empowering institution also influences which journalists shall be permitted at government press briefings and other events. While licensing is sometimes favored by journalists who want to freeze out competition, and gain publicly recognized credentials in a profession, the cost of such government approval is high. The price is paid painfully by the public, which is served by journalists engaged in some degree of self-censorship.

2. As the physical harm done journalists increases, there are understandable attempts to provide mechanisms for protecting them. So far no satisfactory system has been devised. Most plans require identification certifying the carrier to be a journalist. The certifier, under those plans, would be a government agency. That is the equivalent of government licensing of journalists, and it is not acceptable to Western professional associations. The most efficacious arrangement so far is the "hot line" established by the International Committee of the Red Cross. On word that a journalist is in trouble, the ICRC notifies its representative in the appropriate country, and a search is begun. The ICRC has saved several journalists from harm and possible death since the "hot line" was initiated in 1985. The "hotline" phone number—22 734 60 01—should be on the desk of every editor who assigns a reporter to duty in a troubled zone.

3. The right-to-reply seems fair: permit one who believes himself abused or misreported the opportunity to respond or correct a mistake. U.S. newspapers and some broadcasters permit such response in limited letters columns and editorial-reply broadcasts. The "right" to respond, however, has become a government demand, particularly in the Third World, and especially directed to independent news media in the West. That "right" could be widely used to flood the news media with governmental statements. The principle, however, opens the privately owned press to government control. The government of Malaysia has repeatedly demanded space in *Time* magazine, *Far Eastern Economic Review* and the *Wall Street Journal*, and economically penalized all three severely for what it regarded as inadequate responses. The "right" has been declared unconstitutional in the United States. It is legally supported and limitedly operable in France. In Argentina, the right to reply is sustained in defending one's character, but not to project an idea. In most of the world there can be no such right because the news media are under varying degrees of government control, and no words are likely to be printed or broadcast that do not conform to the government's "right." That factor makes the demand for universal recognition of the right to reply as essentially a governmental ploy directed at the largely free, independent news media. There, some replies are carried, but certainly an insufficient number to reflect all the divergent news in a diverse society. It is only in such free countries, however, that *any* right of reply, however limited, or however optional for the medium, exists.

4. "Prior consent" is the key term in international debates over direct-broadcast-by-satellite television (DBS). DBS anticipates cross-border transmissions directly into the receiving sets of another country. The technology already makes possible undirected cross-border telecasts from neighboring states. Indeed, several East-bloc countries already receive unscheduled telecasts from nearby Western countries. Malaysia and Singapore are concerned about one another's cross-border broadcasts, and Canada is concerned over the great flow from the United States. In debates at the United Nations on cross-border flows, only the U.S. has voted in favor of unregulated telecasts into other countries. One specialist writes that in the field of international communications the "vessel of sovereignty" is "leaking, and in some instances may even be sinking."

The Soviet bloc has set three conditions for DBS: (1) prior consent: transmissions across borders would be illegal without the express approval beforehand of the receiving state; (2) state responsibility: every nation would bear responsibility for the telecasts originating within its borders —an impossibility under the U.S. First Amendment; and (3) content regulation: certain specific subjects would be banned, a list obviously open to political control or manipulation.

The potentialities are too great for international communications to have DBS founder because of political considerations. The prior-consent requirement is not an onerous element. It is far less restrictive than the present blanket rejection of DBS which invites jamming of incoming messages and, at worst, shooting down satellites used in DBS transmissions. The U.S. should drop its rejection of prior consent, and join other Western countries in resisting content regulations and state responsibility as the remaining hurdles to DBS. Broad guidelines for cross-border broadcasts can be set forth without accepting program clearance or censorship.

5. At the London conference Voices of Freedom '87, called to challenge the censors worldwide, the declaration approved in January 1987 by 150 participants called for the creation of a "fund against censorship" to support legal challenges of restrictions; compilation of a list of lawyers experienced in such cases; creation of a "censorship hot line" or a clearing house for obtaining help; organization of an "early warning system" on restrictive press laws and other measures; public service advertisements to spotlight such abuses; and dispatch of missions to places where news is being suppressed. Since that declaration, censorious governments have expanded their restrictions. The World Press Freedom Committee (WPFC), which organized this conference, and other agencies worldwide which participated, have used most of these tactics to challenge the censor. But the need is great for financial support of the six projects listed in the Declaration of London.

6. Disinformation has been the patent of the Soviet Union's KGB. It has floated outrageous forgeries intended mainly to separate the allies from the United States and especially turn Third World countries into anti-American outposts. Soviet disinformation, using newspapers and magazines in developing countries, has accused the U.S. of creating and spreading the AIDS virus, assassinating Indira Gandhi, and conspiring to overthrow Third World leaders, among the recent ploys. The U.S. briefly seemed to borrow the disinformation patent when middle-level officials floated information known to be false about Muammar Qadhafi. Such activity demeans a democratic society. And Americans, moreover, are not adept at that game. They should expose Soviet disinformation wherever found, but leave the scurrilous tactic to the professionals in Moscow.

VI. Role of the press

1. The function of the press—newspapers, magazines, radio, television, and now electronic publishing—is different in every country. In the nations where diversity is encouraged, the role of "the press" differs with each medium and with every outlet within the medium. All share one objective, however: to convey to their audience the world as the press sees it that day. The public understands what is happening, nearby or

distant, generally because the news media report it. That is an awesome responsibility. In only one-quarter of the countries is that responsibility left to independent journalists. Elsewhere, governments mainly assume the responsibility both of making life and death decisions, and reporting on those acts—the government's way—to the public.

Governments that do this say the press cannot be trusted. Oddly, the publics in democratic states are also skeptical of the press. In the U.S., with the freest press, opinion polls regularly place journalists near the bottom of professions held most in public esteem. Yet no other profession or business displays its mistakes and shortcomings as publicly as the news media (letters to editors and corrections columns reveal errors every day). One editor puts it frankly, and in perspective: "You can trust the press the way you can trust doctors and preachers and lawyers— you can trust the press if you really believe that all hysterectomies are necessary, the Ten Commandments are obeyed, and everyone in prison is innocent."[1]

Yes, everyone makes mistakes. But the press, wielding power second only to government, imposes its mistakes on the society at large. And, the polls suggest, readers and listeners object not only to mistakes, but also to the arrogance displayed by some journalists, and implicit in some editorials; the intrusion in private lives and businesses; the callousness, particularly of the television camera, during moments of tragedy; the doggedness beyond reportorial necessity that distorts an issue; the countless ways of masking a bias that nevertheless influences audience response; the failure to acknowledge the major errors of editorial judgment; the sensational treatment given a report; and the challenging of personal integrity. These shortcomings are far more serious than abortive relationships with a physician, preacher or lawyer, because a whole society is put at risk when the credibility of the press is reduced. The power of government enlarges, as if by some seesaw effect, when the power of the news and information media diminishes. Governments know that. Some cynically try to undermine, or more subtly encourage the public to challenge the press. That may lead to government controls over the press, and a far more dangerous time for everyone.

There is greater need today than ever for reasonable, intelligent critiquing of the performance of the mass media in the United States. The several magazines of journalistic criticism are useful. So are the occasional books on this subject. During its short life, the National News Council was constructive. (Unfortunately, it was never given the resources or, most important, the cooperation of major newspapers and broadcasters to make possible a credible self-analysis and corrective.) The ombudsmen on some newspapers play very important roles. Every newspaper outlet should have one and they should regularly critique one another. The occasional, sophisticated critique of the press—including television—on CBS's "Sunday Morning" program is the best example of honest self-examination in the news media. It is also inspiringly good viewing for the audience. Journalistic humility, as well as self-analysis, is needed.

2. News from the developing countries is, at best, sparse in the West. That is a major bone of contention for Third World critics of the Western news media. But, then, foreign news generally occupies only a small part of the daily news in most newspapers. That is a significant loss—to Americans. They should know more about the world, especially since the ISDN linkages will soon open many more places and subjects to individual scrutiny. It is a pity that Americans, blessed with diverse mass media even now, remain generally provincial. The Third World news agencies, while still mainly purveyors of "communique news," should be encouraged. They provide some indication of events in the developing countries, even if only as projected by governments. One may hope that these agencies will improve, particularly when they accept the reality that few of their reports are picked up by other Third World government media, let alone outlets in the West.

Government news, however, is news; in many places, it is the only game on the block. U.S. journalists should more frequently examine the output of the Nonaligned News Agency (NANA), Pan African News Agency (PANA), Caribbean News Agency (CANA), Inter Press Service (IPS), Asia-Pacific News Network (ANN), and some of the ninety-odd national news agencies run by governments. The lesson of CANA is important for journalists in the West and the Third World. Its directors include independent media owners as well as government representatives in the Caribbean. CANA has Reuters among its antecedents. CANA is a dependable news service, used widely by many different outlets in the Caribbean, and occasionally quoted as authoritative by the global news services. Though Western journalists are properly reluctant simply to relay accounts from government agencies, they have traditionally done so in Moscow and Beijing, rewriting as necessary for perspective and balance.

3. It should also be recalled that there are many medium-sized national and regional news agencies that have good records for credibility. They often cover their areas better than the global competition. The Italian (ANSA), German (DPA) and Middle East (MENA) services are examples.

4. The Sharon and Westmoreland libel trials did more than place *Time* and CBS in the dock. The cases raised questions about fairness in reporting, and the use of libel laws to test fairness and penalize unfairness. The news media have become too powerful to receive immunity from criticism or even prosecution when the act of transmitting actionable information has become egregious. Yet the penalty, and even the court

process itself, can be unacceptably damaging. Years ago I proposed demonetizing libel suits (Chapter VI). Litigation should be greatly streamlined. Special "libel courts" should be created with specialists in media law. With the introduction of no-money, no-fault libel, the financial burdens for both complainant and defendant would be greatly reduced. The plaintiff would be awarded money sufficient to compensate for actual monetary loss. The ruling on the substance of the complaint would provide psychic and moral satisfaction for whichever party is declared correct. Reputations would thus be protected, and the press would not face costly penalties, perhaps life-threatening to the institution. Credibility, not dollars, should be the most sought-after objective in a libel case.

VII. Opening to the Soviet bloc

Inevitably, when a long-closed society opens its curtains to the world, journalists will flock to see what is going on. The Iron Curtain has been raised across middle Europe, before the Soviet Union, and at the shores of the China Sea. U.S. magazines, newspapers and television correspondents have conducted lengthy interviews with Mikhail Gorbachev. These revealed all—and no more—than the Soviet leader intended to place before the Western publics. Some themes on the Gorbachev melody have been played by astute Soviet officials. The same themes are played repeatedly by Soviet journalists, really civil servants, appearing to act as American or Western journalists—inquisitive, interrogative, investigative. They are, in fact, part of the several-tiered, orchestrated projection of Soviet (Gorbachev party) policy. The fourth tier are the men and women "in the street." They are carefully selected counterparts of U.S. "common men and women." The Soviets participate equally on numerous television bridges linking the U.S. and the USSR. These bridges have had mixed effects. Some are blatantly propagandistic. They provide no inkling of the daily struggles and attitudes of the Soviet participants. There is a certain condescension implicit. The Americans are expected to bare their souls and describe the inequities and difficulties in their country. The Soviet citizens are not to be pressed to reveal their basic problems lest the spirit of goodwill is somehow disturbed, or McCarthy-style anti-communism is suspected. TV bridges can be useful, but they should be constructed so that both ends of the bridge are mounted with equal candor, and without positive or negative biases.

VIII. Preserving cultural differences

There is merit to the claim that news and information flows predominantly from Western mass media to receivers in the Third World. The sheer volume of daily news traffic from the four Western news systems—AP, UPI, Reuters, Agence France-Presse—is greater than all other Western,

East-bloc, IPS, Third World, and national-agency flows combined. But the content is not necessarily biased for or against particular political or national interests. To be sure, the subject coincides with the known interests of the majority of the publics receiving the flow—the mainly Western audiences. But, again, once the selection of subjects by audience interest is conceded, the content of the reports is relatively objective and balanced.

There is, nevertheless, more to be said about cultural products such as film, audio and video cassettes, books and music. These, too, tend to flow north to south. The reasons are complex, and do not meet the charge of imposing cultural values on poorer, smaller societies. The receivers tend to invite the inflow. One may be concerned, not with the cries of "cultural imperialism," which are largely politically motivated, but with the reality. Indigenous cultures *should* be preserved. They *are* subjected to overpowering competition. Rock music alone is dominating most societies, North, South, West and even in the East. That is not to suggest censoring rock, but rather taking steps to preserve traditional, indigenous cultural forms of all kinds.

We are entering a crucial period for cultural preservation. Not only are new communications technologies suppressing older cultural forms, but they are eliminating some as well. This is particularly ironic as we enter the Age of ISDN, when limitless technical linkages—by word, sound, picture—make possible the preservation of traditional forms for masses of individuals to experience. Never before were so many people in distant places able to hear the tribal music of small African peoples, or see art of the Amazon, or view the work of the Australian aboriginals.

To preserve many dwindling cultures, anthropologists and other specialists, as well as ministries of culture, should make a concerted effort to plan a systematic listing of the most threatened cultures and cultural forms. This is an essential program which Unesco is specially equipped to undertake. Heritage and monument preservation has been among Unesco's principal accomplishments. An agenda should be prepared to determine some priority for extensive recording and photographing of the threatened cultural forms. Funds should be sought from governments, foundations, and others to produce accurate documentaries of the songs, dances, art, writing and other forms. All of that can ultimately be shared worldwide in the Age of ISDN. But that is only a beginning. Culture is also the interaction by peoples of diverse backgrounds to the science, the industry, the agriculture, the journalism of each day's grist. That cultural interaction will require the most inclusive linkages in the Age of ISDN.

IX. Assorted reminders

1. The U.S. Agency for International Development (AID) should clearly

place communications development at the top of its priorities—even above "human needs" for awhile. It should be clear now that information of all kinds—if shared with the public—can improve the quality of life, and assist development.

2. Government policies should be reexamined frequently to discover whether, in seeking to advance desirable social, political or national security objectives, they are hindering the competitiveness of American workers and industries in international markets. This concern is particularly applicable in the fields of communications and information.

3. At every stage in the development of ISDN, ombudsmen should monitor the new technologies and their deployment to make sure they are being connected to advance, not reduce, the democratization of the society. We must understand as ISDN develops whether, indeed, our social and political institutions are sufficiently capable of functioning as democratic instruments in the new communications environment.

4. The U.S., in insisting on the free flow of information for news reporting and communications, should separate those elements from the debates over cultural products (films, cassettes, etc.). Indeed, news should also be regarded as distinct from "information," defined as material for social and economic development. News is too important to be regarded as a tool of domestic politics or international geopolitics. When news is labeled "information" and perceived as a polemic, critics call it either "developmental journalism" or "cultural imperialism." Both sets of critics thus regard such "news" as propaganda; not objective, balanced reporting. There are other defenses for the cultural sphere once the vital news flow is defended as a separate and distinct value. The slippery slope should not be feared: by such separation, the U.S. will not concede the need for controls or trade-offs regarding cultural products. But they call for a different defense.

5. Public radio and television in the U.S. serve an important function. Their value should not be gauged by the nightly ratings they draw. Instead, they should be supported for the quality of programming they provide in several fields, quality which the commercial outlets usually do not match. Independent broadcasting, far from being threatened by the presence of public channels, is challenged to match their variety and quality. One may even take some pride, as a democrat, in the political penchant of public broadcasters to favor social and political causes not generally supported by the national administration.

* * *

IN MANY OF these recommendations the element of the slippery slope is invoked. The U.S. government is asked to take some domestic actions which are traditionally regarded as out of character. The action may put the proponent on the slope, at the end of which lies a violation of the First Amendment. But reasonable men have taken such risks before—when they created federal information services under the Morrill Act, or set up Comsat as a satellite monopoly, and then shared control with more than 100 governments. All of that worked well, and the United States and the world are better for it.

U.S. support for communications in the Third World, even if governments at first dominate the technology and content, may similarly prove advantageous even though the initial risk is greater than in supporting a democratic system.

Epilogue

Will ISDN Facilitate Peace and the Human Imagination?

Had the telephone system reached its present perfection previous to 1861 the Civil War would not have occurred.
　—Arthur Pound, *The Telephone Idea,* 1926

THE AGE OF ISDN—the universal networking of diverse information networks—will be unlike any previous human era, *not solely because it is the consequence of unprecedented technical advances,* but no less because two dominant and opposing modern ideologies will be cast aside, and the human imagination, not dogma, facilitated. The two ideologies driving social change in this century either regard *growth* (a capitalist thesis) as the key to highest human achievement, or, its socialist opposite, *revolution,* as the road to utopia. Both theories are myths to be debunked, says the Rutgers sociologist, Peter L. Berger, in his seminal book, *Pyramids of Sacrifice.*[1] The Age of ISDN may well explode both social theories, and help develop a new synthesis.

Development imposes severe human costs, and is driven by growth mainly for the exploitation of resources, writes Berger. "In many Third World countries these costs are prohibitive," he says; the result has been "noncapitalist policies." He acknowledges that capitalism has generated unprecedented productivity and institutions favoring individual freedom, but "these achievements must be weighed against the costs," he says.

Socialist revolutions have also imposed great human costs, Berger notes, and any "egalitarian distribution of the good things of life" ("goodies" still assumed when Berger wrote in 1974, but now largely dismissed), must also weigh the high costs, says the author. He supports critics of capitalism who "reject policies that accept hunger today while promising affluence tomorrow." He also defends critics of socialism who "reject policies that accept terror today on the promise of a humane order tomorrow."

He calls for solutions to problems that accept *neither* hunger *nor* terror. Such solutions can only come, says Berger, from "cognitive participation"—knowledgeable participation by *individual* citizens in the definitions of the problems, and in the process of making decisions. Most social-policy decisions, he says, "must be made on the basis of inadequate knowledge," and often lead to high human cost: physical deprivation and suffering. Human beings, Berger adds, "have the right to live in a meaningful world." That requires an assessment not only of costs but of a "calculus of meaning."

The Age of ISDN, I believe, will provide ever widening and meaningful citizen participation, along with the individual's prompt access to the ideas and agenda for social action as well as change. The diversity of information that will be readily available through ISDN should reduce the price individuals and nations must pay for modernity. For ISDN can provide the history of misdirected development and productivity, as well as the record of social and political failures in the name of egalitarianism or utopianism. ISDN also supports "smallness" in the development of communicating instruments. That is the philosophical as well as technological antithesis of growth as the standard. Most important, ISDN places the irreducible number *one*—the individual person—at the center of the communicating universe.

The intelligence of ISDN will facilitate human intelligence and its application to what Berger calls "intermediate structures"—forms of association that lie between the modern state and the mass of "uprooted individuals" found in all modern societies. Such structures, in Berger's terms, will "cut across the capitalist/socialist dichotomy."

* * *

ARTHUR POUND, QUOTED above, would likely regard sending a live instantaneous telecast from, say, the top of Mount Everest to a living room in Manhattan, as enhancing the possibility of peace. In May 1988 British mountain climbers did just that. They ascended Everest carrying with their dehydrated rations a small portable satellite communications earth station. An even more significant telecast, using a similar transportable earth station, came from Moscow during the Reagan-Gorbachev summit. More extraordinary than telecasting from Mount Everest was broadcasting live from Moscow, without the traditional mediation of Soviet censors or wary technicians. The possibilities of expanding international understanding, and perhaps relieving cross-border tensions, may be increased by such independent communications. These are "possibilities" hedged by a "perhaps." But these possibilities are worth probing.

The use of satellite transportables carried by a single journalist makes some governments nervous, says Gavin Trevitt, spokesman for Inmarsat (the International Maritime Satellite Organization). Some people see the flyaways, barely ten years old, as a threatening sign of the future. "It brings up the issue of national sovereignty," says Trevitt. The Soviet Union argued that way for years, yet it did permit a transportable to operate in 1988. That was a good omen—far better than Worldwatch's warning in 1983 that greatly expanded information capabilities in the U.S. and USSR tended to overwhelm those in command, and increase the temptation to launch a nuclear attack.[2] Too much cross-border information, said Worldwatch.

Yet quite the opposite—superpower peace for forty years—has resulted from the expansion of American information flows associated with transnational corporations. A society committed to free enterprise, at home and abroad, has concentrated for four decades on ways to stabilize the international military as well as economic system. Far from imposing a "naked and arbitrary power" on the world (C. Wright Mills's term), the U.S. "power elite" has facilitated global trade for transnationals of Scandinavia, the Netherlands, Japan, West Germany, and the burgeoning economies of several former Third World countries, as well as the United States itself.

The same economic and financial development opportunity was present for the Soviet Union and its allies, but their dogmatic ideology undermined even relatively healthy economies such as Czechoslovakia and East Germany, and reduced the bloc to the level of the sluggish Soviet economy. No small aspect of this condition, despite the rhetoric of the Cold War, was the decision by the Soviet Union and the United States to deploy their reconnaissance satellites. These twenty-four hour windows of surveillance enabled both countries to read even an adversary's newspaper from more than 100 miles in space. By that gentlemen's agreement, both countries evolved the important tradition of allowing adversarial satellites to pass over their own territories with impunity. Thus, for years, both sides have had a good understanding of the economic as well as the military capabilities of the other. This reconnaissance may have forecast a still more important decision in the Kremlin. By agreeing to allow uninhibited U.S. surveillance the Soviets, in effect, dropped a fundamental part of their ideology: They concluded that war was no longer the road to the global revolution they still seek. This decision, for the present at least, favored the information revolution. Without a public revelation, the Kremlin may even have allowed some fresh ideas to enter the obsolescent ideological debates over Marxism. Such debates, however, still have not filtered down through the public information systems.

The inevitable linkage of Soviet information systems to the worldwide ISDN presages the opening of all of Soviet society—not just the military elite—to massive information flows. That is not likely to happen soon. The full impact of ISDN will not be felt, even by presently free countries, for decades. And the present glasnost policy in the USSR is still a centrally regulated management tool, not a freeing of the society. But ISDN packs a real potential for extending inter-nation comity.

The 1988 presidential campaign, sadly, was an opportunity lost to the American people to gain a better understanding of the choices facing the United States on the world scene. There was no discussion of the bankruptcy of Soviet policies, or the coming political and financial creation of a West European Empire—and the implications of both for America. Nor was there even a glimmer of the information revolution that will further inundate a population already "informationalized" but not educated to many dilemmas, not the least the fundamental choices posed by ISDN. "A popular government without popular information or the means of acquiring it is but a prologue to a farce or tragedy or perhaps both," said James Madison.

It may be that our news media are merely reflections of the society they serve. "Freedom of the press," said Lord Windelshaw, "is a state of affairs as well as a state of mind." Press freedom is fundamental to political freedom, but the *quality* of information available to citizens enables a society to plan coherent, consistent policies to meet rapid changes in the world. The most vital policies demanding national attention will be the flow and quality of information itself.

No nation, large or small, free or not, will escape the Age of ISDN. Only the most oppressive ruler would want to keep his people off the information line. To deny one's constituency the maximum available information will not only retard the development of the nation as well as its citizens, but reduce the possibility for amicable relationships with other nations. The free exchange of information within a country encourages rule by rationality. And that makes military adventurism less likely.

It is not historic coincidence that the Soviet Union, in practicing a limited glasnost, finds its adventure in Afghanistan more difficult to sustain. The global response—particularly Third World reaction—suggests that widespread *information,* at home and abroad, about the Soviet invasion has added to an already heavy cost. If information systems of many kinds become pervasive in the USSR, many fears born of secrecy will not arise. Information, extensive and rapidly accessible, could then become a deterrent to rash action. It may be more effective, ultimately, than the nuclear deterrents now in place, and still needed. Perhaps after some years of normalization of information, domestically and internationally, in matters far removed from weaponry, Cold War suspicions may cease. This would be based primarily on broad areas of information exchanges in fields of general culture, natural sciences, business and education. Few exchanges so far—except at the high levels of the sciences—begin to meet these objectives. There is little real probing of the vast domestic terrain by Soviet citizens. Only after they have opened their own society to such observation will real cultural exchanges with America be meaningful. Then there may follow the public sharing and discussions of information on arms and defense policies beyond that needed for continuing, highly technical arms-reduction agreements by the U.S. and the USSR.

Some critics of ISDN are not as optimistic. They fear the homogenization of cultures through cross-border flows. Or they share Orwell's fear that interlocking technologies will further centralize information controls. While acting to democratize, we should keep in mind that engines of information have been used—and could be again—for totalitarian surveillance and control, and cultural domination. Yet the new technologies are inherently individual-oriented. They require the personal mastering of the tools, and the ability to conceptualize and improvise—hardly the domain of the censor. Just as vast computer systems have been shown susceptible to man-induced "viruses," so the normal functioning of the new technologies can place political as well as information systems on the defensive. It will be essential to build such democratizing elements into all the integrated systems coming along. They should, indeed, be subject to the proper influences of individuals and small groups.

The telecommunications sage, the late Arthur C. Clarke, made the optimistic point in 1983.[3] He saw journalists and transmitters becoming totally independent of national communications systems in ten years. "The implications of this are profound," said Clarke, "and not only to media news gatherers who will no longer be at the mercy of censors or inefficient (sometimes nonexistent) postal and telegraph services. It means the end of closed societies and ultimately...the unification of the world." Clarke rejected the notion that "many countries wouldn't let such subversive machines across their borders." He countered: "They could be committing economic suicide, because very soon they would get no tourists, and no businessmen offering foreign currency. They'd get only spies, who would have no difficulty concealing the powerful new tools of their ancient trade." He said the old debate about the free flow of information "will soon be settled—by engineers, not politicians. (Just as physicians, not generals, have now determined the nature of war.)"

That is not to suggest that journalists or data processors will get assigned roles as peace mongers. Western journalists have long rejected any such task assigned by national or international directives. But there may well be a surrendipitous peace dividend resulting from the great interaction and interdependence which information linkages provide.

It may be recalled that Unesco was created in 1946 not solely to promote knowledge as a discrete end, but also a means of advancing peace.

William Benton, the senior American delegate, joined with Leon Blum and Julian Huxley in insisting that more education, for peace or any other purpose, could only be achieved through "the modern instruments of mass education"—the press, radio and motion pictures.

The implications of the maximizing of information materials through ISDN are enormous, particularly for peace.

The widely diverse information channels will become available to average citizens because the computers and the data they deliver will be essential for the maintenance of an information-era society. Without such accessibility the economy will deteriorate, and with it social stability, and even national security.

Once the gates are opened to information networks, the flows are likely to move more freely both inside and between countries. Oppressive governments will not readily opt for a more open society. But even they will be reluctant to appear instantaneously on international television with pictures of a massacre underway. As Clarke warned, "It will be useless to shoot the cameraman. His pictures will already be safely seen 5,000 kms away, and his last image may hang you." To be sure, exposure of political abuse by visiting journalists can be irritating, but also valuable. "Many rulers might still be in power today, or even alive," Clarke reminds us, "had they known what was really happening in their own country."

Technological changes must be faced not only by governments which now monopolize or influence most of the information media within their borders. Private owners, particularly of newspapers and press chains, will also be prime movers in the vast networking to come. How can ISDN become a democratizing force if major links of the networks are to come under private monopoly control, and, particularly, if electronic systems in the U.S. and Europe are subject, as newspapers now are, to massive acquisition by a few proprietors?

International buy-outs of major newspapers, magazines, television networks, and cable systems are already under way. Most of the newspapers in the UK are now owned by people living outside the country. The most important newspaper, *The Times*, and the most popular paper, *The Sun*, are both owned by Rupert Murdoch, an Australian living in New York. He has also bought television stations in three countries, and is rapidly moving into satellite communications, films, books and the linking of all these media. Clearly, Murdoch understands the promise of the future, and is endeavoring to put his lock on a portion of it. So far, it cannot be said that his newspapers or other media take greatly different editorial positions than the same outlets showed under different owners. No significant internationalist viewpoint is demonstrated in these media. Indeed, all seem to feature domestic crises as saleable newsbreaks.

At the press chains in the U.S. similar freedom is generally given the local paper's editors. New magazines are coming on the scene as new technologies steadily increase the competition. As Murdoch acquired another television channel, he had to divest a newspaper in the same city. At the same time, a new cable system came on line to diversify that city's competitive offerings still further.

Such diversification may be expected as new networks are added to ISDN—myriad interactive media, as well as direct-service channels. The number and variety should increase the competition, rather than lead to monopolization. And if market forces are stymied by financial buy-outs, as a last resort government regulation should maintain the free flow of ideas.

The liberal democratic model of the press—print, broadcast or digital—has the same imperatives, whether applied to a single newspaper or broadcaster, or a chain of media mega-properties crossing national borders. The press is a *public* intelligence service. Whoever the owner, the principal receiver of information is the public. The public must be adequately informed of matters of relevance to citizens, and of the range of choices available to both public and officials. The press must check the abuse of power by any sector of the society—including the press itself. The press must also provide the exchange of ideas between the government and the citizenry.

These implicit tasks of the press are not assigned by government but by the social contract which a democratic society has with its citizens. By accepting citizenship in that society, all citizens assume some social responsibilities—again, not assigned by the political entity of government, but the overriding power of a free society itself. For the press, this means the responsibility to pursue truth as a self-righting process based on the competition between differing facts and opinions. Fairness and balance are required. Competing sides should receive the most effective projection of their respective views. Indeed, there should be special care to enable holders of unpopular views to gain a forthright hearing. A monopoly of information by any interest group should be strenuously resisted, and bias recognized for its deforming characteristics.

Social responsibility of the press also implies a positive imperative: the conveying of information which the public *should* have, whether or not at that moment there is a visible or an economic demand for such information. The categories of such "silent demand" may include unsensational news and information about real life in many developing countries, the growing concern about population growth and ecological crises, health hazards from foods and lifestyles, and issues of incipient domestic or international conflict or its resolution. Many editors respond to this imperative by saying their audience is simply not interested, and so they spike such news and opinions. Yet a good writer can produce a moving human interest story about reforestation today—often an unspectacular subject—rather than wait until floods kill 1,000 people in that place, and create a one-day horror report. Such coverage by "process" rather than "hard news" reporting may help prevent the deluge, and meet the journalist's implicit social responsibility.

The greatest protection for a free society—and ultimately for today's least-free nations which will one day come into the modern era—is diversity. The availability of different viewpoints, even differing reports of the same event, reflects reality. The Rashomon principle—different viewers perceiving the same message differently—is at work in every edition of the press and every televised report. Bias is not the principal charge; rather, the greatest dereliction is the failure to reflect diversity.

The Age of ISDN will provide countless opportunities to plug into "truths" of many kinds. The West Europeans are working hard now to decide what they will offer their twelve publics after 1992. Mikhail Gorbachev is planning to transform Soviet society in ways we do not really understand. The Japanese have set in place extraordinary information links, and are probably developing artificial intelligence faster than are American technologists. The Chinese are beginning to understand they must reorder their restrictive society to accommodate the information age.

But the United States, the grandfather of the information society, has no plan for the crucial time just ahead. This is an age of great discontinuity, especially for Americans used to leading and dominating. Hereafter, leadership—economic as well as political—must be consensual among nations, not overwhelming. We must be able to operate in systems and formats of ambiguity, and not become paralyzed. We must act, yet collaborate. Indeed, not only will American provincialism be dangerous, but conceiving of any U.S. economic policy as intrinsically "domestic," will hereafter be misleading and unrealistic.

Will the U.S. treat ISDN as another plaything—just video games on a vast scale? Or an opportunity to increase citizen literacy, both in the printed and electronic words? Should Shakespeare find his way regularly into U.S. interactive media? What about the daily availability of encyclopedic information, as well as music videos and sports?

The least-free nations, the majority in the world, will have to learn, with James Madison, that a freer press must be tolerated for the greater good. Even rulers of partly free nations, where "guided journalism" is actively practiced, will discover that "some degree of abuse is inseparable from the proper use of everything, and in no instance is this more true than in that of the press." Madison added, "It is better to leave a few of [the press's] noxious branches to their luxuriant growth than, by pruning them away, to injure the vigor of those yielding the proper fruits."

Epilogue

Madison concluded, "And can the wisdom of this policy be doubted by any who reflect that to the press alone, checquered as it is with abuses, the world is indebted for all the triumphs which have been gained by reason and humanity over error and oppression?" The United States was very much a "developing nation" when Madison uttered these words.

Such great press responsibility should evoke not journalistic arrogance, but a willingness to recognize one's own fallibility and publicly correct mistakes and unbalanced reportage. Charges of imbalance and other criticisms should not be turned aside, but welcomed as useful gauges of the press's own effectiveness and credibility. Journalists should not "whore the immunity which they demand and (in America) largely receive," says Herb Greer.[4]

Nor should the response to error or arrogance in the press be governmental controls, though the Reagan administration, like its predecessors, used executive powers to try to manage news and newspersons. The slippery slope of undemocratic control, while often apparent in such attempts, was not far traversed. Some new communications implements make surveillance and control easier. But other new tools offset and defeat their counterparts. And for every bureaucrat committed to censor, there are others, in and out of government or the press, ready to monitor abuses and reveal them.

Indeed, the First Amendment has withstood repeated challenges in recent years. Abuse by intelligence agencies of domestic political surveillance has been somewhat reduced. The Supreme Court has refused efforts to limit First Amendment protection to political questions. The Court still rejects the notion of prior restraint (censorship before publication) but may grant it temporarily while considering a particular case. While acknowledging the citizen's right to access of information, the Court has not clearly tackled this very difficult issue. It is precisely here that ISDN—networks of networks—can provide limitless choices and easy access to diverse information.

In the Age of ISDN, Americans can contribute immeasurably to personal progress and world peace. But the nature of general education and national politics will have to change. The media systems have already become the most powerful channels of mass education. It will no longer do for these media to emphasize violence in fiction and nonfiction. Such presentations either hold the audience to a fear-laden image of life, or encourage withdrawal from the world. The new media have the responsibility of projecting with no less conviction and intensity the realities of the environment, both ecological and political.

That responsibility does not require the creation of false expectations—of heroes or power where there are none, or villains or threats where they do not exist—but the projection instead of less frenzied use of pictures, language and music to excite the mind, not only the emotions, to focus on ideas, not merely images.

The traditional ignoring of history, and the failure of the mass media to place current events in deep perspective—all of this provides insufficient clarification of the *process* by which today's events can be explained, and tomorrow's actions contemplated. Edmund Burke wrote: Americans will have to learn the complex mathematics of the new politics. The uses of a free man's rights are, he said, "in balance between differences of good, in compromises sometimes between differences of good and evil, and sometimes between evil and evil. Political reason is a computing principle: adding, subtracting, multiplying, and dividing, morally and not metaphysically, or mathematically, true moral denominations."

Today's new worldview requires a far greater use of reason, than simply of bald emotion, of rational assessment of realities, not propagandistic responses to polemics. Even if simplistic tactics and arguments were believed necessary to win office, they can no longer suffice once the buck stops at the new president's desk. For too long, domestic politics have governed or overly influenced U.S. policies in much of the world. That can no longer be afforded. With ISDN, the world finally becomes that global village we have been told to expect. In such a time, Americans may come to understand how broad U.S. interests, world peace and stability can be subverted by ill-conceived domestic U.S. policies that afflict volatile regions abroad. Not only is a "kinder, gentler America" needed, but one which can rationally frame creative, long-term goals and with consistency realize them.

The new communications networks, consequently, can become a major debunker of the "know nothing" strain in American politics. The mass media, and the ever more diverse information systems have the obligation, then, to help educate an America whose policies should match the complexities and opportunities of the new age—the post-Communist era, the Age of ISDN.

The educating and debunking will not come automatically from the powerful new communications systems. They are facilitators of the information flow, and should not become controllers of policy or destiny. They will provide options, but humans must make the choices. Even when machines choose, they will only be following programs created by humans. Technology can provide information about more and more, but unless human wisdom is applied the increasingly complex problems of the world will remain little altered. The new communications technologies are not panaceas. They are, rather, infinitely faster ways to use the creativity and imagination of the human mind; no less, no more. Thus employed, the Age of ISDN will be a great boon.

Yet, communications technology cannot guarantee that a system will be accessible to all, or convey ideas adequately diversified. Communications tech can also run amok, as it did in the wiretaps at Watergate. Communications tech can also correct improvident use of technology, as did the subsequent television hearings of that escapade. In the decades ahead, there will be many opportunities to misuse the new technology by the application of political power. It is essential, therefore, to begin now to raise to public attention and debate the formulation of a national policy for the Age of ISDN. This is needed to protect the U.S. citizen as the nation enlarges the freest, most information-oriented system anywhere, and to assure that America is technologically, politically and culturally linked not only to its Western friends and the East bloc, but to the billions of individuals in developing countries who clamor for two-way communication.

To do so, we should recognize that all the communications technologies, new or old, are merely facilitators of human intelligence. The information systems should be no less diversified than the universal family. Every member has a special way of observing the present and imagining the future. The interplay of such views and imaginings is the distinct human attribute. The coming Age of ISDN will facilitate the diversified exchange of human intelligence, focusing the mind outward from neighborhood and nation to the horizon beyond.

Moral
"Newsgathering Barred"—becomes news

The New York Times

CHINA HARD-LINERS SEND TROOPS TO BEIJING; PARTY CHIEF IS OUT, NEWSGATHERING BARRED

ON A SUNNY day in mid-May 1989 in Tiananmen Square, Beijing, 1,000,000 Chinese challenged the Communist regime, suggesting for the first time in forty years what a (democratic) People's Republic might be.

Simultaneously, great prodemocracy demonstrations massed in Shanghai and scores of other cities across the vast land of China. Overseas Chinese supported the democracy movement in Hong Kong, San Francisco and New York.

The movement for several years had been demanding an end to corruption, and especially the right to freer expression and a freer press. In the veiled and sophisticated language of their politics, Chinese at every level debated the central role of Chinese journalism in the democratization of their country. Spurts of freedom were permitted journalists, followed by regressions. When the students took Tiananmen Square, however, broadcasters across China picked up the account, as did international television, its eye focused "live" on the historic scene.

In the tradition of censors everywhere, the frightened authorities ordered state broadcasters to prevaricate, and pulled the plug on foreign television satellites. That merely signified the government's own fears. More than that, the China Spring demonstrated the massive use of information power to assemble resisters and challenge state power. That nonviolent act, whatever the short-term consequences, is of historic proportions.

The students of Tiananmen Square are among the pioneers of information power in the coming Age of ISDN—the networking of diverse information networks. At play in China, as in Panama, Nicaragua, the Philippines, Poland, Hungary and elsewhere, were fax machines, copiers and cellular telephones. They were used by the new revolutionaries and by foreign and domestic journalists. Such diversified flows of news and information will speed democratization—everywhere.

Appendices

Appendix A
Journalism Morbidity Table—1988

These statistics are inclusive through 31 December. These record only the physical and psychological harassment of journalists and their media. The figures do not reflect other forms of official and unofficial editorial censorship, and diverse methods of economic and political pressuring of the mass media. The statistics are a clue, however, to those official actions which generate self-censorship by journalists.

	1988	1987	1986	1985	1984	1983	1982
Journalists killed	38[a]	32[b]	19	31[c]	21	14	9
Kidnapped, Disappeared	14	10	13	13	5	4	11
Arrested, Detained	225	188	178	109	72	80	145
Expelled	24	51	40	9	22	19	23

Other 1988 Statistics

Journalists wounded: 28 in 7 countries
Journalists beaten: 40 in 6 countries
Journalists otherwise assaulted: 50 in 12 countries
Death threats and other threats to journalists: 43 in 9 countries
Journalists' homes raided: 11 in 6 countries
Journalists' homes destroyed: 1 in 1 country
Charges filed against journalists: 48 in 6 countries
Films or manuscripts confiscated: 82 cases in 13 countries
Press credentials withdrawn: 7 in 2 countries
Journalists refused credentials: 7 in 4 countries
Journalists harassed: 46 in 10 countries
Passport withheld or expulsion threatened: 2 in 2 countries
Closed newspapers or radio stations: 40 in 12 countries
Banned publications or radio programs: 31 in 10 countries
Bombed or burned newspapers or radios: 8 in 8 countries
Occupied newspapers or radios: 7 in 4 countries
Radio station destroyed: 1 in 1 country

These harassments totalled 465 cases in 70 countries in 1988. The year before there were 436 cases in 57 countries. This year, as before, the figures greatly underestimate both the number of cases and the individuals involved. Some single cases here involving the closing of media facilities affect scores of journalists. Many cases are not reported, though journalists are increasingly aware that maltreatment of the messenger by governments and others is aimed primarily at all citizens. The fate of journalists, therefore, should be considered of interest and importance to everyone.

---

a. Killed: In Afghanistan, 5; Algeria, 1; Brazil, 2; Camaroon, 1; Chad, 1; Colombia, 4; Ethiopia, 1; Greece, 1; Guatemala, 1; Honduras, 1; India, 4; Iran, 1; Mexico, 4; Pakistan, 1; Peru, 3; Philippines, 3; Soviet Union, 1; Thailand, 1; Turkey, 1; Vietnam, 1;
b. Corrected from last year's table with later information.
c. In 1985, 16 of 31 murdered journalists were killed in the Philippines.

Appendix B
National News Agencies
*Independent of Govt. Control/Ownership

Africa
Angola	ANGOP
Benin	ABP
Burundi	ABP
Cameroon	CAMNEWS
Cen. African Rep.	ACAP
Chad	ATP
Congo	ACI
Ethiopia	ENA
Gabon	AGP
Ghana	GNA
Guinea-Bissau	ANG
Ivory Coast	AIP
Kenya	KNA
Liberia	LINA
Madagascar	ANTA
Malawi	MANA
Mali	AMAP
Mozambique	AIM
Nigeria	NAN
Rwanda	ARP
Senegal	APS
Seychelles	SAP
Sierra Leone	S.L. News Agency
Somalia	SONNA
Tanzania	SHIHATA
Togo	ATOP
Uganda	UNA
Zaire	AZAP
Zambia	ZANA
Zimbabwe	ZIANA

Middle East
Algeria	APS
Democratic Yemen	ANA
Egypt	MENA
Iraq	INA
Israel	*ITIM
Jordan	JNA
Kuwait	KUNA
Lebanon	NNA
Libya	JANA
Mauritania	WAMS
Morocco	MAP
Qatar	QNA
Saudi Arabia	SPA
Sudan	SUNA
Syria	SANA
Tunisia	TAP
United Arab Emirates	WAM
Yemen	SABAA

Asia
Afghanistan	BIA
Australia	*AAP
Bangladesh	*APIS, *BSS
China	XINHUA China News
India	*UNI, *PTI
Indonesia	ANTARA
Iran	IRNA
Japan	*KYODO
Korea (S.)	*YONHAP
Korea (N.)	KCNA
Laos	KPL
Malaysia	BERNAMA
New Zealand	*NZPA
Pakistan	*APP, *PPI
Philippines	PNA
Thailand	TNA

Europe
Albania	ATA
Austria	*APA
Belgium	*BELGA
Bulgaria	BTA
Cyprus	CNA
Czechoslovakia	CTK
Denmark	*RB
Fed. Rep. of Germany	*DPA
Finland	*STT-FNB
France	*AFP
German Democratic Rep.	ADN
Greece	*ANA
Hungary	Magyar Tavirati Iroda
Italy	*ANSA
Netherlands	*ANP
Norway	*NTB
Poland	PAP
Portugal	*NP
Romania	AGERPRES
Spain	*EFE
Sweden	*TT
Switzerland	*ATS
Turkey	AA
United Kingdom	*PA, *REUTERS
USSR	TASS
Yugoslavia	TANJUG

Latin America
Argentina	*TELAM SIA y P
Bolivia	*ANF
Brazil	*AE, *EBN
Ecuador	ECUAPRESS, *ED
Nicaragua	ANN

North America
Canada	*CP
United States	*AP, *UPI

International agencies
GNA	(Gulf News Agency), Bahrain
IINA	(International Islamic News Agency), Saudi Arabia
*IPS	(Inter Press Service), Italy
OPECNA	(OPEC News Agency), Austria
PANA	(Pan-African News Agency), Senegal

* * *

90 countries with national news agencies. Of these 28 are independent, 62 are government-owned.

Listing is based on information provided by Unesco, current to 1986.

Appendix C

Letter of Violeta Barrios de Chamorro

Señor Presidente de la República
Comandante Daniel Ortega
Managua, 7 June 1988

I BRING BEFORE you my strongest protest against the injurious campaign mounted against me by the Sandinista Television System—the informative medium of your state Party, which has reached vulgar and dangerous extremes never before seen in our nation.

The most recent campaign of this propagandistic arm of Sandinismo, also reproduced by your Party newspapers, *Barricada* and *El Nuevo Diario*, publicly shows the sacred remains of my husband, ridden with gunshot wounds, next to photographs or videos of myself, taken out of context, and which show me as appearing grateful for my husband's assassination.

Señor Ortega: When you came to my house [on 19 September 1987], pressured by the circumstances of the moment, to offer me a shady agreement for the reopening of *La Prensa*, not only did I refuse it but with Don Rodrigo Madrigal Nieto, the foreign minister of Costa Rica, as a witness, I clearly told you that *La Prensa* would either be published without censorship or it would not be published at all. In addition, I refused to accept as a form of indirect pressure "the responsible exercise of journalism" —a phrase that your negotiators wanted to impose on me in order to insinuate that a prearranged agreement had been devised for the reopening of *La Prensa*.

You know very well that *La Prensa*, in its new stage of life, survived a brutal closing for fourteen months, arbitrarily imposed by your government, and you also know that it will continue to develop an undeviating journalism in behalf of the people's interests, without ignoring the many vices that distinguish your government's actions.

If you, in spite of this, permitted *La Prensa*'s reopening as a result of Esquipulas II [signed in Guatemala 7 August 1987] and not from any good will, then you should know that you will have to tolerate responsible criticism of your actions and those of your functionaries—as is the case in every democratic country—a state of democracy that was the goal of the Guatemala accord that you signed guaranteeing compliance with its articles.

But the vacillations and duplicities of your government in regard to this compliance compel us to speak the truth: your government has demonstrated that it does not want to honor the accord you signed, and it is my obligation and *La Prensa'*s to reveal this to both the Nicaraguan people and the civilized world.

I would also point out that it has been your government, and not *La Prensa*, that negotiated with the Nicaraguan Resistance in spite of having frequently insisted that it would never do so. And since among the delegates of the Nicaraguan Resistance, Comandante Enrique Bermúdez, former colonel in the National Guard, has also arrived to negotiate— it is exclusively your government that has accepted him as a legitimate interlocutor.

If these negotiations are not favorable for the political interests of your state Party, the blame will belong to your government—not to me nor *La Prensa*, which limits itself only to tell the truth. In this case, the truth, whether you like it or not, consists in showing that your government, in addition to ruining the country, has dedicated itself to the liquidation of all rebellion, without correspondingly restoring any of the basic human rights of the Nicaraguan people.

Your government, besieged by disaster and inefficiency and by the incompetence of its functionaries, is obligated to comply with Esquipulas II, and you, your functionaries, and state employees all know it.

All Nicaragua knows that you and your government have ruined the country and that, consequently, compliance with the peace accord can no longer be delayed—an accord that requires the restoration of the Nicaraguan people's human rights and their liberties, long confiscated.

Even the Soviet Union and its satellites have grown weary of this tremendous disaster, and it is widely known that they are no longer committed to your political and military survival.

In regard to all the above, neither *La Prensa* nor myself is to blame, much less the memory of my husband Pedro Joaquín Chamorro—a memory that should be, above all, sacred to you, the Sandinistas, beneficiaries of his death.

The exhibition of my husband's body—destroyed by assassin bullets whose origin I am uncertain of even today—transmitted and publicized through the infamous media of your state Party, together with images of myself suggesting my satisfaction with his death, constitutes a cowardly and repugnant act that can only be explained by the miserable morality that distinguishes your government.

It is one thing to attempt to hide from the people and your own Party members the weakness of you and your government—a weakness that might even lead to your compliance with Esquipulas II—forced by the terrible reality that we are living through today. But it is something else to attempt to camouflage this reality, exposing through the state media the desecrated body of my husband—a man declared "Martyr of Public Liberties" by your own law—and insulting and slandering his widow, who as both a widow and woman deserves a minimum degree of respect. In any country this would be called villainy, infamy, and ignominy: all characteristics of your immoral government—a government that I helped and supported when I was carried away with the strong emotions that characterize a woman, actions which I repent and abhor as I do my sins.

As a Nicaraguan I work for peace. As a woman I unite myself with the widows and mothers of Nicaragua who suffer so much because of your government. And as director of *La Prensa*, I promise you that I will continue denouncing and combating the acts of your government, until you and your government silence me with the brutal club or the assassin's bullet.

Power, the Press and the Technology of Freedom

Appendix D
Categories of Press Freedom
Degree of Change in 1987

Most Free

Great Improvement	Some Improvement	No Change	Some Deterioration	Great Deterioration
Dominican Rep.	Belgium	Antigua	India	
Malta	Bolivia	Austria		
Portugal	Finland	Barbados		
	Granada	Canada		
	Iceland	Colombia		
	Jamaica	Costa Rica		
	Korea (So.)	Denmark		
	Nigeria	Dominica		
	Trinidad	France		
		Germany (Fed. Rep.)		
		Italy		
		Japan		
		Kiribati		
		Luxumbourg		
		Nauru		
		Norway		
		Papua New Guinea		
		Peru		
		Philippines		
		St. Kitts-Nevis		
		St. Lucia		
		St. Vincent		
		Solomon Islands		
		Spain		
		Turkey		
		Western Samoa		

Least Free

Guatemala	Albania	Afghanistan	Paraguay	Bangladesh
	China	Cameroon	Tanzania	Fiji
	Czechoslovakia	Chile		Indonesia
	Egypt	Cuba		Malaysia
	Guyana	Ghana		Panama
	Kenya	Jordan		South Africa
	Liberia	Senegal		Sri Lanka
	Nepal	Togo		
	Nicaragua	Zimbabwe		
	Poland			
	Soviet Union			
	Sudan			
	Taiwan			
	Uganda			
	Vanuatu			
	Zaire			

Appendix E

The M'Bow Correspondence

This author engaged Unesco's Director-General Amadou Mahtar M'Bow in correspondence from 4 August to 31 October 1984. That exchange produced a response which this writer regarded as a change of basic policy although one Unesco official maintained it is only a "clarification." The reader may judge. In any event, the correspondence reveals the sensitive nature of Unesco's communications controversies.

Mr. M'Bow had invited me to participate at Unesco, Paris, 16-20 July, in the consultation group organized to advise him on reforms in public information activities. The group of nineteen was chaired by Mohamed Heikal, former minister of information of Egypt. From the outset, Mr. Heikal and I agreed that the main purpose of the consultation should be to discuss those basic matters that influence the global perception of Unesco—the organization's programs and debates. The majority at the consultation, however, felt we should devote the week to recommending improvements in Unesco's public information office. The final report included an appendix expressing my belief that member-states should demonstrate restraint in proposing communications programs, if Unesco is not to suffer divisive, harmful reactions.

On 4 August 1984, I met privately in New York City with Mr. M'Bow to express my concern over the specific issue of the protection of journalists.

My 8 August letter to Mr. M'Bow stated:

It is unfortunate, I believe, that the understandable desire of many developing countries to participate in the Information Revolution has been termed a demand for some still undefined New World Information and Communications Order. The yearning is valid and understandable; [Yet] the term NWICO is regarded as threatening statist control of news and information media. I also know some member-states do indeed seek that objective; most do not. And such an objective has never been approved at Unesco.

Yet some debates and programs at Unesco have lent credence to such Western fears. This despite the fact that substantial steps were taken at the last General Conference to dispel such fears.

Now, however, a disturbing sign has appeared.

I learned in Paris that a meeting was held in Geneva last month to plan a seminar in Mexico next March that will discuss, among other matters, the protection of journalists. That and other related subjects also on the agenda seem to repeat the pattern of the acrimonious meeting at Unesco in February, 1981. Many of the same nongovernmental organizations are sponsoring this...And the use of identification cards for journalists—a form of licensing—already is being discussed as part of the new as the past meeting.

I regard the proposed Unesco co-sponsorship of the meeting in Mexico and elsewhere on the subject of "protection of journalists," among other matters, as certain to end any chance of the United States remaining inside Unesco for years to come. I know I need not spell out the reasons, and I also know the organizational explanations for having convened the recent meeting in Geneva to plan this meeting in Mexico. All of that notwithstanding, I have absolutely no doubt that the mere planning of such a meeting with Unesco participation will provide a completely predictable negative reaction in the United States, the United Kingdom, and perhaps elsewhere. In that renewed combatative climate, all the earnest efforts to improve the functioning of the organization will be submerged in a new emotionalized attack that will easily carry the day.

You have asked me to provide my best judgment on matters of public information. I said repeatedly at the consultation that no organization, certainly not Unesco, can separate the content of its programs and debates from the public image it projects. Wholly deserved or not—I believe by now you know my balanced view on this matter—Unesco is perceived in the West as using the "protection" issue to advance press controls. No Western government can ignore that issue when perceived as a challenge to basic rights. And at this moment, such a challenge would be regarded as a distinct rejection of Western concerns.

Unesco will not avoid attack because another international agency may claim to run the "protection" segment of a trilateral meeting. Unesco will be charged with major responsibility for the entire meeting, and this will be seen as continuing an objectionable pattern that helped bring Unesco to its present crisis.

Director-General M'Bow replied 29 September:
First let me re-state my position and of course that of Unesco, which has been and will continue to be that questions of the working conditions of journalists, including the protection of journalists, are of direct concern to the members of that profession, and it is not for us but for the journalists themselves and their nongovernmental professional organizations to discuss and deliberate on them. Intergovernmental Organizations such as Unesco can only be involved in such deliberations at the request of the professional organizations concerned.

The meeting in question will eventually be convened in 1985 by FELAP.

At present, to my knowledge, this has been supported by the following organizations:
- International Federation of Journalists (IFJ);
- International Organization of Journalists (IOJ);
- International Catholic Journalists Union (ICJU);
- Union of African Journalists (UAJ).

The interorganization meeting which was held at the ILO headquarters in Geneva in June 1984, requested that ILO, International Committee of the Red Cross and Unesco prepare the following information documents:
a) ILO: Selected problems related to employment
b) CICR: Protection and safety of journalists
c) Unesco: Implementations of new technologies.

This meeting also expressed the wish that Unesco, the International Labour Organization and the International Committee of the Red Cross co-sponsor the proposed 1985 conference. However, no formal request to this effect has been made.

Let me assure you that I am already giving the question you raised in your letter my personal attention and that I shall take the concerns you have expressed into consideration when it comes to any action relating to Unesco's involvement or participation in the proposed meeting.

I did not consider that reply adequate. The text of the letter was read to me from Paris on 2 October and telefaxed to New York a few hours later. That same day, my response to Mr. M'Bow by telefax was as follows:

You listed four organizations of journalists which support the meeting scheduled for 1985. These four NGOs do not represent the entire field of international organizations of journalists. At a meeting in Washington, September 28-29, 1984, for example, representatives of journalists organizations from several continents discussed with apprehension the proposed conference in Mexico City.

Two responses are already available: The magazine *Editor and Publisher* reflecting thousands of newspapers in the United States, advised in the headline over its only editorial this week, "Don't Relax on Unesco." The editorial was devoted entirely to the Mexico City conference in which "protection" would be discussed. The editorial warned that this meeting shows that "licensing of journalists, an imposed code of conduct, rules for the content of news...are all still on the Unesco agenda."

Similarly, the American Newspaper Publishers Association passed a special resolution, September 19, saying it is "alarmed at Unesco's participation in and support for a new meeting now scheduled in Mexico City on 'working conditions and security of journalists.'" The ANPA stated it "would welcome signs (Unesco) will move from such controversial and confrontational activities, to constructive programs supporting Unesco's original commitment to a free flow of information."

I deeply regret this compounding of Unesco's problems vis-a-vis the news media in many countries at the very moment when substantive changes in administrative procedures and programs are being debated by the Executive Board. My letter of August 8 was written precisely to help avoid this development.

Even now—or should I say, particularly now?—I believe a salutary step should have been taken.

You have restated your and Unesco's position: intergovernmental organizations can be involved in such deliberations as the working conditions of journalists only at the request of the professional organizations concerned. It follows, then, that you cannot be involved in such deliberations if some professional organizations advise you not to become involved, even while others take the contrary position. Since there is no unanimity among professional organizations on this subject, Unesco should leave it entirely to the professional organizations themselves to discuss and possibly resolve matters of working conditions of journalists.

You know my position on communications issues generally. I believe that Unesco has the responsibility to examine many aspects of the communications revolution that affect every country and every citizen. It is particularly appropriate for Unesco to help developing countries expand their communications capabilities, and their access to information in many fields, and to international dialogues in general.

I sincerely hope you will act promptly to make it clear that Unesco will neither be involved nor participate in the proposed meeting in Mexico.

My 2 October reply was available in Paris on the 3rd, when Mr. M'Bow addressed the Executive Board on this subject. He moved a bit closer to my recommendation that he make a clear-cut denial of participation in the forthcoming "protection" conference. But he allowed an ambiguous loophole: "we have so far received no official request" for Unesco participation. His remarks were greeted with a mixed response in Paris.

Mr. M'Bow's letter to me of 20 October was a far clearer commitment to distance Unesco from even the discussions of "protection" (which, in the past, have been linked to licensing, monitoring and penalizing of journalists).

Mr. M'Bow wrote:

On 5 October 1984, I made a statement at the present session of the Executive Board (document 120 EX/INF. 6 prov.) where I referred, among other things, to this matter, saying that:

"The programme and budget approved for 1984-1985 by the General Conference does not foresee the organization of such an activity and therefore contains no corresponding budgetary provision. Several leading journalists' unions announced last July, however, the convening in Mexico City in 1985 of a world conference on the working conditions and security protection of journalists, in view of the considerable danger to which pressmen are exposed in the course of their work. Three of the associations concerned—the International Federation of Journalists (IJF), the International Organization of Journalists (IOJ), and the Federation of Latin-American Journalists (FELAP), met last week to prepare this world conference. Unesco did not attend the meeting.

"It is true that the international and regional journalists' associations concerned have indicated their desire that the world conference be placed under the joint auspices of Unesco, the International Labour Organization (ILO) and the International Committee of the Red Cross (ICRC), but we have so far received no official request.

"I should like to take this opportunity to recall that Unesco's attitude so far in this field has been that questions concerning the working conditions of journalists should be settled by the journalists themselves. The Organization confines itself to providing support, to the full extent that its resources allow, and as is customary, to professional organizations which have submitted a request through the usual channels. I personally hope that, on a matter which is essential to journalists' free exercise of their profession and their physical safety, the professional associations will come to an agreement among themselves."

You stated in your letter that, since there is no unanimity among the professional organizations on the subject of the proposed conference, Unesco should leave it entirely to the professional organizations themselves to discuss and possibly resolve the matters involved. Indeed, this is precisely the position taken by Unesco.

I should however point out that we have been receiving requests from several nongovernmental organizations inviting us to intervene in one way or another. Some requested us to intervene in specific cases (US Committee to Protect Journalists), others wanted Unesco to co-sponsor with them meetings on the subject (International Press Institute), a third party requested Unesco to finance studies (International Federation of Journalists), and yet another solicitied our participation in and contribution to meetings on the subject (Dutch Association of Journalists).

Furthermore, I have personally been approached on many occasions to contact a number of governments, so that journalists could perform their duties in better conditions. As you know, these contacts have always been carried out with complete discretion and without public announcement.

At present, Unesco finds itself in the middle of a controversy because of a decision taken by a number of professional organizations to convene a conference on the subject and because they expressed their wish that Unesco, together with another organization of the United Nations family (ILO) and the International Committee of the Red Cross, should cosponsor this conference. I have been informed by my collaborators that when the subject came up at a meeting at the ILO office in Geneva, the Unesco delegation made it quite clear that all parties concerned with the working conditions of journalists should participate in any undertaking in this field. Publishers, broadcasters and news agencies were mentioned in particular. Following this meeting, this was repeated in writing in letters sent to the Secretaries-General of FELAP, IFJ and IOJ, the members of the preparatory commission of the conference. It was also the Unesco delegation to the Geneva meeting which proposed that before a conference takes place, the commission would do well to hold one of its meetings in the United States so that contacts could more easily be established with organizations there.

You will thus realize that any decision with regard to the preparation of the conference, let alone the items on its agenda, rests in the hands of the professional organizations concerned.

On this occasion, I would like to express my concern over the alarming conditions under which journalists work in many parts of the world. I received with great distress the report of a recent meeting of the Association of Foreign Correspondents in Mexico, chaired by the CBC correspondent, which disclosed that 154 journalists were killed and 52 disappeared in Latin America alone in the course of the last ten years. Last week, during a television interview I was asked what Unesco could do to help in the movement to free a French journalist at present prisoner in Afghanistan.

Any suggestion you may have on a possible Unesco action in this connection would be appreciated. As for the proposed conference in Mexico, the views I have expressed in this letter will be made known to all the orgaizations concerned; unions of journalists, employer organizations, as well as other international organizations concerned with communication problems.

This statement breaks new ground in these ways:

1. It accepts the premise that in the absence of "unanimity" among professional organizations Unesco should leave the matter "entirely" to the journalists themselves to discuss and possibly resolve.

2. It properly points out that Western journalists continue to ask Unesco to aid journalists in trouble, and Mr. M'Bow has quietly sought to do so. Such efforts have brought neither him nor Unesco the public acknowledgement of the news media.

3. Unesco did not attend the second planning meeting for the proposed "protection" conference.

4. Mr. M'Bow, in effect, suggested that American and other Western news media have virtual veto power over the content of future international communications conferences that Unesco would finance and participate in. He urged that a planning meeting (for the Mexico conference) be held in the United States with the participation of "publishers, broadcasters, and news agency representatives." These would be Westerners, and not likely to approve an agenda or content for any conference they find objectionable. That would provide a de facto veto for Unesco's support and participation in the conference, if it is subsequently held without Western approval.

In the legalistics of Unesco language and protocol, this represents a significant sign of progress—not only on the limited, though vital, "protection" issue, but for all the related questions in the communications sector.

The controversial "protection" meeting was not held as scheduled. But this would not influence the U.S. decision to withdraw from Unesco.

QUESTIONNAIRE

On Impediments to the Free Flow of Information in the News Media of the World in 1987

I. IN THE FIELD OF JOURNALISM AND THE NEWS MEDIA IN YOU COUNTRY, WAS 1987 A YEAR OF
(Please check one box only)

1-great improvement [ ]

2-some improvement [ ]

3-no change [ ]

4-some deterioration [ ]

5-great deterioration [ ]

II. OWNERSHIP OF THE MEDIA IN YOUR COUNTRY
(Please place one number on each line [A to E], a total of 20 numbers)

	A Number of Newspapers	B Number of Magazines	C Number of Radio Sta.	D Number of TV Channels	E Number of Nat'l. News Agencies
1-State-Owned	___	___	___	___	___
2-Privately-Owned	___	___	___	___	___
3-Shared---Public and Private	___	___	___	___	___
4-Other Forms of Ownership	___	___	___	___	___

III. CHECK THE KIND AND DEGREE OF INFLUENCE OVER THE EDITORIAL CONTENT OF THE MEDIA
(Please place a check in one of three columns under A to D, on all the lines [1 to 6]; a total of 24 checks)

Degree of Influence

Kinds of Influence	A-Newspapers Strong Some None	B-Magazines Strong Some None	C-Radio Stations Strong Some None	D-TV Channels Strong Some None
1-Political Influences	__ __ __	__ __ __	__ __ __	__ __ __
2-Economic Influences	__ __ __	__ __ __	__ __ __	__ __ __
3-Social/Racial Influences	__ __ __	__ __ __	__ __ __	__ __ __
4-Commercial Influences	__ __ __	__ __ __	__ __ __	__ __ __
5-Ideological Influences	__ __ __	__ __ __	__ __ __	__ __ __
6-Judgments by Editors/Journalists	__ __ __	__ __ __	__ __ __	__ __ __

Appendices

IV. CHECK THE DEGREE OF FORMAL GOVERNMENTAL INFLUENCES ON THE MEDIA
(Please place one check in one of three columns under A to D, on all the lines [1 to 8], a total of 32 checks)

Degree of Influence

Kinds of Influence	A-Newspapers Strong Some None	B-Magazines Strong Some None	C-Radio Stations Strong Some None	D-TV Channels Strong Some None
1-Information Ministry				
2-The Military				
3-Government Press Club				
4-Newsprint Controls				
5-Price Setting				
6-Licensing of Media				
7-Licensing of Journalists				
8-Restrictive Legislation				

V. CHECK THE DEGREE OF INFORMAL GOVERNMENTAL INFLUENCES ON THE MEDIA
(Please place a check in one of three columns under A to D on all lines [1 to 4]; a total of 16 checks)

Degree of Influence

Kinds of Influence	A-Newspapers Strong Some None	B-Magazines Strong Some None	C-Radio Stations Strong Some None	D-TV Channels Strong Some None
1-Favor Friendly Journalists				
2-Penalize Critical Journalists				
3-Threats to Harm Journalists				
4-Threats to Deprive Journalists of Employment				

VI. CHECK THE DEGREE OF NEWS MANAGEMENT ACTUALLY PRACTICED BY THE GOVERNMENT
(Please place a check in one of the columns under A to D on all lines [1 to 6]; a total of 24 checks)

Degree of Management

Kinds of Management	A-Newspapers Strong Some None	B-Magazines Strong Some None	C-Radio Stations Strong Some None	D-TV Channels Strong Some None
1-Limiting Access to Government Information				
2-Selective Leaking of Info.				
3-Dissemination of False Info.				
4-Dissemination of Favorable Information				
5-Denunciation of Journalists or Media by Officials				
6-Restrictions on Size of Print or Broadcast Audience				

3

VII. CHECK THE DEGREE TO WHICH JOURNALISTIC CONTENT GENERALLY---REPORTS OR COMMENTARIES---HAVE BEEN ACTUALLY INFLUENCED BY THESE PHENOMENA
(Please place a check in one of three columns under A to D on all lines [1 to 5]; a total of 20 checks)

Effective Influence

Effective Event	A-Newspapers Strong Some None	B-Magazines Strong Some None	C-Radio Stations Strong Some None	D-TV Channel Strong Some None
1-Murder of Journalists				
2-Imprisonment of Journalists				
3-Harrassing of Journalists				
4-Banning or Closing of Media				
5-Confiscation of Product or Equipment				

VIII. CHECK THE FORMS OF CONTROL OVER INTERNATIONAL NEWS OFTEN PRACTICED BY GOVERNMENT

1-Jamming of Incoming Broadcasts [ ]

2-Barring or limiting of Foreign Publications to General Public [ ]

3-Channeling Incoming Foreign News Through a Government News Agency [ ]

4-Channeling Outgoing Domestic News Through a Government News Agency [ ]

5-Restricting the Movement and Access to Information by Foreign Journalists [ ]

6-Expelling foreign journalists [ ]

IX. CHECK THE DEGREE TO WHICH MEDIA CONTENT ACTUALLY IS INFLUENCED BY
(Please place a check in one of three columns under A to D, on all lines [1 to 8], a total of 32 checks)

Degree of Influence

Kind of Influence	A-Newspapers Strong Some None	B-Magazines Strong Some None	C-Radio Stations Strong Some None	D-TV Channels Strong Some None
1-Owner's Order (if privately owned)				
2-Pressure on Media by Special Interests				
3-Advertisers' Interests				
4-Mergers or Enlarging of Networks or Press Chains				
5-Pressure from Competitive Media				
6-Imposing Security Restrictions				
7-National Development Needs				
8-Self-restraint by Journalists				

Appendices

X. IN 1987, THROUGH OCTOBER 31, HOW MANY INSTANCES OF THE FOLLOWING HAZARDS HAVE BEEN RECORDED IN YOUR COUNTRY?
(Place a number in each box in which this was an incident)

1-Journalists Killed [ ]

2-Journalists Shot [ ]

3-Journalists Beaten; Clubbed [ ]

4-Journalists Kidnapped; "Disappeared" [ ]

5-Journalists Arrested [ ]

6-Death & Other Threats to Journalists [ ]

7-Journalists Harrassed [ ]

8-Banned Newspapers or Radio Stations [ ]

9-Bombed, Fired Media Plants [ ]

10-Journalists Charged, But Not Jailed [ ]

11-Journalists Expelled [ ]

12-Journalists Barred Entry to Country [ ]

13-Media Plants Raided; Equipment Impounded [ ]

14-Publications Taken Over [ ]

15-Publication License Removed [ ]; Removal Threatened [ ]

(This questionnaire is copyright by Freedom House, New York © 1987. No part or the whole may be reproduced except by the copyright holder.)

Finally, using the above questionnaire as an outline, please write 500-1,500 words on

(1) the normative role of the news media in your country and the degree of credibility of the media in the eyes of the public;

(2) the degree to which the flow of news and information is subject to impediments from outside and inside the media; and

(3) your estimate of how journalists and the news media in your country fared in 1987, as compared to 1986 and/or years past. If there was a major improvement or deterioration in the news flow in 1987 please describe the cause(s) and effects.

Power, the Press and the Technology of Freedom

Appendix G
News Media Control by Countries

	Least free	Most free		Least free	Most free		Least free	Most free
s-	Afghanistan		o	Gambia		-	Oman	
s-	Albania		-	Germany (E)		+	Pakistan	
c-	Algeria		s+	Germany (W)		s-	Panama	
-	Angola		s-	Ghana				
s+	Antigua & Barbuda		c+	Greece		s+	Papua New Guinea	
c+	Argentina		s+	Grenada		s-	Paraguay	
+	Australia		so	Guatemala		s+	Peru	
s+	Austria		-	Guinea		s+	Philippines	
+	Bahamas		-	Guinea-Bissau		s-	Poland	
-	Bahrain		s-	Guyana		s+	Portugal	
s-	Bangladesh		c-	Haiti		-	Qatar	
s+	Barbados		o	Honduras		-	Romania	
s+	Belgium		c-	Hungary		s+	St. Kitts-Nevis	
+	Belize		s+	Iceland		s+	St. Lucia	
-	Benin		s+	India		s+	St. Vincent	
-	Bhutan		s-	Indonesia		-	Sao Tome & Prin.	
so	Bolivia		-	Iran		-	Saudi Arabia	
o	Botswana		-	Iraq		so	Senegal	
c+	Brazil		+	Ireland		-	Seychelles	
-	Brunei		c+	Israel		-	Sierra Leone	
c-	Bulgaria		s+	Italy		co	Singapore	
-	Burkina Faso		s+	Jamaica		-	Somalia	
-	Burma					s-	South Africa	
-	Burundi		s+	Japan		s+	Spain	
-	Cambodia		s-	Jordan		so	Sri Lanka	
			s-	Kenya		so	Sudan	
s-	Cameroon		-	Korea (N)		+	Suriname	
s+	Canada		s+	Korea (S)		-	Swaziland	
-	Cape Verde					c+	Sweden	
-	Central Afr. Rep.		-	Kuwait		c+	Switzerland	
-	Chad		-	Laos		-	Syria	
			-	Lebanon		s-	Tanzania	
s-	Chile		-	Lesotho		o	Thailand	
s-	China (Mainland)		s-	Liberia		s-	Togo	
so	China (Taiwan)		-	Libya				
s+	Colombia		s+	Luxembourg		s-	Tonga	
-	Congo		-	Madagascar		s+	Trinidad & Tobago	
			-	Malawi		-	Tunisia	
s+	Costa Rica		so	Malaysia		co	Turkey	
-	Cote d'Ivoire					s-	Uganda	
s-	Cuba		o	Maldives		s-	USSR	
+	Cyprus (G)		-	Mali		-	United Arab Emirates	
+	Cyprus (T)		s+	Malta				
			-	Mauritania		c+	United Kingdom	
s-	Czechoslovakia		+	Mauritius		+	United States	
s+	Denmark					+	Uruguay	
s+	Dominica		o	Mexico				
s+	Dominican Rep.		-	Mongolia		s+	Vanuatu	
+	Ecuador		-	Morocco		+	Venezuela	
			-	Mozambique		-	Vietnam	
so	Egypt		s+	Nauru		-	Yemen (N)	
o	El Salvador					-	Yemen (S)	
-	Equatorial Guinea		so	Nepal				
			+	Netherlands		-	Yugoslavia	
c-	Ethiopia		+	New Zealand		s-	Zaire	
so	Fiji		s-	Nicaragua		-	Zambia	
			-	Niger		s-	Zimbabwe	
s+	Finland							
s+	France		s+	Nigeria				
-	Gabon		s+	Norway				

The Table

This table shows the countries with the *most free* print and broadcast media (shown with +), *least free* media (-), and intermediate (0). The graph depicts the most free with the longest lines.

Of the 159 countries shown, 57 are in the *most free* category (36 percent of the nations); 83 are among the *least free* (52 percent); and 19 are in the *intermediate* group (12 percent).

The countries marked with an (S) are included in the book's Press Freedom Survey; those with a (C) are described only in the book's country reports.

Appendices

Appendix H
Government Press Ownership, Licensing, Censorship, "Guidance"

Afghanistan 2
Albania 2
Algeria 2
Angola 2
Antigua & Barbuda 2
Argentina 3, 4
Australia
Austria
Bahamas
Bahrain 1

Bangladesh 1, 5
Barbados
Belgium
Belize
Benin 1

Bhutan 5
Bolivia 4
Botswana 2
Brazil 3, 4
Brunei 1

Bulgaria 2
Burkina Faso 2
Burma 1, 3
Burundi 2
Cambodia 2

Cameroon 2
Canada
Cape Verde 2
Central Afr. Rep. 2
Chad 2

Chile 1, 3
China (Mainland) 2
China (Taiwan) 3, 5
Colombia 4
Congo 2

Costa Rica 4
Cote d'Ivoire 2
Cuba 2
Cyprus (G) 3, 4
Cyprus (T)

Czechoslovakia 2
Denmark
Dominica
Dominican Rep.
Ecuador 4

Egypt 5
El Salvador

Equatorial Guinea 2
Ethiopia 2
Fiji 5

Finland
France
Gabon 2
Gambia
Germany (E) 2, 3

Germany (W)
Ghana 2, 3
Greece
Grenada
Guatemala 4

Guinea 2
Guinea-Bissau 2
Guyana 3
Haiti 4
Honduras 4, 5

Hungary 2
Iceland
India
Indonesia 3, 5
Iran 1

Iraq 2
Ireland
Israel 1, 3
Italy 4
Jamaica 3

Japan
Jordan 3, 4, 5
Kenya 3
Korea (N) 2
Korea (S) 3

Kuwait 1, 3
Laos 2
Lebanon 4
Lesotho
Liberia 4, 5

Libya 2
Luxembourg
Madagascar 1
Malawi 1, 3
Malaysia 3, 5

Maldives 3
Mali 2
Malta
Mauritania 2
Mauritius

Mexico
Mongolia 2
Morocco 1
Mozambique 2
Nauru

Nepal 1, 3, 4
Netherlands
New Zealand
Nicaragua 2, 4
Niger 2

Nigeria 3
Norway
Oman 2
Pakistan 3
Panama 1, 4

Papua New Guinea 3
Paraguay 1, 3
Peru 4
Philippines
Poland 1

Portugal
Qatar 5
Romania 2
St. Kitts-Nevis
St. Lucia

St. Vincent
Sao Tome & Prin. 2
Saudi Arabia 1, 3
Senegal 3, 4
Seychelles 2

Sierra Leone 1, 3
Singapore 3, 5
Somalia 2

South Africa 1
Spain

Sri Lanka 5
Sudan 3, 4
Suriname
Swaziland 1
Sweden

Switzerland
Syria 2
Tanzania 2, 3
Thailand 3, 4
Togo 2

Tonga 1
Trinidad & Tobago
Tunisia 1, 4
Turkey 5
Uganda 1

USSR 2
United Arab Emirates 4, 5
United Kingdom
United States
Uruguay

Vanuatu 1
Venezuela 4, 5
Vietnam 2
Yemen (N) 1
Yemen (S) 2

Yugoslavia 2
Zaire 4, 5
Zambia 2
Zimbabwe 2

Numbers beside countries show:

1 Independent press subject to censorship: 25 countries (16 percent).
2 Government owns all or most of the print and broadcast media: 159 (32 percent).
3 Government licenses independent press: 31 (20 percent).
4 Government licenses journalists: 25 (16 percent).
5 Government "guides" the press: 17 (11 percent).

Of 159 countries, 115 (72.3 percent) practice one or more of these controls.

Appendix I

The London Information Forum of the Conference on Security and Cooperation in Europe (CSCE) 18 April to 12 May 1989

AS A MEMBER of the United States delegation to the CSCE Information Forum I delivered three formal statements as the context for making a series of proposals which will be part of the agenda of the follow-up conference in Helsinki in 1992.

Excerpts from those statements follow:

[T]he role of independent journalism is essential for compliance with the [CSCE's] Vienna Concluding Document and Article 19 of the Universal Declaration of Human Rights. The document and the declaration speak of facilitating "the freer and wider dissemination of information of all kinds" and ensuring that "individuals can freely choose their sources of information."

The emphasis on the individual citizen surely is not accidental. For the right of the individual to choose information is inescapably linked to the right of the journalist to publish or broadcast. And to ensure that information "of all kinds"—the document's own words—are accessible to citizens there must inevitably be unofficial as well as official news and information produced and distributed.

Given the momentum of glasnost [in the USSR], we are encouraged to believe that independent journalism will finally be permitted. But recent incidents are troubling. There is, for example, no independent journalist on the Soviet delegation to this conference, though several Eastern European delegations have independent journalists. Yet there are several hundred unofficial publications now appearing in the Soviet Union. They operate with great difficulty. They cannot receive regular supplies of paper or other printing materials. They face harassment and interference in the distribution of their products. Surely, these alternative journalists and other writers deserve formal recognition and assistance rather than rejection.

[Journalists all over the world will watch] whether the Soviet Union in its forthcoming press law finally approves independent, unofficial journalism. Access to independent journalism is a right of the citizens in all countries committed to the CSCE...

[T]he widest global impact, occurring in by far the shortest time-span and affecting all of humanity most profoundly, will be the Information Revolution.

It may be technically known as the Age of ISDN—the Integrated Systems of Digital Networks—the universal networking of networks of all kinds: news, information, data, history, culture, entertainment; by sound and picture; conveyed by small and large technologies mainly over telephone lines linking people to people everywhere.

This revolution will empower the individual citizen as never before, in all societies whatever their social or political structure. The Age of ISDN will be a democratizing force motored by technology, but not restricted by the application of technology. We need not fear an Orwellian outcome in which communication machines either subvert the words of men and women or hold them hostage to the technocrat at some central authority.

ISDN should be seen as a vast array of switching points worldwide. Those switches will direct the flow of all kinds of news and information from one point to another no matter how distant. But that flow must be unimpeded by political or bureaucratic controls. Put positively (as I believe we can these days) we, in all countries, must develop the political will to allow technology to drive social change without impediments.

Some broad guidelines may be useful. There should be:

1. A commitment to provide the basic technology that will put everyone, everywhere on line. That means providing telephones and terminals within easy access and at affordable cost. The International Telecommunication Union (ITU) has set the goal of having everyone on earth near a telephone by early in the 21st century. This includes all Third World countries. Surely all CSCE nations can meet that objective.

2. Universal compatibility of ISDN services is a human right. Small and large communication instruments should be able to speak to one another no matter where they are created or deployed. Only open architecture of computers, television receivers (including high definition TV), and, most important, telephones can provide universal access to all digital systems. This means inter-regional as well as regional standardization and compatibility. Commercial or political competition should not block access to communications from other regions, either by incompatible technology or tariffs. To provide diverse news and information at lowest cost worldwide, for example, there should be developed an integrated packet of video, audio, and data services having universal application. All news suppliers, particularly those originating in the developing world, should have ready access to such integrated packets. This is in compliance not only with the [CSCE] Vienna Concluding Document, but all fundamental declarations and covenants in the field of human rights.

3. Technology should drive politics. The philosophy of openness and of the individual's right to communication—certainly to the extent envisioned in the Age of ISDN—will be an extension of present political

and civil rights in some countries, and more innovative in other countries. No nation, however, can avoid making social, political, and even legislative adjustments to the new age. But these adjustments should be mainly to guarantee openness and diversity of content, and avoidance of monopoly control of the carriers. While legislation may be needed to deregulate present governmental monopolies of communication, legislation should be limited only to ensure that new monopolies, of whatever form, do not appear. There should be deregulation to free the flow of content and regulation only to ensure the maintenance of diversity.

4. Not only technology, but human rights should drive the politics of the Age of ISDN. I have noted earlier at this conference that the Vienna Concluding Document of the CSCE called for information "of all kinds" to be accessible to all citizens. That means unofficial as well as official information; independent views and criticisms as well as governmental dicta and indoctrination. That calls for independent, non-governmental journalists and information processors. No nation can fulfill its commitment to the CSCE Process without enabling independent journalists and information processors to work in freedom without prior- or post-censorship, and without fear of reprisal or worse.

To improve the free flow of information among CSCE participating States, I propose that each Government:

1. Will comply with the CSCE commitment for access to information "of all kinds" by removing restrictions on independent journalists and other information processors.

2. Should, where necessary, introduce legislation to guarantee openness and diversity of information flow, and regulate only to maintain the diversity of content.

3. Remove restrictions on the importation of equipment by journalists for cross-border transmissions of audiovisual information and data.

4. Remove licensing requirements for the use of satellite television reception dishes by organizations and private individuals.

5. Simplify the procedures for television and radio news organizations to obtain a permit, and limit the need for assignment of PTO (Public Telecommunications Operator) personnel, for each use of satellite up-link fly-away terminals.

6. Reduce administrative obstacles to greater use of satellite and cable television, open architecture in computers and other communication devices, open system interconnective networks, and trans-border flow of data.

7. Encourage development of globally accepted common standards for telecommunications and computers, so as to enhance greater access to information in all CSCE countries by universally compatible integrated packets of video, audio and data services.

8. In accordance with the goals of the International Telecommunica-

tions Union, all CSCE countries should provide nearby telephone service to all of their citizens by the year 2010.

Earlier, here, I stressed the commitment of the CSCE countries to allow individuals as well as organizations to "distribute information of all kinds." That commits CSCE countries to permit independent journalists and other information processors to function openly.

I emphasize [now] the commitment in the same paragraph (VCD number 34) to allow independent individuals to *reproduce* information material of all kinds.

The act of copying, either by printing or by photocopy, is essential to the further commitment to exchange and permit an unimpeded flow of information.

Yet there are countries represented here which prevent the use of copier machines, computer printers and other copy facilities, except for official purposes. And some states represented here license the typewriter to prevent even the minimal exchange of information uncontrolled by central authority.

I am, therefore, formally submitting this proposal for the consideration of the conference and for relaying to Helsinki in 1992:

With the increasing demand for copying written and photographic material in small numbers as well as for desktop publishing, it is essential to insure the right of individuals and organizations to purchase such equipment and use it without official licensing or controls. I, therefore, propose that CSCE countries:

1. Remove administrative and criminal restrictions and penalties leveled against independently obtaining, possessing, reproducing, publishing and distributing printed and photographic materials.

2. Permit private ownership, use of and access to typewriters, word processors, copying machines and related instruments.

3. Respect intellectual property rights in the use of such reproducing facilities.

NOTES

Preface
Pages 1-4

1. The two-volume 1,445-page edition of *Big Story: How the American Press and Television Reported and Interpreted the Crisis of Tet 1968 in Vietnam and Washington* by Peter Braestrup, published by Westview Press (1977), is available only from Freedom House. The second abridged, one-volume edition: New Haven: Yale University Press, 1983 (paper).

2. Raymond D. Gastil, editor and principal author. The annual has been published since 1978.

Introduction
Pages 5-19

1. Lewis Mumford, "Authoritative and Democratic Technics," *Technology and Culture*, 5 (Winter 1964) page 7.

2. Decree of 1586, quoted in John Shelton Lawrence and Bernard Timberg, *Fair Use and Free Inquiry*, Norwood, NJ: Ablex, 1980.

3. "Sunday Morning," CBS network, 25 September 1988.

Chapter I
The Word is Power
Pages 21-51

1. New York *Herald Tribune*, 2 October 1951.

2. See Daniel Bell, *The Coming of Post-Industrial Society*, New York: Basic Books, 1976.

3. Simon Nora and Alain Minc, *The Computerization of Society*, Cambridge: MIT Press (English reprint) 1980, page 70. This book, with its English introduction by Daniel Bell, was written for the Inspection Generale des Finances at the request of the President of France. The French report: *L'Informatisation de la Societe*, Paris: La Documentation Francais, 1978.

4. Ibid., page 72.

5. Lincoln Steffens, *Autobiography*, New York: Harcourt, Brace & Co., 1931, page 311.

6. Mario Cuomo, "A Brief on the Freedom of the Press" (mimeo), before the New York Press Club, 25 November 1986.

7. Herb Greer,"The Ethics of Fallible Journalism," *Encounter*, December 1986.

8. Herbert I. Schiller, *Communication and Cultural Domination*, White Plains, NY: International Arts and Sciences Press, 1976; *Who Knows: Information in the Age of the Fortune 500*, Norwood, NJ: Ablex, 1981; *Information and the Crisis Economy*, Ablex, 1984.

9. See Ben A. Bagdikian, *The Information Machines: Their Impact on Men and the Media*, New York: Harper & Row, 1971.

10. Joseph Kraft, "The Imperial Media," *Commentary*, May 1981, page 42.

11. S. Robert Lichter, Stanley Rothman, Linda S. Lichter, *The Media Elite: America's New Powerbrokers*, Bethesda, MD: Adler & Adler, 1986.

12. S. Robert Lichter, Daniel Amundson, Richard Noyes, *The Video Campaign: Network Coverage of the 1988 Primaries*, Washington: American Enterprise Institute for Public Policy Research, 1988, page 108.

13. See C. Anthony Giffard, *Unesco and the Media*, New York: Longman, 1989.

14. Joseph C. Goulden, *Fit to Print: A.M. Rosenthal and His Times*, Secaucus, NJ: Lyle Stuart Inc., 1988, pages 147-8.

15. J. Herbert Altschull, *Agents of Power*, New York: Longman, 1984, pages 194-5.

16. Op. cit., Lincoln Steffens, page 515.

17. Walter B. Wriston, keynote address at annual conference of the International Institute of Communications, Washington, DC, 13 September 1988.

18. A claim made by Yoshizo Ikeda, president of Nippon Hoso Kyokui (NHK) at the annual conference of the International Institute of Communications, Washington, DC, 14 September 1988.

19. Ibid, IIC conference, 12 September 1988.

20. Hamid Mowlana is professor of international relations and director of the international communications program at the American University of Washington.

21 The *Statesman*, Calcutta, 5 September 1988.

22. For a lucid account see, William Dorman and Mansour Farhang, *The Press and Iran*, Berkeley, CA: University of California Press, 1987.

23. From Nielsen Media Report, interview with the author 14 February 1989.

24. Thomas Carlyle, *On Heroes, Hero-Worship and the Heroic in History*, London: Chapman & Hall, 1889, page 134.

25. This anecdote is recalled by Ronald Koven.

Chapter II
Revolution Without Losers
Pages 53-91

1. *Documents in American Foreign Relations*, 1964, New York: Harper & Row, 1965 page 373.

2. *IPI Report*, organ of the International Press Institute, London, August 1988, page 5.

3. George Gerbner, "The Challenge Before Us," in *Communication and Domination: Essays to Honor Herbert I. Schiller*, Norwood, NJ: Ablex, 1986, page 234.

4. Ibid., Gerbner, page 239.

5. Ibid., James D. Halloran, page 243.

6. Ibid., Halloran, page 248.

7. Jonathan Fenby, The *International News Services,* New York: Schocken Books, 1986.

8. William L. Rodgers, "Effects of Cable and Radio Control on News and Commerce," Annals of the American Academy of Political and Social Sciences, 112, March 1924, page 250.

9. Kent Cooper, *The Right to Know,* New York: Farrar, Straus & Cudahy, 1956, page 153.

10. Quoted in Margaret A. Blanchard, *Exporting the First Amendment,* New York: Longman, 1986, page 23.

11. Samuel DePalma, *United Nations Affairs,* 17 June 1949, RG 59, NA: Box 2255, 501.BD Freedom of Information/6-2949.

12. Herbert I. Schiller, Journal of Communications, summer 1987, page 156.

13. Mustopha Masmoudi, *The New World Order for Information,* Secretariat of State for Information, Tunisia, February 1977, page 8 (pamphlet).

14. See C. Anthony Giffard, *Unesco and the Media,* New York: Longman, 1989.

15. Kaarle Nordenstreng, *The Mass Media Declaration of Unesco,* Norwood, NJ: Ablex, 1984, page 14.

16. U.S. Department of State, U.S./Unesco Policy Review, February 27, 1984.

17. Op. cit. Nordenstreng, page 31.

18. *New York Times,* 16 October 1983, page A-11.

19. Leonard R. Sussman, An Approach to the Study of Transnational News Media in a Pluralistic World, number 18 in the series prepared for the International Commission for the Study of Communication Problems (MacBride Commission), Paris: Unesco, April 1978.

20. Colleen Roach, *Journal of Communication,* Autumn 1987, page 36.

21. Miklos Haraszti, *The Velvet Prison,* New York: New Republic Book, 1988, pages 144-5, translated from the Hungarian by George Konrad.

22. Report on the Situation of Human Rights in Chile, Organization of American States, Washington, DC, 1985.

23. These four models of the press, and the critiques of each appeared originally in Wilbur Schramm and Frederick Siebert, *The Four Theories of the Press: The Authoritarian, Libertarian, Social Responsibility and Soviet Communist Concepts of What the Press Should Be and Do,* Chicago: University of Illinois Press, 1963.

24. Mary Anne Fitzgerald, *African Report,* Center for Strategic and Intelligence Studies, Washington, DC, March-April, 1987.

Chapter III
The Too-Easy Equivalence of Liberty
Pages 93-97

1. George Orwell, "The Freedom of the Press," *New York Times Magazine,* 8 October 1972, page 12.

Chapter IV
Fear of Information Power in "Socialist" Countries
Pages 99-124

1. *Editor and Publisher,* 25 June 1988.
2. Everett Carl Ladd, *Christian Science Monitor,* 17 June 1988.
3. Politicheskiy Slovar, 4th edition, *Politizdat,* 1987; translated by David Berley.
4. Mikhail Gorbachev, *Perestroika: New Thinking for a Communist Country and the World,* New York: Harper & Row, pages 19, 21, 23, 31, 32.
5. Vladimir Solovyov and Elena Klepikova in *Newsday,* 6 March 1988.
6. *New York Times,* 6 February 1988.
7. *Moscow News,* number 27, 10-17 July 1988.
8. *Moscow News,* number 20, 22-29 May 1988.
9. *Moscow News,* number 4, 31 January-7 February 1988.
10. *Pravda,* May 5, 1988.
11. John D. H. Downing, "Trouble in the Backyard: Soviet Media Reporting on the Afghanistan Conflict," *Journal of Communication,* Spring 1988, page 5.
12. Told to Edward Gorman, the *Times* of London, published in *IPI Report,* March 1989.
13. See Peter Braestrup, op. cit., *Big Story.*
14. Angus Roxburgh, *Pravda: Inside the Soviet News Machine,* New York: George Braziller, Inc., 1987, page 270.
15. Op. cit. *Times,* 6 February 1988, page 52.
16. Jadwiga Pastecka, "The Socialist Approach," in *The Right to Communicate: A New Human Right,* Dublin, Ireland: Boole Press, 1983.
17. *China Daily,* 18 July 1985.
18. *China Daily,* 23 June 1988.
19. The survey was conducted by the Journalism Research Institute under the Chinese Academy of Social Sciences and the Capital Society of Journalism. Of the 472 respondents, 235 (about 50 percent) were Communist party members, 80 (17 percent) belonged to "democratic" parties, 118 (25 percent) had no party affiliation, and 4 (1 percent) were Communist Youth League members.
20. Milton Hollstein, *Editor and Publisher,* October 22, 1988.
21. Leonid Treyer, *Moscow News,* number 46, 20-27 November 1988.

Chapter V
Power and Press in the Developing Countries
Pages 125-157

1. See Robert L. Stevenson, *Communication, Development and the Third World,* New York: Longman, 1988.

2. *Sudan Times*, 12 June 1988.
3. *Sudan Times*, 21 June 1988.
4. *Sudan Times*, 12 June and 4 July 1988.
5. *Sudan Times*, 16 November 1988.
6. *Sudan Times*, 18 November 1988.
7. Botswana, Grenada, Tuvalu, Cyprus, Dominica, Kiribati, Papua New Guinea, St. Kitts-Nevis, Tonga.
8. See the most recent Comparative Survey of Freedom in the yearbooks, *Freedom in the World: Political Rights and Civil Liberties,* published by Freedom House.
9. M. G. G. Pillai, *The Statesman,* Calcutta, 22 December 1987.
10. William A. Rugh, *The Arab Press,* Syracuse: Syracuse University Press, 1979.
11. Leonard R. Sussman, *Mass News Media and the Third World Challenge,* Washington, D. C.: SAGE Policy Paper, CSIS, Georgetown University, 1977, page 39.
12. Tran Van Dinh and A. W. Singham, *Non-Aligned Movement in World Politics,* Westport, CN: Lawrence Hill & Co. 1977, page 30.
13. C. Anthony Giffard, report prepared for the International Association for Communication Research; published at University of Washington, Seattle (mimeo), 1983.
14. Sarath Amunungama, associate secretary-general, Worldview International Foundation, address to Institute for International Communication, Washington, D. C., 14 September 1988.
15. *World Military Expenditures and Arms Transfers,* U.S. Arms Control and Disarmament Agency, 1985, page 3.

Chapter VI
Count the Ways of Censorship
Pages 159-188

1. Quoted by Samir Atallah of *Al Mostagbal* at the London conference, Challenging the Censors, January 1987.
2. See Nomavenda Mathiane, *South Africa: Diary of Troubled Times,* New York: Freedom House, 1989.
3. See R. Bruce McColm, *To License A Journalist?,* New York: Freedom House, 1986.
4. Sean MacBride, et al, *Many Voices, One World,* report of the International Commission for the Study of Communication Problems, Paris: Unesco, 1980, pages 236-237.
5. Nelson Etukudo, ed., *Issues & Problems in Mass Communications,* Calbar, Nigeria: Development Digest Ltd., 1986; page 1.
6. Ibid, Etim Anim, "Editing a Government Newspaper," page 19.

Chapter VII
How the News is Reported and Distorted: A Survey of State vs. Information Power in 74 Countries
Pages 189-204

1. See "Freedom in the World—1989," chart, *Freedom at Issue*, January-February 1989, page 47.

2. Leonard R. Sussman, "Communications: Openness and Censorship," in *Freedom in the World: Political Rights and Civil Liberties; 1987-1988*, Raymond D. Gastil, New York: Freedom House, 1988, page 135.

Chapter VIII
The Information-Power Struggle: 88 Country Reports
Pages 207-340

1. S. B. Majrooh and S. M. Y. Elmi, *The Sovietization of Afghanistan*, Peshawar: Printing Corp. of Frontier, Ltd., 1986; chart no. 1, pages 39-44.

2. *IPI Report*, June/July, 1988.

3. *Columbia Journalism Review*, August 1988.

4. Francisco Goldman in "Some Tales of La Libertad de Prensa," *Harper's*, August 1988.

5. *New York Times*, 29 July 1988.

6. *Laws and the Soviet State*, 1938.

7. Associated Press, W1414, 11 September 1988.

Chapter IX
The Reagan/Press Conflict
Pages 341-355

1. "Protection of Classified Information," Joint Hearing, House Subcommittee on Civil Service and Subcommittee on Civil and Constitutional Rights, December 6, 1985, Washington: GPO, 1987, page 37.

2. *Washington Post*, 30 March 1967.

3. Associated Press, W1182, 28 November 1988.

4. Dick Adler, *Los Angeles Times*, 14 May 1974.

5. American Newspaper Publishers Association, 28 April 1961.

6. Op. cit. Lichter-Rothman, *The Media Elite*.

7. See Mark Hertsgaard, *On Bended Knee: The Press and the Reagan Presidency*, New York: Farrar, Straus & Giroux, 1988.

8. See *Battle Lines: Report of the Twentieth Century Fund Task Force on the Military and the Media;* background paper by Peter Braestrup, New York: Priority Press Publications, 1985.

9. *New York Times* editorial, 18 September 1988.

Notes

10. *New York Times,* 20 March 1985

11. Michael J. Robinson, Maura Clancy and Lisa Grand, "With Friends Like These...," *Public Opinion* 6, June/July 1983, page 24.

12. David L. Paletz and K. Kendall Guthrie, "The Three Faces of Ronald Reagan," *Journal of Communication,* Autumn 1987, page 7.

13. Joe S. Foote, "The Rating's Decline of Presidential Television," Radio-Television Department, Southern Illinois University [date].

14. Walter Karp, "Liberty Under Siege," *Harper's,* November 1985.

**Chapter X
Networks of Freedom
Pages 357-417**

1. The letter, addressed "To the capitalists of the Electric Telephone Company," appears in Ithiel de Sola Pool, *Forecasting the Telephone: A Retrospective Technology Assessment,* Secaucus, NJ: Ablex, 1983, page 21. The late Prof. Pool was the most prominent exponent of the uses of telephony and other "small" communications systems as a democratizing force in human relationships. The retrospective analysis of the telephone here is drawn from Dr. Pool's book.

2. Associated Press, W1373, 20 September 1988.

3. Parker W. Borg, "Building a Flexible Framework for New Information Services," Current Policy No. 1123, U. S. Department of State, 19 October 1988.

4. Zbigniew Brzezinski, *Between Two Ages: America's Role in the Technetronic Era,* New York: Viking Press, 1970.

5. Shahwar Junaid, *Communications Media and Statecraft,* Rawalpinida: Publishing Consultants, 1988.

6. "Science Technology and the Constitution," Office of Technology Assessment, Congress of the United States, 1987.

7. Herbert I. Schiller, *Who Knows: Information in the Age of the Fortune 500,* Norwood, NJ: Ablex, 1981, page 100.

8. George Shultz, "The Shape, Scope and Consequence of the Age of Information," Department of State Policy Paper number 811, 21 March 1986.

9. George Kroloff and Scott Cohen, "The New World Information Order," report to the Committee on Foreign Relations, U.S. Senate, November 1977, mimeo.

10. "International Information Flow: Forging a New Framework," report number 96-1535, 11 December 1980, for the Committee on Government Operations, House of Representatives.

11. Matthew Nimitz, Under Secretary of State for Security Assistance, Science and Technology, at the OECD, Paris, 6 October 1980.

12. Report to the President by the staff of the Domestic Council's Committee on the Right of Privacy, Washington: Government Printing Office, 1976, page vi.

13. "Science, Technology and the First Amendment," Office of Technology Assessment, United States Congress, Washington: U.S. Government Printing Office, January 1988.

14. Felix Rohatyn, "On the Brink," *New York Review of Books*, 11 July 1987, page 3.

15. Final Draft, U.S. Development Communications Assistance Programs, a report by the Senior Interagency Working Group for Communications Development Assistance chaired by Ruth Zagorin for the Office of the Coordinator for International Communication and Information Policy, Department of State, November 1984.

16. George Shultz letter, 21 September 1983, to Senator Charles H. Percy, chairman of the Senate Foreign Relations Committee.

17. Sarath Amunugama, speaking at the IIC, Washington, D.C., 14 September 1988.

18. Record of the 237th meeting of the UN Committee on the Peaceful Uses of Outer Space, UN Doc. A/AC.105/PV.237, at 3 (1982).

19. Rod McShane, *Intermedia*, January 1988.

20. "Shakeup in Illicit Market Due as Turkish Homevid Goes Legit," *Variety*, 25 February 1987.

21. *The Democratic Journalist,* January, 1988, organ of the International Organization of Journalists, Prague.

22. See Rolf T. Wigand, "Integrated Services Digital Networks: Concepts, Policies and Emerging Issues," *Journal of Communication*, Winter 1988.

23. G. Russell Pipe, *Transnational Data Report*, January 1989, page 19.

24. Associated Press, W1223, 30 December 1988.

25. John Diebold, concluding address, European Economic Community Conference on ESPRIT, 20 May 1988, mimeo.

26. "Commercializing High-Temperature Superconductivity," OTA report brief, June 1988.

27. John Diebold, letter to *New York Times*, 11 November 1988.

28. Associated Press, W854, 21 April 1988.

29. Francois de St. Phall, "And Now, the Tiger of Europe," op ed page, *New York Times*, 24 March 1988.

30. G. Russell Pipe, "Perspectives in Transnational Data Regulation," independent consultant, Amsterdam, Netherlands, undated, mimeo.

31. Background Report/133, 7 August 1987, Radio Free Europe/Radio Liberty.

32. *Science,* vol. 242, 25 November 1988, page 1118.

33. *Miami Herald,* 6 May 1984.

34. Warren Unna, "India Not to Get Latest U.S. Supercomputer," *The Statesman*, Calcutta, 28 March 1987.

35. Associated Press, W865, 22 April 1988.

36. Wilson P. Dizard, Jr., *The Information Age*, New York: Longman, 1982.

37. Michael Palmer and Jeremy Tunstall, "Deregulation and Competition in European Telecommunications," *Journal of Communication*, Winter 1988.

38. *Bulletin*, International Federation of Journalists, February 1988.

39. Rupert Murdoch quoted by Aidan White, general secretary of the International

Federation of Journalists at the L'Observatoire de l'Information conference, Montpellier, France, 13 October 1988.

40. Philip H. Abelson, deputy editor, *Science,* 28 October 1988.

Chapter XI
Don't Fear the Slippery Slope, and Other Recommendations
Pages 419-437

1. Lynn Ashby, editor of the *Houston Post,* in *Editor and Publisher,* 23 July 1988.

Epilogue
Will ISDN Facilitate Peace and the Human Imagination?
Pages 439-449

1. See Peter L. Berger, *Pyramids of Sacrifice: Political Ethics and Social Change,* New York: Basic Books, 1974.

2. *New York Times,* July 31, 1983.

3. *Chronicle of International Communications,* July 1983.

4. Herb Greer, "Fourth Estate," letter, *Encounter,* December 1988.

Index

Abel, Elie, 185
Action of National Information Systems (ASIN), 154
advertising, 202, 391-92
Afanasyev, Viktor, 113-14
Afghanistan, 208-10
  invasion of, 111-12
Africa, media in, 184-86
Agence France-Presse, 30, 63
Agent Identities Bill, 347
Aggarwala, Narinder, 142
Albania, 210-11
Algeria, 212-14
Altschull, Herbert, 38-41
American Society of Newspaper Editors, 65
Amin, Idi, 151-52
Amunugama, Sarath, 386
Anim, Etim, 184-85
Animal Farm (Orwell), 95-96
Antigua, 214-15
Arab countries (*see* Moslem countries)
Arab Project for Communication and Planning (ASBU), 151
Argentina, 215-16
Arms Export Control Act, 346
Asia-Pacific News Network (ANN), 151
Asiavision, 44, 422
Associated Press, 30, 64-65
AT&T, 375
  breakup of, 48
Austria, 216-17

Bacon, Francis, 21
Bagdikian, Ben, 34
Bandung conference, 139
Bangladesh, 217-18
Bao Xingjian, 83
Barbados, 218-19
Barrett, Edward W., 69
Barron, Jerome, 179
Beckett, Denis, 177
Belgium, 219-20
Bell, Alexander Graham, 357-58
Bell, Daniel, 21, 59

Bentham, Jeremy, 160
Benton, William, 2, 68-70, 444
Berger, Peter L., 439-40
Big Story: How the American Press and Television Reported and Interpreted the Crisis of Tet 1968 in Vietnam and Washington, 2
Binder, Carroll, 70
Blum, Leon, 444
Bogart, Leo, 89
Bolivia, 220-21
Botswana, 135
Boulding, Kenneth, 362
Braestrup, Peter, 2
Brazil, 221-22
Bulgaria, 117, 222-23, 388
bureaucrats:
  vs. journalists, 100
  Soviet, 106-9
Burke, Edmund, 448
Burnham, Forbes, 143-44
Buthelezi, Mangosuthu, 176-77

Cable News Network, 50, 392
cable television, 50, 90, 376
Cambodia, 49
Cameroon, 223-25
Campbell, Frank, 142-46
Canada, 58, 225-27
capitalism, vs. Marxism, 439-43
Caribbean News Agency (CANA), 151, 433
Carlyle, Thomas, 50-51
Carter, Jimmy, 372
Casey, William, 346
CBS, 177-78
CBS Records, 391
CCITT, 396-97
censorship:
  historical background of, 10, 12, 63-65
  recommendations for ending, 428-31
  self-, 203
  types of, internationally, 159-88

487

and writing between the lines, 81-83
Chafee, Zechariah, 354
Chamorro Cardenal, Jaime, 99
Chernobyl, 104-5
Chile, 83, 227-30
China, People's Republic of:
  censorship in, 56, 160
  communication technology, 44, 63, 404
  economy, 6
  freedom of the press in, 134-35, 230-33
  reformist communism in, 83, 118-23, 451
  unofficial media in, 118-22
China Daily, 118-21
China Spring, 118-22, 451
Chirac, Jacques, 51
Clark, William, 346
Clarke, Arthur C., 443, 444
classified information, 341, 345-47, 350-52, 381
Clawson, Kenneth, 342
Cohen, Scott, 368
colegios of journalists, 181-83
Colombia, 57, 161, 233-35
common carriers, 373-75, 396-97, 424-26
communication, power of, 21-51
communication industry:
concentration of power in, 33-34
  foreign investment in, 391-92, 413
  subsidies to, 421-22
  U.S., 408-9
communication policy:
  in developing countries, 6-7, 125-57, 409-11
  in Marxist countries, 99-124
  U.S., 22-23, 423-24, 446-49
  U.S., recommendations, 423-36
communication revolution:
  and democracy, 9-13, 48-51, 416
  and freedom, 7-9
  resistance to, 8-13, 16-19
communication systems:
  effect on nation-states, 53-55
  foreign, assistance to, 78, 156-57, 383-88
  ownership of, 58-60, 366-67
  regulation vs. deregulation, 26, 378-79, 396-97, 424-26
communication technology, 5-19, 358-417
  openness vs. protectionism, 394-95
  state vs. private ownership of, 82-83, 364-65, 396-97, 421
Communist countries (*see* Marxist countries)
computers (*see also* data processing)
  increasing use of, 8-9
  regulation of, 405-7
  standards for, 360
  super-, 357, 408
  worms and viruses, 380
Constitution, U.S, and freedom of the press, 32-33, 101-2, 372-73
consumerism, 156
Cooper, Kent, 64, 67
Corcoran, Farrel, 110-11
Costa Rica, 135, 181-83, 235-37
C-SPAN, 50, 353
Cuba, 237-38
cultural imperialism, 390-93
Cultural Revolution (China), 118
culture:
  popular U.S., 390-93
  traditional, protection of, 389-93, 434-35
Cuomo, Mario, 32
Czechoslovakia, 115, 117, 238-40

Daily Telegraph (Alton, IL), 177
data bases, online, 374, 405
data processing (*see also* computers) regulation of, 405-7
Deakin, James, 21
defense spending, U.S., 397-99
democracy:
  communication revolution and, 9-13, 48-51, 416
  vs. freedom, 134-35
Denmark, 240-41
DePalma, Samuel, 70
deregulation:

of communication systems, 26, 378-79, 396-97, 424-26
of television, 90, 354
Deutsch, Karl, 394
developing countries (*see also* Third World)
 aid to, for communication technology, 78, 156-57, 383-88
 communication policy in, 125-57
 communication systems in, 6-7, 409-11
 economy, 6-7
 future of ISDN in, 16-17
 media's impact on, 43-44, 47-48
 modernization in, 127-29
 news coverage of, 142-55, 433
 and NWICO, 71-77
 political systems, 137-38
 vs. West, 137-42
developmental journalism, 142-46
dictatorships, information suppressed by, 159-88
Diebold, John, 401-3
diplomacy, secret, 55-56
direct-broadcast-by-satellite television (DBS), 430-31
disinformation, Soviet, 174-75, 431
Di Yi, 120-21
Dizard, Wilson P., Jr., 409
Dominica, 241
Dominican Republic, 241-42
Dougan, Diana Lady, 381-82
Dounaev, Vladimir, 100
Downing, John D. H., 111-12
Drozd, Vladimir, 105
Dzhirkvelov, Ilya, 174

Eastern Europe, glasnost in, 115-18
Eastern liberals, 343
East Germany, 115, 116
Eban, Abba, 55-56
Egypt, 242-43
electronic publishing, 373-75
El Salvador, 112, 161
Ershov, Andrei, 18
Esprit program, 404-5
Estonia, 105
ethics of journalism, 36-37

Ethiopia, 243-49
Ethnos (Greek newpaper), 174-75
Etukudo, Nelso, 184
European Economic Community:
 economic policy, 23, 404-7
 future of ISDN in, 15-16, 58, 396-97
 technology, 358-61, 401-2
Export Administration Act, 350
export restrictions, 407-9

Fairness Doctrine, 352-53, 421
Falwell, Jerry, 349
Far Eastern Economic Review, 179-80
Fascell, Dante B., 390
Federal Communications Commission, 352-53, 371, 381
Fenby, Jonathan, 63
Fiji, 249-50
financial information, 43
Finland, 250-52
First Amendment (*see* Constitution, U.S)
Fitzgerald, Mary Anne, 90
Fleming, Anne Taylor, 175
Foreign Agents Registration Act, 348
foreign investment, in U.S. communications industry, 391-92, 413
foreign policy, television and, 48-50
France, 23-24, 252-53, 375-76
Franklin, Benjamin, 10
freedom:
 communication revolution and, 7-9
 vs. democracy, 134-35
 Freedom House survey of countries, 38, 169, 189-206
 networks of, 357-417
Freedom House:
 and freedom of the press, 2-3, 113
 freedom surveys, 38, 169, 189-206
 journalism morbidity statistics, 57
 press-freedom survey, 3, 189-206
Freedom in the World: Political Rights and Civil Liberties, 3
Freedom of Information Act, 346-47
freedom of the press:

Constitution, U.S., and, 32-33, 101-2, 372-73
Freedom House and, 2-3, 113
Freedom House survey of, 3, 189-206
and freedom of communications systems, 424-26
international guarantees of, 66-81
legal decisions re, 12-13, 347-49
restraints on, 32-33
Frontline (South Africa), 176-77

Gainza Paz, Alberto, 21
Gandhi, Indira, 148
Gandhi, Rajiv, 47
gazetting, 179-80
Gergen, David, 343
German Democratic Republic (*see* East Germany)
German Federal Republic (*see* West Germany)
Ghana, 253-55
Gideonse, Harry D., 2
Giffard, C. Anthony, 149, 151, 153
glasnost, 45, 63, 93-97, 102-18
  fear of, 122-24
Glavlit, 63, 113-15
Gorbachev, Mikhail, 17, 72, 89, 103-4, 107-9, 113, 434
Gostelradio, 45
Gould, Jack, 2
government (*see also* nation-states)
  journalists vs., 100, 419-20
  and media, 2-3, 5-7, 12-13, 21-51, 54-57, 83-91, 189-206, 341-55, 416, 419-37
  ownership of media, 81, 129-34, 184-86, 411-12
government information, 44-48, 159-88
government publications, U.S., discontinuance of, 345
Greece, 255-58
Greer, Herb, 32, 447
Grenada, invasion of, 258-59, 349
Grigoryants, Sergei, 104, 115
Gromyko, Andrei A., 108
Group 77 (*see* Nonaligned Movement)
Guatemala, 260-61
guided journalism, 83-89
Guyana, 143-44, 261-62

Haber, Anton, 175
Haiman, Franklyn S., 179
Haiti, 166, 262-63
Haldeman, H. R., 342
Hammarskjold, Dag, 55
Haraszti, Miklos, 81
Havas, Agence, 63, 64, 65
HDTV, 6, 398-403
He Qiu, 118-19
Hertsgaard, Mark, 344
high-definition television (*see* HDTV)
Honduras, 176
Hook, Sidney, 3
HTS (*see* superconductivity)
Hull, Cordell, 65
Hume, Brit, 36
Hungary, 116, 171-72, 263-65
Hustler magazine, 349
Hutchins, Robert, 69
Hutchins Reports (Commission on Freedom of the Press), 69
Huxley, Julian, 444

IBM, 23-24
Iceland, 265
imperialism, cultural, 390-93
India, 47, 63, 148, 265-67, 408
Indonesia, 75, 267-68
industrial policy, U.S., 399-404, 423-24
information:
  balanced flow of, 139-42
  barriers to flow of, 368-69
  about business, regulation of, 405-7
  free flow of, 66-81, 139-42
  about government, restriction of, 345-52
  government control of, 44-48, 159-88
  vs. news, 155
information policy, U.S., 368-80
  international, 381-95
information revolution, 14-15, 42-44
  international effects of, 53-91

potential of, 439-49
a revolution without losers, 53-91
information systems (see communication systems)
Integrated Systems of Digital Networks (see ISDN)
Inter American Press Association, 166
international communications:
  barriers to, 368-69
  ITU rules, 397
  new patterns in, 368-69
  regulation of, 393-95
  U.S. policy re, 381-95, 426-27
international community, information revolution's effect on, 53-91
International Federation of Journalists, 182, 412
international journalism, 203-4
International Program for the Development of Communications, 78-81, 151, 384, 386
International Telecommunications Union, 18, 155, 370, 397
Inter Press Service, 149-55
investigative journalism, 55-56, 341-52
Iran, 45-47, 48
Iran-Contra affair, 48-49, 346
Irwin, Will, 25
ISDN:
  future of, 5-6, 15-19, 57-58
  vs. NWICO, 9, 25
  phasing in of, recommendations, 427-28
  potential of, 358-76, 439-49
  problems and dangers of, 413-16, 443
  technology of, 395-98, 404-5
  worldwide linkages of, 23-25
Islamic revolution, 46-47
Israel, 173, 268-70
Italy, 270-73
ITT, 409
Jamaica, 273-74
Japan:
  communication systems, 23
  freedom of press in, 274-75
  journalism in, 173
  technology, 6, 62, 398-401
Jefferson, Thomas, 26
Johnson, Bennett, 399
Johnson, Lyndon, 2, 112, 342
Jordan, 275-76
journalism:
  as a calling, 50-51
  developmental, 142-46
  economic restrictions on, 183-84
  ethical, 36-37
  guided, 83-89
  international, 203-4
  investigative, and official secrets, 55-56, 341-52
  laws affecting, 175-83, 376-80
  training programs, 173-74, 386-88
  in Washington, D.C., 341-55
  Western, 86-89
journalists (see also media)
  on dangerous missions, advice to, 166-68
  vs. government, 100, 419-20
  licensing of, 181-83, 196, 419-20, 429
  violence against, 57, 161-68, 199-201, 429

Keen, Sam, 110-11
Kennedy, John, 343
Kenya, 168, 276-78
KGB, 104
Khomeini, Ayatollah, 48
Kiribati, 278
Kissinger, Henry, 55
Kleinwachter, Wolfgang, 393
Korea (see South Korea)
Kozhukov, Mikhail, 111-12
Kraft, Joseph, 34
Kravchenko, Leonid P., 45
Kroloff, George, 368

labor unions, 27-28
Langley, Samuel, 357
Lasswell, Harold D., 2
Latin America, licensing of journalists in, 181-83
League of Nations, 66
leaking, 345-46, 350-51

Lear, John, 1-2
Lebanon, 167-68
Lee Kuan Yew, 179-80
Lenin, 105, 112, 113
Lewis, Anthony, 344
libel laws, 176-79, 349, 433-34
liberals:
  Eastern, 343
  in media, 34-36, 343-44
Liberia, 280-82
libertarianism, criticism of, 85-89
licensing, of journalists, 181-83, 196, 419-20, 429
Lichter, S. Robert, 21
Lichter-Rothman studies, 34-36
Lippman, Walter, 342
Lubin, Isador, 70-71
Luxembourg, 282

MacBride Commission, 71, 76, 80
MacLeish, Archibald, 68-69
Madison, James, 442, 446-47
Malaysia, 84-89, 136, 282-83
Malta, 283-84
Malwal, Bona, 131-34
Marxism, vs. capitalism, 439-43
Marxist analysis of media, 38-41, 59
Marxist countries:
  communications policy in, 99-124
  future of ISDN in, 9, 17-18
  guided journalism in, 85-86
  news coverage of, 434
  reformism in, 83, 118-23, 451
Masmoudi, Mustapha, 71-72
mass culture, television and, 60-62
Mass Media Declaration, 66, 72, 74-75, 78
Mathiane, Nomavenda, 177
Mayor Zaragosa, Federico, 77, 140-41, 388
M'Bow, Amadou Mahtar, 79
McColm, R. Bruce, 182
McLuhan, Marshall, 51
McNeil, Hector, 69
media (*see also* journalists)
  control of, in developing countries, 125-57
  control of, in Marxist countries, 99-124
  criticisms of, 33-37
  economic interests of, 29, 38-41
  government and, 2-3, 5-7, 12-13, 21-51, 54-57, 83-91, 189-206, 341-55, 416, 419-37
  government ownership of, 129-34, 184-86
  government vs. private ownership, 80-81, 411-12
  liberal biases in, 34-36, 343-44
  Marxist analysis of, 38-41, 59
  monopoly ownership of, 31-33, 48, 59, 444-45
  ownership of, and news content, 201-3, 344
  power of, 28-44, 54-55, 186-88
  in the power structure, 38-41
  public perception of, 431-32
  responsibility of, 1-3, 431-34, 445-46
  self-criticism by, 432-33
  "small," 155-57, 410-11, 415-16
  underground, 81-83
  watchdog role of, 37-38, 54
media elites, 34-37
Meese, Edwin, 350
Mejia, Alfredo, 166
Mexico, 57
military secrets, 349
Mills, C. Wright, 441
Minitel, 405
Mitterand, Francois, 404
modernization, in Third World, 127-29
Mohamad, Mahathir, 84-89
money markets, international, 43
Montesquieu, 134
Moscow News, 113, 114-15
Moslem countries, 45-47, 136-37
Mowlana, Hamid, 46
Mpassi-Muba, Auguste, 152
Mumford, Lewis, 10
Murdoch, Rupert, 444-45

Nation, 40
National News Council, 432
National Public Radio, 421
national security, protection of, 350-52

nation-states:
    communication systems as affecting, 53-55
    power of, 54-55
Nauru, 284
neocolonialism, 137-38
Nepal, 285-86
networks of freedom, 357-417
network television, 50, 353-54
Newell, Gregory, 79
news:
    definition of, 31, 56
    vs. information, 155
    management of, 198
news agencies:
    alternative, of developing countries, 125, 146-55, 168-69, 433
    government-run, 168-69
    history of, 63-65
    Unesco and, 150-51
    Western, 30-31
news coverage:
    biased, 34-37
    of developing countries, 142-55, 433
    by news services, 30-31
    by television, 48-50
    of wars, 111-12
newspapers, 193, 374-75, 425-26
new world information and communication order (*see* NWICO)
New York Times v. Sullivan, 177
Ngatara, Ludovick A., 129
Ng'weno, Hilary, 129-31
Nicaragua, 175, 286-89
Nigeria, 135, 185, 289-90
Nintendo, 8
Nixon, Richard, 342, 353, 371
nomenklatura, 106-9
Nonaligned Movement, 71-72
Nonaligned News Agencies Pool, 72, 146-49, 152, 169
Nordenstreng, Kaarle, 74, 75
Norway, 290-91
NWICO:
    developing countries' demand for, 125, 138-42

vs. ISDN, 9, 25
    proposed, 71-77

Oettinger, Anthony, 409
Office of Telecommunications Policy, 371-72
Ogonyok, 113
open standards, 360
optical fibers, 404
Orwell, George, 59, 83, 95-96, 413

Palmer, Michael, 409
Pan-African News Agency (PANA), 151-52
Panama, 83, 291-93
Papua New Guinea, 293-94
Paraguay, 167, 294-96
Pastecka, Jadwiga, 117-18
Pelton, Ronald, 346
perestroika, 45, 102
Peru, 75, 297-300
Phillipines, 57, 168, 300-301
Pierce, Charles S., 21
Pillai, M. G. G., 136
Pipe, G. Russell, 397
Poland, 116, 170-71, 301-4
Pool, Ithiel de Sola, 59, 373
Portugal, 304-5
Posner, Vladimir, 94, 110-11
post, telegraph and telephone systems (PTTs), 393
postindustrial revolution, 22
Pound, Arthur, 439
power:
    of media, 28-44, 54-55, 186-88
    of nation-states, 54-55
    nature of, 27-28
    the word as, 21-51
power structure, media in, 38-41
Prague Spring, 117
President, U.S. power of, 27
presidential elections, media coverage of, 35
press, freedom of (*see* freedom of the press)
press, the (*see also* media)
    extended definition of, 372-76
prior consent, 430-31
productivity, U.S., 398-99

prophets, Old Testament, 54
protectionism, 359, 394-95, 398
  European, 405-7
Public Broadcasting System, 354, 421
public/private sector cooperation, 399-404, 422-23

QUBE system, 412-13
radio, 127, 155, 193
Reagan, Ronald, 13, 25-26, 35, 73, 341-55, 367, 421, 447
regulation (*see* deregulation)
reply, right to, 179-81, 430
research & development, in communications, spending on, 404-5
Reuters, 30, 64, 65
revolution, high social cost of, 439-40
Roach, Colleen, 80
Robinson, Michael, 35
Rockefeller, Nelson, 372
rock music, 391, 435
Rogers, Walter S., 66
Rohatyn, Felix, 379
Roosevelt, Eleanor, 67
Roosevelt, Franklin, 342
Roosevelt, Theodore, 41-42
Rosenthal, A. M., 36-37
Rostow, Eugene, 371
Rousseau, 134
Rugh, William, 137
Rumania, 116
Rusk, Dean, 53

Sadat, Anwar al-, 44
Safire, William, 346
St. Kitts-Nevis, 305
St. Lucia, 305
St. Vincent, 305-6
Sakharov, Andrei, 107, 109, 115
Sakharov, Nicholai, 113
satellite technology, 126, 146, 363-64, 366-67, 371, 377, 430-31, 440-41
Savio, Roberto, 150-51, 153
Schiller, Herbert, 33, 59
Schmidt, Stephen B., 181-83
Schramm, Wilbur, 371

Schultz, George, 25-26, 349, 367, 385
scientific information, 350-51
secret diplomacy, 55-56
secrets, official, exposing of, by journalists, 55-56, 114-15, 341-52
Selyunin, Vasily, 101
semiconductors, 402
Senegal, 306-7
Sesinyi, Andrew, 21
Sharon, Ariel, 177-78
Shaw, David, 343
Shevardnadze, Edward, 77
shield laws, 421
Shukla, V. C., 148
SIG report on international communications policy, 382-86
Singapore, 136, 179-81, 307-8
Sisulu, Zwelakhe, 168
slippery slope, fear of, 3, 89-91, 355, 419-37
"small" media, 155-57, 410-11, 415-16
Smyser, Richard D., 33
socialist countries (*see* Marxist countries)
software (*see* computers)
Solidarity, 116
Solomon Islands, 308
Solzhenitsyn, Alexander, 175
Sony, 391
sources, protection of, 348
South Africa, 63, 175, 176-77, 308-15
South Korea, 13-14, 44, 172-73, 278-80
Soviet Union:
  censorship in, 63, 113-15, 160
  communications policy, 102-15
  disinformation by, 174-75, 431
  economy, 6, 62
  freedom of press in, 315-20
  future of ISDN in, 17-18
  media in, 38-40, 45
  political changes in, 441-43
  political system, 11, 39-40, 93-97, 107-9
  public opinion in, 105-7
  technology, 82-83, 360-61, 407

494

and Unesco, 74-75
  vs. United States, 93-95
Spain, 320-21
Sri Lanka, 321-22
Stahl, Leslie, 352
Stalin, 93-96, 114, 118
standards, open, 360
states (*see* nation-states)
Steffens, Lincoln, 29, 41-42
stock market trading, 413
Strategic Defense Initiative, 367
Strzyzewski, Tomasz, 170
Sudan, 63, 131-34, 322-24
supercomputers, 357, 408
superconductivity, 402-3
Sussman, Leonard R., 142-43, 145
Sweden, 324-25
Swinton, Stan, 63
Switzerland, 325

Taiwan, 325-27
Tanzania, 327-28
technology:
  of communication, 5-19, 358-417
  economic aspects of, 364-65
  export restrictions, 407-9
  sharing of, 411
  training in, 417
telenovelas, 393
telephone system, 155, 357-58, 387, 408, 425-26
television, 2, 10-11, 15, 193
  cable, 50, 90, 376
  cross-border, 430-31
  deregulation of, 90, 354
  and foreign policy, 48-50
  HDTV, 6, 398-403
  international, regulation of, 389-90
  and mass culture, 60-62
  networks, 50, 353-54
  news coverage, 48-50
terminals, computer, free access to, 15-16, 405
Thailand, 136
The Discovery Channel (TDC), 50
Third World (*see also* developing countries)
  cultural values of, 389-93, 434-35
  Western journalism distrusted by, 126-27
Time magazine, 177-78
Tinofeyev, Leo, 83
Togo, 329-31
Tonga, 331
Trevitt, Gavin, 441
Trinidad and Tobago, 331
Truman, Harry, 370
Tueni, Ghassan, 137, 167-68
Tunstall, Jeremy, 409
Turkey, 135, 331-33, 392-93
Tusa, John, 188

Uganda, 333-34
underground media, 81-83
Unesco, 66, 370
  administration of, 77-81
  developing countries in, 138-42
  founding of, 68-70
  and news services, 150-51
  Soviet Union and, 74-75
  U.S. policies re, 140-42, 426-27
  U.S. withdrawal from, 36, 70-81, 383
United Kingdom, 334-36
  censorship in, 63
  withdrawal from Unesco, 77
United Nations:
  and freedom of the press, 67-81
  U.S. attitude re, 79-81
United Nations Development Program, 156-57
United Nations Educational, Scientific and Cultural Organization (*see* Unesco)
United States:
  challenges to, re ISDN, 365-76
  communication industry, 391-92, 408-9, 413
  communication policy, 22-23, 423-36, 446-49
  foreign trade, 397-98
  future of ISDN in, 15-16
  industrial decline of, 398-99
  industrial policy, 399-404, 423-24
  information policy, 368-80
  international communication policy, 381-95, 426-27

media in, 2-3, 38-42
vs. Soviet Union, 93-95
technology, 14, 398
and Unesco, 36, 70-81, 140-42, 383, 426-27
United States Information Agency, 387, 420
Universal Declaration of Human Rights, 68
UPI, 30, 65
U.S. Telecommunications Training Institute, 386, 388

Value Added Networks (VANS), 360
Van Bennekom, Pieter, 167
Vanuatu, 336
VCRs, 392-93
Versailles, Treaty of, 66
video games, 8
Vietnam war, 2, 112
Voice of Freedom '87 conference, 431

Wang Ying, 120
wars, news coverage of, 111-12
Washington, D.C., journalism in, 341-55
watchdog role of press, 37-38, 54
Wei Jingsheng, 118

West, vs. developing countries, 137-42
Western journalism:
criticism of, 86-89
news agencies, 30-31
Third World distrust of, 126-27
Western Samoa, 336
West Germany, 253
Westmoreland, William, 177-78
Whitehead, Clay, 353
Wick, Charles, 44
Wilson, Woodrow, 5, 64
Windelshaw, Lord, 442
Wolff Agency, 64
World Administrative Telephone and Telegraph Conference (1988), 397
Worldnet, 44
World War II, 65-67, 95
Worldwatch, 441
Wriston, Walter, 43
writing between the lines, 81-83

xerography, 410-11
Xinhua, 120

Yakovlev, Yegor, 115
Yang Baikui, 134-35

Zaire, 337
Zimbabwe, 338-40

Freedom House Books
General Editor: James Finn

Yearbooks

Freedom in the World: Political Rights and Civil Liberties,
Raymond D. Gastil; annuals from 1978-1989.

Studies in Freedom

Escape to Freedom: The Story of the International Rescue Committee,
Aaron Levenstein; 1983.

Forty Years: A Third World Soldier at the UN,
Carlos P. Romulo (with Beth Day Romulo); 1986. *(Romulo: A Third World Soldier at the UN,* paperback edition, 1987.)

Today's American: How Free?
edited by James Finn & Leonard R. Sussman, 1986.

Will of the People: Original Democracies in Non-Western Societies,
Raul S. Manglapus; 1987.

Perspectives on Freedom

Three Years at the East-West Divide,
Max M. Kampelman; (Introductions by Ronald Reagan and Jimmy Carter; edited by Leonard R. Sussman); 1983.

*The Democratic Mask: The Consolidation
of the Sandinista Revolution,*
Douglas W. Payne; 1985.

The Heresy of Words in Cuba: Freedom of Expression & Information,
Carlos Ripoll; 1985.

Human Rights & the New Realism: Strategic Thinking in a New Age,
Michael Novak; 1986.

To License A Journalist?,
Inter-American Court of Human Rights; 1986.

The Catholic Church in China,
L. Ladany; 1987.

Glasnost: How Open? Soviet & Eastern European Dissidents; 1987.

Yugoslavia: The Failure of "Democratic" Communism; 1987.

The Prague Spring: A Mixed Legacy
edited by Jiri Pehe, 1988.

Romania: A Case of "Dynastic" Communism; 1989.

Focus on Issues

Big Story: How the American Press and Television Reported and Interpreted the Crisis of Tet-1968 in Vietnam and Washington,
Peter Braestrup; Two volumes 1977;
One volume paperback abridged 1978, 1983.

Afghanistan: The Great Game Revisited,
edited by Rossane Klass; 1988.

Nicaragua's Continuing Struggle: In Search of Democracy,
Arturo J. Cruz; 1988.

La Prensa: The Republic of Paper,
Jaime Chamorro Cardenal; 1988.

The World Council of Churches & Politics, 1975-1986,
J.A. Emerson Vermaat; 1989.

South Africa: Diary of Troubled Times
Nomavenda Mathiane; 1989.

The Unknown War: The Miskito Nation, Nicaragua, and the United States,
Bernard Nietschmann; 1989.

*Power, the Press and the Technology of Freedom
The Coming Age of ISDN*
Leonard R. Sussman; 1989

An Occasional Paper

General Editor: R. Bruce McColm

Glasnost and Social & Economic Rights
Valery Chalidze, Richard Schifter; 1988.

PN 4748 .D44 S87 1989

	DATE DUE	
MAR - 4 1992		